Hidden World

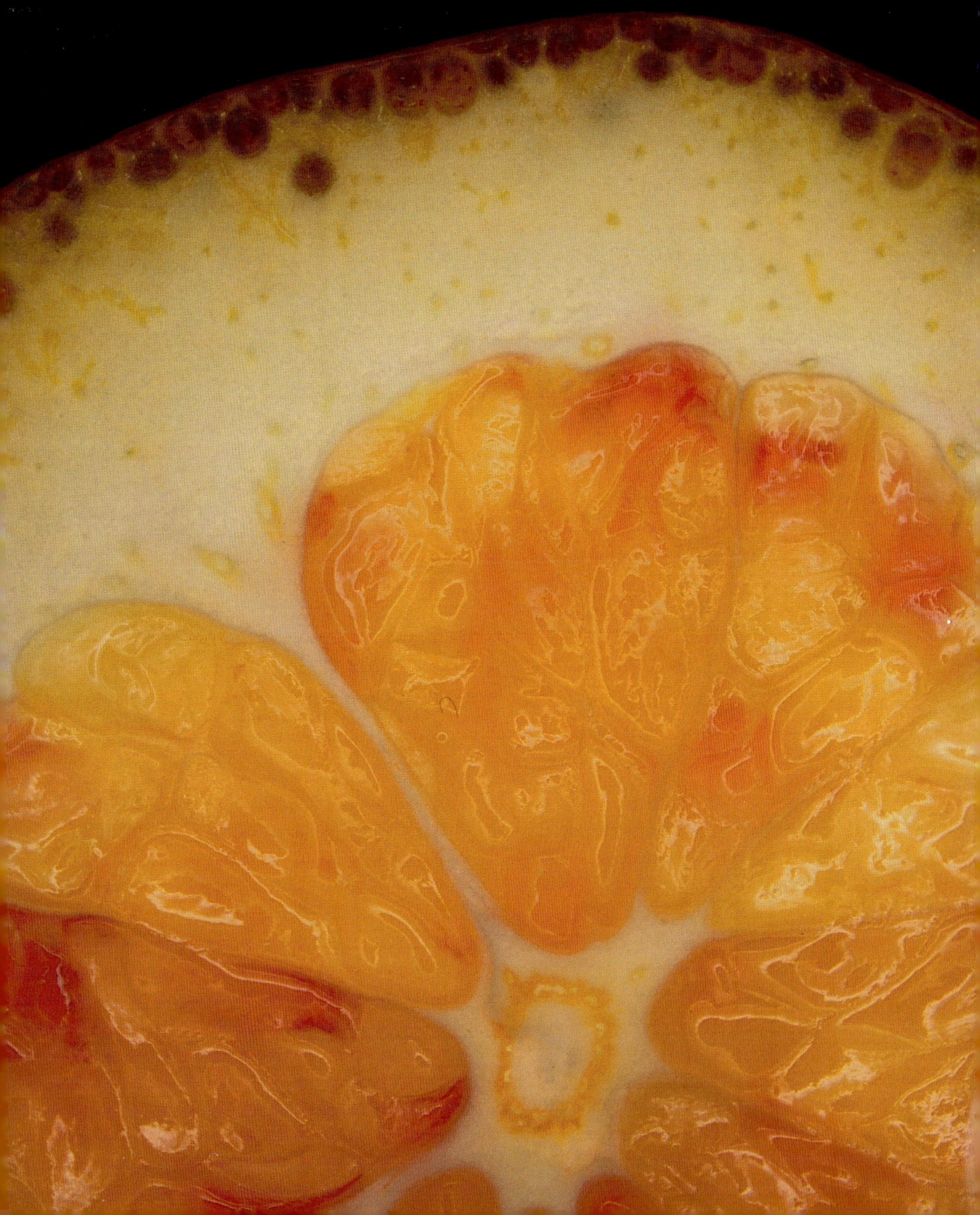

Hidden World

The Survival Systems of Plants

KEVIN TARNER AND RACHEL HUGHES

The University of Georgia Press Athens

Athens, Georgia 30602
www.ugapress.org

Designed by Erin Kirk
Set in Minion Pro and Cinta
Printed and bound by Versa Press

The paper in this book meets the guidelines for permanence and durability of the Committee on Production Guidelines for Book Longevity of the Council on Library Resources.

Most University of Georgia Press titles are available from popular e-book vendors.

Printed in the United States of America
29 28 27 26 25 P 5 4 3 2 1

Library of Congress Cataloging-in-Publication Data

Names: Tarner, Kevin, author. | Hughes, Rachel (Ecology lecturer), author.
Title: Hidden world : the survival systems of plants / Kevin Tarner and Rachel Hughes.
Description: Athens : The University of Georgia Press, [2025] | Includes bibliographical references and index.
Identifiers: LCCN 2024056045 | ISBN 9780820370026 (paperback)
Subjects: LCSH: Plant physiology. | Plants—Adaptation.
Classification: LCC QK711.2 .T364 2025 | DDC 581—dc23/eng/20250312
LC record available at https://lccn.loc.gov/2024056045

This book is dedicated to
Marvin and Phyllis Hughes.
Their tireless passion for the
environment has been an
inspiration to us both.
We thank you, we love you,
and we don't know what
we'd do without you.

Contents

Part Two Survival against Natural Forces

Acknowledgments

Writing a book is a unique undertaking involving some combination of passion, learning, sharing, and sacrifice. This book would not have been possible without the love and support of our families and friends. First, we would like to thank Phyllis Hughes for lending us her artistic skills and thoughtful commentary. Her husband, Marvin Hughes, also was kind enough to give us his input and edits. Both have been essential in sparking our passion for the natural world.

We have a collection of friends to thank for their support and assistance. First, fellow plantsmen and photographers Paul Barden and Lars Lohn lovingly gifted us the DSLR that shot many a photo found in this book. Additional photography has been kindly shared by photographer, plant scientist, and friend Mason McNair. For assistance in the procurement of various organisms for photography, we would like to thank our friends Håkon Jones, Brandon Coker, Mark Hopkins, Matt Seader, Dr. Soraya Bertioli, and Katherine Hardigree, as well as the members of Dr. Melissa Mitchum's nematology lab.

The support, training, and access to equipment we received at Georgia Electron Microscopy at the University of Georgia was invaluable for much of this work's close-up photography. Beth Richardson deserves a special mention for her assistance and advice. We also appreciate the expertise and troubleshooting provided by Dr. John Shields, Mary Ard, Dr. Eric Formo, and Dr. Tina Salguero.

We'd also like to thank Jane Kobres for inspiring us to take this idea to a publisher and for believing in the merits of a book on plant defense. Additionally, this book would never have come to fruition without the support of the UGA Press. We appreciate that support very much.

Finally, thanks also to our friends Amy Bohon, Aaron Calvin, Dr. Wendy Zomlefer, Dr. David Giannasi, Steven Hughes, Taylor Harrell, Matt Diephouse, Tyler Thompson, and Dr. Amanda Nash for listening to us, supporting us, and being all-around awesome people.

Hidden World

Introduction

Every plant around us is kept alive by an often overlooked and largely imperceptible hidden world. Though typically unnoticed by humans, this complex system supports and protects plant life around the globe. We often view plants as static, immobile organisms that exist in their environment and are completely susceptible and submissive to whatever fate befalls them. However, we have come to learn that this is rarely the case. These sun-soaked oxygen producers have an entire evolutionary arsenal of survival adaptations to ward off threats to their well-being and existence. These adaptations are omnipresent, hypervigilant, and responsive, and they take on a variety of forms and use a variety of techniques.

A plant's preference in life would be to grow verdant and leafy, reproduce prolifically, and harmlessly utilize the sun's warming rays in perpetuity. Plants instead find themselves at the base of a ravenous food web full of hungry animals and microorganisms ready to attack and consume them. If that weren't challenging enough, plants must also contend with natural forces that frequently impact their success. Unlike many animals, plants cannot run from nature's merciless pressure and must weather disturbance such as wildfires, drought, cold, storms, and floods. Additionally, individual plants must successfully obtain resources in a sea of other plants and photosynthesizing organisms. Backed against a wall of munching herbivores, competing plants, and environmental stresses, plants have been forced to evolve ingenious but not always apparent solutions to these problems. To survive, plants have developed sophisticated perception systems that are acutely attuned to the world around them. These systems are then coupled with systems of chemical communication techniques as well as chemical and physical defenses in a bid to survive. Every plant cell is packed full of genetic armament to help the plant win as many battles as possible.

The necessity for plant defense has created a complex and secret landscape of strategies for signaling, responding to, and producing the physical and chemical defenses found throughout the plant kingdom. Like green sentinels, plants vigilantly watch the weather, the seasons, and the organisms around them. They report what they discover to each other. All the while, plants also gear up for war. These adaptations have allowed the plant kingdom to journey successfully through natural history and triumph over endless cataclysms.

These changes have greatly affected the success of plants, but humans have also been impacted by these changes in our passage through time. We have coexisted with and even co-opted plant defenses, resulting in tremendous benefits for our species. Humankind has lived alongside of, utilized, and exploited the survival systems of plants in a variety of ways. These defenses and adaptations are so intimately parallel to our daily lives that we often bump into them unwittingly. We encounter plant defenses when we cook vegetables, add spices to our food, brew coffee, and smell plants when we use them for decorations or when we extract or interact with them. Botanical defenses form the basis of enormous commercial industries, scientific breakthroughs, and technological advancements. The latex in rubber, the wood fiber in paper, the botanical origins of explosives and movie film, chemical medicines such as morphine, and recreational products such as caffeine and nicotine all originated as plant defenses. We borrow from the plant world structures and molecules to use in the fields of nanotechnology, engineering, chemistry, and pharmacy. We will explore all of these borrowings in greater detail as we move through this book.

One reason we fail to see these encounters as a brush with plant defenses is that many are harmless and inconsequential daily events. Another is that the chemicals and foods in our homes are long separated from their botanical origins by a global chain of production, processing, and transport. Additionally, we may never notice some of the most powerful plant defenses, such as pigments, flavors, and scents, because they target other threats. These warning shots are not intended for us to perceive. In general, however, all plant defenses pack a specifically tailored wallop to something somewhere. What we hope to teach our readers is the ability to read plants. As we discover this realm, we will glimpse the intricacy, awe, trickery, diversity, and successfulness of the plant survival systems operating everywhere around us.

Sometimes, however, humans are the target of this defensive lineup. In many instances, we interact with plants as an herbivore would and thus are perceived by the plants as an immediate threat. Often, we may inadvertently or intentionally make the first strike against a plant's defense systems. Damaging, chopping down, or munching a plant jolts us into an evolutionary war for survival millions of years in the making. While plants would prefer to be left alone, their defense systems rally to their aid when they perceive even the slightest threat or attack.

Ultimately, the purpose of plant defenses is to teach us a lesson: cease and desist or risk the consequences. Defenses often begin as slaps on the wrist, but some plants are prepared to ratchet up their efforts to lethal levels. Harmful encounters that result in poisonings and intoxication are indications that plant defenses are ever present. An ignored warning such as a bitter taste or an unpleasant smell or reaction can lead to severe health complications.

Not all plant defenses involve direct aggression, however. A plant might emit a smell that communicates a warning to other plants or recruits predators of a plant's pests. Spicy tastes from wasabi and chilies act as botanical "do not disturb" signs to our taste and pain receptors. Wood and bark act as armor to damage, disease, desiccation, wind, and fire. Plant cells and surfaces are brimming with complex and otherworldly fortifications. Awareness of these omnipresent tactics operating around us creates a surreal feeling akin to entering another dimension. While the strategies plants use to survive are amazing, they can be subtle and surreptitious. As a result of our own lack of awareness, we often overlook the complexity of this leafy world, which can lead us to undesirable consequences.

We are only truly safe if we are on plants' good side: helping disperse their offspring in fruit and seeds, propagating them as crops, pampering them as houseplants, or depositing water and nutrients in our landscaping and gardens. Humans can be both a friend and a foe to the plant kingdom.

Symbols of this unseen world are everywhere, and this book teaches you how to find them. We hope that by the time you have finished this book you will be able to identify these symbols each time you go outside. You will begin to pick up on this complex network by noticing the shape of a tree or the color of a leaf. You will never look at plants the same way again, whether they are floating, towering, spiny, or toxic. We hope this journey inspires an appreciation for the complexity and finesse found within every plant.

CHAPTER 1

The Hidden World within Plants' Lives and Our Lives

The Basics of Plant Survival

Why Do Plants Need Defenses?

Plants make food for other organisms and are an excellent source of energy. As such, they are frequently attacked and eaten for the resources inside their cells. Plants are at the bottom of the food web, and for good reason. They manufacture their own sugar, which is then propelled upward through the other trophic levels. Each trophic level in an ecosystem comprises organisms with similar functions in the food web (i.e., primary energy producers such as plants, followed by herbivores, carnivores, and apex predators). All other levels, humans included, have to consume or absorb another organism to gain energy. Plants accomplish this sugar generation through a process known as photosynthesis. Since they create their own energy, plants are known as autotrophs. Heterotrophs, by contrast, must glean food from the organisms they digest, absorb, or parasitize; they include herbivores, carnivores, parasites, fungi, and the majority of microbes. During the process of photosynthesis, plants pull in carbon dioxide through the pores on their leaves and absorb water through their tissues; then they combine these elements to produce a sugar (glucose). Sunlight is the catalyst that drives this chemical reaction forward. The end product, glucose, is so important because almost all life-forms use sugar to fuel their lives.

The chemical formula for photosynthesis can look quite intimidating: $6CO_2 + 6H_2O \rightarrow C_6H_{12}O_6 + 6O_2$. However, we'd like you to think of this formula as a recipe. The $C_6H_{12}O_6$ is the chemical makeup of glucose, the sugar that serves as the primary energy source for a plant. Each of the letters or groupings of letters in the formula represents an element from the periodic table: C for carbon, H for hydrogen, and O for oxygen. The subscript numbers tell you how much of each element the plant needs to form the product the plant wants to make. Sitting anchored in the soil, the plant will have the easiest time acquiring these components from water (H_2O) in the soil and carbon dioxide (CO_2) in the air. The number in front of each molecule or element tells you how many elements the plant needs, just as a recipe tells you that you need six eggs or five tablespoons of butter. Lucky for us, this recipe ends up with a surplus of oxygen. This leftover oxygen becomes a waste product released as oxygen gas, which we and many other organisms use for breathing. Accordingly, photosynthesis provides sugars as an energy source for much of the life on the planet, as well as providing an oxygen-rich atmosphere. This recipe keeps all life on this planet running. Plants are one of the main building blocks at the base of food webs around the world.

In addition to manufacturing sugar, plants actively harvest and store water and nutrients from their surroundings, which also makes them into perfect natural sources of water and minerals for other organisms. Along with this rich stockpiling behavior, plants mostly lack mechanisms to physically move to avoid danger. All of the energy, water, and minerals they contain would seem ripe for the taking. As rooted beings, plants also seemingly have no methods to avoid natural cataclysms such as floods, droughts, and fire as well as damage by herbivorous animals or pathogenic microbes. However, plants have actually been developing solutions to these problems for millions of years. Plants' ability to produce energy and sequester resources, combined with their immobility, is a basic premise underlying the thousands of unique defensive solutions that plants have evolved. When everything wants to eat you and you cannot run away, you have to get crafty. Plants' solutions to these challenges include physical defenses such as spines, epidermises, and trichomes, as well as chemical defenses such as adjustments to shape and color and chemical signaling to themselves, other plants, and helpful creatures.

The Structure of Plants

Before we jump into these defenses, let's get acquainted with plants, both how they look and how they function (a science called plant physiology). There is so much plant biodiversity out there that we have to make a few generalizations. Keep in mind, though, that there are usually exceptions to every rule.

Most plants have two parts. Aboveground, the shoot system comprises stems, leaves, and reproductive structures. Belowground, the root system anchors plants to the soil and branches widely to absorb water and mine necessary minerals. The shoot system is responsible for harvesting light energy from the sun to use as a catalyst for photosynthesis. Leaves are green because of a pigment called chlorophyll that helps plants absorb

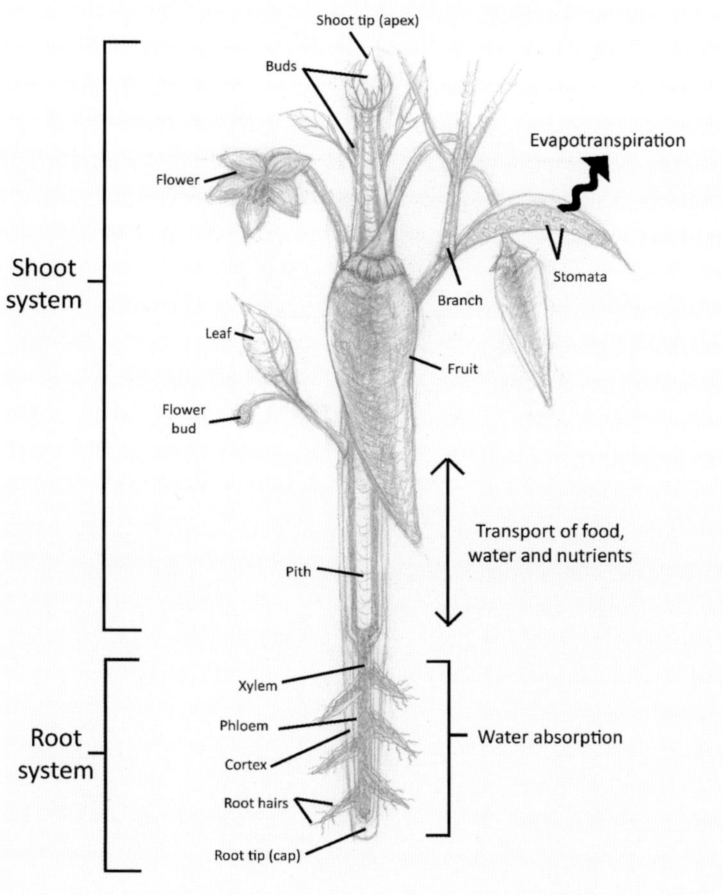

energy from the sun. The energy from sunlight powers photosynthetic reactions. Thus, the aboveground tissue is the site of photosynthesis. Additionally, gas exchange occurs above the soil. Plants harvest carbon dioxide gas and expel oxygen and water vapor as waste products. On leaves, special pores called stomata regulate this gas exchange using guard cells, which surround the pores and can open and close depending on environmental factors.

The aboveground surfaces of many plants include specialized physical defensive structures and materials such as spines, versatile hairs called trichomes, resilient materials such as wood and bark, and a tough, waxy surface called the epidermis. Shoot and root tissues might also produce chemicals that repel, poison, cause pigmentation, communicate messages, or coordinate other defenses.

The majority of plants include a vascular system, which is a plant's internal pipeline for resource distribution—moving around sugars, water, and nutrients. This tubing is made up of tissues called xylem and phloem. Phloem cells send sugars to all parts of the plant, including roots. Xylem conducts water and nutrients absorbed by the roots to all parts of the plant.

A plant relies on basic laws of chemistry to move water and nutrients through its system. Water most typically enters a plant's xylem tissue through the roots. Water is a polar molecule, which means each molecule is attracted to another molecule. The molecules cling together like tiny liquid magnets. This phenomenon is created by two forces called cohesion and adhesion. You know of this attraction if you have ever looked at a droplet of water on your car windshield or at the surface tension in a glass of water.

But how does this chain of water move upward? Stomata, or the pores that control gas exchange with the environment, regulate, among other things, the movement of water out of a plant. When the top molecule of water in a chain reaches this pore, the external heat causes the water molecule to energize and transform into a gas called water vapor. As the molecule transforms states and leaves the plant, it tugs on the chain of water, bringing the chain forward up to the leaf. This tiny tug starts a mass flow of water through the entire plant, reaching successively to water molecules in the roots and drawing new water in from the soil. The process by which this water column moves upward from the roots throughout the plant and then evaporates from the leaves and stomata is called transpiration. Water transpiration keeps plant cells alive. Transpiration provides the water for chemical reactions, structural turgidity of cells, and evaporative cooling for the plant's aboveground tissues. Should the ground be too dry and no water available, the tugging of transpiration leaves the end of the chain without a water molecule to grab. If this continues, the plant eventually wilts. Once it reaches the wilting point, or the point of no return, no amount of water can save it. That's because the chain has moved inside the plant and can no longer rely on polar bonds to adhere to any additional water molecules. This air gap in plant conductive tissues is called an embolism. The effects of an embolism can eventually starve the plant of water for photosynthesis and cooling. The root and shoot systems fuel and balance each other. Damage to or death of one results in proportional damage and imbalance to the other.

A Brief Evolutionary History of Plants

To understand the different types of plants we will be discussing and the defenses they've established, we need to understand where those defenses fit in the evolutionary history of plants on this planet. Plants evolved from other photosynthesizing organisms found in aquatic systems around the world, including algae and cyanobacteria. As these organisms washed up along the banks of rivers, wetlands, and lakes around the world, some of them gained adaptations to live right at these boundaries.

To survive on land, plants face many challenges. Plants have made numerous adjustments to get water, nutrients, and sugar on a terrestrial surface. Algae are able to float around in a soupy nutrient mix of water. These moist photosynthesizers can diffuse nutrients and water across their entire surface. They do not have to worry too much about structural support, as they float in a buoyant medium. Moreover, algae reproduce by having their sperm swim and their embryos develop in this aquatic system. Keep this in mind: everything needs to stay moist. Organisms are made mostly of water and need to keep sex cells and embryos moist. Plants are no different in this regard, and land plants have to figure out ways to stay moist. Finally, land plants have to combat intense solar energy and UV radiation, which threaten their tissues and lead to additional dryness.

The earliest group to try to tackle these issues is known as the bryophytes, a group comprised of the mosses, liverworts, and hornworts. These plants have just barely made the transition to the terrestrial landscape. They do not have true leaves or roots, and they require a good deal of water to survive and reproduce. They reproduce in a way that is a remnant of their algae ancestors as they wait for water to flood over them. Once a plant is fully submerged, the sperm from its reproductive structures swims to the egg of another. Additionally, the plant's embryos require a decent amount of available water to survive. These plants have not addressed any of the challenges of structural support on land or how to move nutrients and water around a larger structure, so they remain short and compact. However, bryophytes are impressive and remain a pioneer species on rock outcrops, newly formed land, and areas of great disturbance.

The next plant group to evolve are seedless vascular plants, which include ferns, horsetails, spike mosses, and clubmosses. These plants have made a huge change: they have developed vascular tissue, the tubing network that moves water, nutrients, and sugar around the plant. Lignin, a polymer that adds strength to a plant's tissues, shows up in this group. Now plants can get taller. They are able to move resources from the roots upward to the leaves and move sugars downward. These plants still rely heavily on water for reproduction, so they remain in damp habitats. This worked out well, though, as much of the planet was covered in wetlands at the time of these plants' arrival.

Two huge changes occur in the next group of plants: the evolution of pollen and the evolution of the seed in the plant group known as the gymnosperms. Gymnosperms, from Greek words meaning "naked seed," are also known as conifers and include pines, spruces, cypresses, larches, hemlocks, redwoods, junipers, cycads, ginkgoes, welwitschias, and more. Let's look a bit closer at these two impressive changes.

We've all come face to face with pollen in the spring. It blankets our cars, buildings, and landscapes. Those of us with an unfortunate tendency toward allergies will start sneezing, wiping our watering eyes, or forming welts. Those without allergies must still face the constant assault of yellow dust everywhere. What is the purpose of that cloud of pollen floating through the air? As noted, every organism must maintain moisture and keep its sex cells moist. Having to rely on flooding or a constant water supply limited plants' ability to move onto land, especially as our planet began to dry. None of that drier land could be colonized unless a new technique arose.

Enter pollen. Each grain of pollen is a sack containing future sperm cells. The sack's membrane keeps the sperm from drying out—thus, it keeps it moist. This is a game changer. Now a plant doesn't need to rely on water for reproduction. However, the plant is going to need some way to get that sperm to the female reproductive structure (a cone in this case) of another plant. What was readily available at this time in our planet's history? Wind! If you've ever seen a pine tree release a yellow cloud at the slightest breeze during those warm spring months, you've seen gymnosperm plant reproduction at work. The reason pine trees make so much of this pollen mess is because they are relying on wind to get that sperm to its contact point. That's quite a gamble. Anchored to the ground, gymnosperms are at the mercy of the strength and direction of the wind. Thus, they must make a ton of pollen to ensure its arrival at the target. Sperm manufacture from the male cones is relatively cheap from a biological perspective, so manufacturing a lot of it is a reasonable way for them to hedge their bets. This evolutionary change allows plants to colonize drier areas that were previously unattainable.

The gymnosperms gained another big adaptive change to go along with pollen: the arrival of the seed. Seeds are another way for plants to keep their offspring moist. These little containers are made up of three parts: the seed coat, the endosperm, and the embryo. You are familiar with a seed coat if you have ever cooked dried beans (an angiosperm—we will get to these soon) and seen the thin skin of the bean shuck off from the rest. This coating provides insulation and moisture protection, once again freeing plants from the shackles of being near water. Remember, embryos, like sex cells, must stay moist. The seed also includes the endosperm or food supply. This is the parent plant's way of packing a little lunch to get its offspring started. It's a harsh world out there. Having some energy to jump-start the process is essential, especially in shaded environments or challenging terrains. Finally, the seed includes the embryo, which is the next generation—the offspring. Thus, the humble seed is an impressive leap forward in the evolutionary history of plants. Seeds and pollen give gymnosperms the ability to colonize previously inaccessible areas of the planet, which will be a major advantage as the earth goes from a landscape of wetlands to a drier climate in the coming millions of years.

The flowering plants were the last group to show up on the planet. These are also known as angiosperms, from the Greek words *angeîon* (container) and *spérma* (seed). Flowering plants gained two additional adaptations: the development of a flower and the development of a fruit. Though sperm is relatively cheap, using wind to move around pollen isn't the most efficient system. Enter flowers: attractive structures that call in pollinators to handle the plant's reproductive requirements. Flowers are made up of the main reproductive structures of the plant as well as some showy elements such as petals and often a nutritious reward. Now, instead of sending buckets of sperm into the air and hoping they find success, plants call in pollinators to hand-deliver the sperm directly to the female reproductive structure of another flower. A great diversity in pollinator relationships has developed over evolutionary time. Plants began offering a sugary reward or, in some instances, visual trickery in order to entice bees, moths, and butterflies as well as beetles, bats, wasps, flies, mammals, and birds. Most commonly, the reward is in the form of nectar. Plants are careful to make their nectar sweet enough to attract a pollinator but not so sweet that their pollinator fills up on one flower. Instead, pollinators move from flower to flower, depositing pollen that inadvertently sticks to their bodies as they feed from a flower's nectar reserve. Though the sugar is more costly to manufacture than excessive sperm, the efficiency of delivery is well worth the cost. Over time, flowers developed a variety of colors, patterns, scents, shapes, and ultraviolet markings specifically tailored to the particular pollinator they are trying to attract. This efficient system allows for the large diversity of plants we see today and helped make angiosperms the largest group of plants on the planet. The next time you give someone a bouquet or enjoy the scent of a lovely flower, take a moment to appreciate these reproductive structures that have given us so much biodiversity.

In addition to flowers, angiosperms developed fruits to assist in distributing their offspring into the world. As the ovary ripens, it develops a dry or fleshy exterior around the seed. The fleshy fruits include things you think of as fruits (cherries, grapes, peaches, watermelons, and the like) as well as things people often forget are fruits (peppers, squash, eggplants, and more). Most of the fleshy fruits encourage distribution through feeding—either by being carried off and consumed, with the seeds discarded, or by being consumed whole, with the seeds taking a ride through the digestive system. The action of the enzymes and acids in the guts of the consumer scars the tough seed coat and prepares the seed to germinate once it has successfully exited the organism. This seed also conveniently finds itself preplanted in a fresh pile of fertilizer in the form of animal waste. Some fruits use other methods of distribution. Dry fruits can take on a variety of forms to utilize wind, water, or animal movement. Burrs, for example, catch in animal fur or skin; the animal then carries the seed far from its origin. Coconuts are capable of floating to a new destination and come prepackaged with some fresh water for the future offspring to enjoy in an otherwise saline environment. Dandelions form a parasol out of their fruit that allows their seeds to float away in the breeze. Fruits allow the offspring to gain some distance from their siblings and parents, reducing competition and increasing the chance that at least a few of them might luck out in an adequate starting environment. This is well worth the cost in creating these structures, which use up key macro- and micronutrients. So the next time you are hanging out in the produce aisle at the grocery store, take a moment to admire this impressive distribution system that we find so tasty and nutritious.

Plant Growth

Once the plant has successfully created the next generation of plants and distributed them into the world, it is up to that little embryo to make it on its own. Young plants are similar to but slightly different in appearance from mature plants. A plant begins as a tiny embryo inside a seed (gymnosperms and angiosperms) or spore (mosses and ferns). Environmental cues initiate a process called germination, the start of growth and differentiation for the embryo within a seed or spore. After this, the nascent embryo is free to develop into a tiny plantlet and begins to produce leaves, stems, and roots. The growth and development of these organs are called organogenesis. To grow, plants first make a tissue called a meristem, which is a special type of undifferentiated cell. Meristem cells have the power to grow up and become anything they want to be in life: roots, leaves, flowers, new stems, seeds, bark, xylem, phloem, guard cells, trichomes, spines, chemical storage cells—you name it! Meristems are generally located at the growing tips of a plant's shoot or root systems. Meristems give plants the power to adapt their growth in response to threats and changing environmental conditions. In primary growth, cells in the meristem divide and begin to differentiate, resulting in the lengthening of a stem or root. Cambium, a layer of live growth tissue between a plant's xylem and phloem, produces partially differentiated cells that become new xylem, phloem, wood, and bark cells. In secondary growth, cells in a plant's cambium divide and differentiate. This is how roots and shoots become wider—if you've ever seen the growth rings on a tree, you've seen secondary growth at work. Wood and bark are products of plant secondary growth, adding significant structural support and protection.

Cell elongation is another type of plant growth. You're familiar with this if you've ever noticed houseplants bending toward the light of a window or seen vines wrapped around a support. Plant cells can grow longer as individuals or in a grouped pattern. Plants control all of their growth using chemicals called hormones, which regulate a variety of processes within an organism and kickstart action within cells or tissues. When plant hormones signal cells on one side of a stem to grow faster or longer than the other side, the stem begins to bend. Through elongation, plants can grow toward interesting things in their environment. Roots might use elongation to grow toward rich sources of water (hydrotropism), nutrients (chemotropism), or oxygen in flooded soil (aerotropism). Roots also use gravity to determine which direction to elongate (gravitropism). Stems might bend toward light sources (phototropism) or respond to physical forces (thigmotropism). For example, vines make thigmotropic growth to climb objects that they physically touch, while wind causes plants to grow stronger by exerting mechanical forces on them. The clandestine detective work plants perform in the outside world includes but is not limited to these tropisms. Plants gather and respond to this information in ways that are optimized for their survival.

CHAPTER 2

How Enemies Attack Plants

Ecology and Plant Pathology

As strong, complex, and dynamic as plants are, they're not without weaknesses. Innumerable ravenous herbivores and microbes swarm around plants constantly, mercilessly seeking even a small chink in their armor. An almost overwhelming array of threats assaults plants every day. A plant's simultaneous priorities might include finding enough water for the day, beating back an onslaught of aphids, preparing for oncoming severe weather, and growing in the best pattern to evade the shade created by competing neighbors. Another plant might be smoldering from a recent wildfire, its few surviving leaves and shoots voraciously descended upon by hungry browsing mammals and a horde of migratory locusts. Every day, for every plant, the story of natural history and its hidden world unfolds. Let's elucidate some of those stories of what a plant might face in its lifetime.

One of the basic tenets of living things is that stress is weakening. This is true of us humans. If we aren't eating well or sleeping well, we might be at greater risk of catching a cold or getting an infection. When plants are stressed and weakened by natural forces, herbivory, or diseases, they more easily fall prey to further attacks. The exhausting, constant pressures they face may stretch their resources thin, open wounds and entry points not otherwise available, shut down defense production, or otherwise test their limits. Additionally, defense production is a costly investment for plants. Physical defenses, a category that includes cell walls, spines, trichomes, wood, bark, and seed coats, intensively tax a plant's carbon supplies. The opportunity cost of dedicating this carbon toward defenses occurs at the expense of creating energy for growth, reproduction, and photosynthetic tissues such as leaves. Chemical defenses such as toxins are equally taxing, depleting a plant's nutrient reserves of elements such as nitrogen and sulfur. Should any of these resources fall into short supply, a plant's ability to manufacture and maintain its defense systems is crippled.

Environmental stress is one of the most common ways an attack on a plant can begin. Drought, heat, and cold stress can limit the ability of plants to maintain their turgor pressure or close off openings such as the stomata. (Turgor pressure is the internal force generated by a cell's internal water supply. The water wants to leak outside the cell naturally through a variety of means, but the cell wall holds it back. Therefore, the plant cell wall is said to be under turgor pressure and forced outward. There is less water within a wilting cell, so the pressure on the cell wall is weak, leading it to become wilty or floppy.) Nutritional stress and shading can limit defense production just when it may be needed to fend off an outbreak of caterpillars. Snapped branch wounds remaining after a hurricane might become host to a new colony of wood-boring beetles. The list of possibilities is endless.

Wounds are a common way for pests and diseases to enter a plant. Attacks on plants easily beget further attacks. Sometimes, numerous species of different organisms might simultaneously vie for opportunities to colonize a plant's wound. Depending on the nature of the plant's injury, physical defenses such as cell walls, spines, seed coats, and bark might be broken, rendered ineffective, or breached. For example, the feeding site of a microscopic, plant-parasitic nematode lies under the soil. The piercing wound exacted by the nematode's stylet might easily become the next infection point for virulent, root-rotting fungi and bacteria or the egg-laying site for root-feeding insects. When the needle-like mouthparts of aphids or other pests pierce a plant's cell to feed, they might be simultaneously inoculating the cell with phytoplasmas, bacteria, or viruses that live inside their guts and saliva.

Even when plants are healthy and unstressed, they still have natural entry points that must be closely monitored. All plants must exchange gases with the outside environment, and to do so they have pores in their leaves (stomata) and pores in their woody tissues (lenticels). These pores must open and close throughout the day in order for the plant to survive and meet its photosynthetic and other biological needs. Some plant diseases specialize in finding these open entryways, forcing themselves through them and initiating infections.

The horizontal brown stripes on this black cherry (*Prunus serotina*) trunk are called lenticels. They are breathing pores in plant stems that help permit gas exchange. Unfortunately, they are also an entryway for plant diseases and insects.

Piercing-Sucking Feeding

Piercing-sucking feeders feed on plants by piercing plant cells and sucking out the contents. Each herbivore species employing this method is selective and preferential in its diet and may target individual cells or the xylem or phloem tubes of the vascular system. This type of attacker has specialized mouthparts that can penetrate plant tissue and withdraw food. In general, piercing-sucking insects have a pointed, beak-like mouthpart called a rostrum. Inside this protective sheath resides a pair of stylets. (Arachnids, which include mites, have stylets but no rostra.) Stylets are like hypodermic needles, sharp and capable of easily driving into and through plant tissue. When the stylets are pressed together, two tubes form. The first tube is connected to salivary glands. It deposits the attacker's saliva, which contains glues (to cement the bug to the plant while feeding), lubricants, or chemicals called effectors, which deactivate or manipulate plant cellular defenses. The second is the feeding tube, which is connected to the guts and capable of withdrawing plant sap.

The feeding of these attackers not only injures plant tissues and opens wounds in protective barriers such as the cuticle and cell wall but also

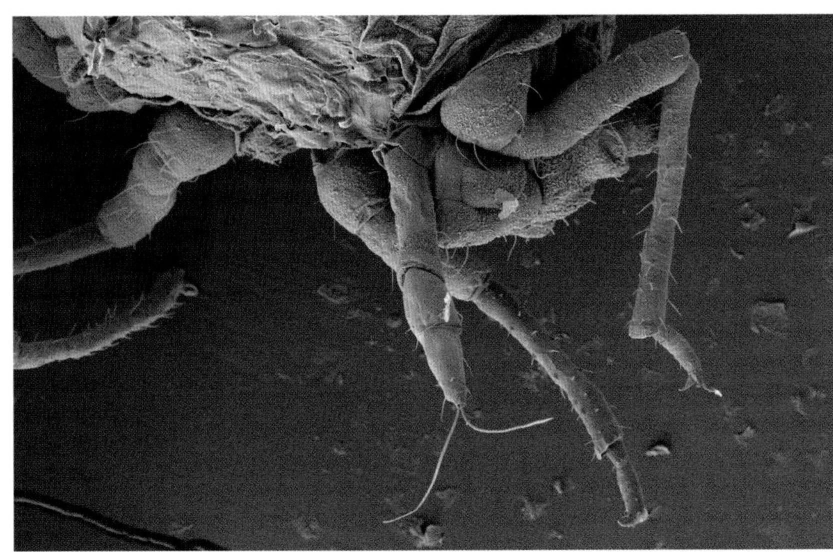

Seen here is an aphid specimen under a scanning electron microscope. The beak-like mouthpart seen here is the rostrum, while the two fine, needle-like projections from inside are called the stylets. One stylet secretes saliva, while the other withdraws food.

Seen here is an aphid family feeding from a daylily leaf. Visible are a winged adult male, a wingless adult female, and a nymph, or baby. The dried exoskeleton in the photo has likely been left behind from a molt. Interestingly, these cast-off skins, called exuviae, commonly lie among live aphids in a colony. They often act as decoys, absorbing attacks from aphid predators and parasitic wasps.

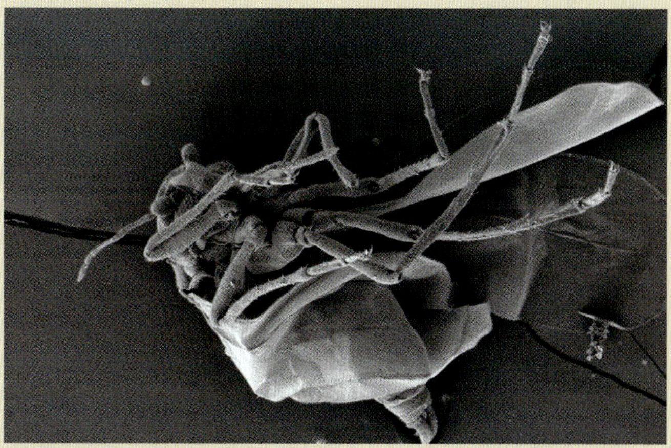

Seen here is a whitefly, captured under a scanning electron microscope. Whiteflies are named after the powdery white coating on their wings, which protects them from the sticky, sugar honeydew they secrete. Because their wings are so waxy, slick, and waterproof, they are also fairly resistant to pesticide sprays.

Scale insects, seen here feeding from a Meyer lemon tree, are herbivores with piercing-sucking mouthparts that withdraw nutrient-rich plant sap. More than eight thousand species of scale herbivorize plants. Juvenile stages of the insects are mobile and colonizing. Adult females settle down, become immobile, and begin growing armor. This brown, thick, and protective shield or scale of wax gives this insect its name. The insects are feeding so heavily on this plant that the proper development of the Meyer lemon fruit is severely impacted.

deprives plants of water, energy, and nutrients needed for growth. Piercing-sucking feeders can additionally carry and vector plant diseases such as phytoplasmas, viruses, and bacteria, which travel around in the bug's guts and saliva. The defecations of piercing-sucking feeders, called honeydew, are sticky and rich in excess sugars. As pests deposit honeydew on plant leaves, further damaging microorganisms such as sooty molds and mildews can colonize plant surfaces. You might have experienced honeydew indirectly if you've ever parked a vehicle underneath a tree for a while. Thousands of tiny, daily feedings and depositions from aphids and other piercing-sucking insects can easily generate a dark, sticky, and, eventually, moldy buildup on items underneath a tree.

Piercing-sucking feeders include aphids, mealybugs, scale, whiteflies, thrips, mites, shield bugs, plant hoppers, psyllids, and cicadas.

Chewing, Cutting, and Scraping

Chewing and cutting insects use a sharp and powerful set of opposing mouthparts called mandibles to feed on plants. Mandibles are heavy-use appendages that damage plant tissue such as leaves, cells, wood, and seeds by chewing and slicing through it. A second set of jaws, the maxillae, chew and crush the food further for processing by the insect's gut. A wide variety of insects feed using a chewing technique, including the larvae of butterflies and moths (otherwise known as caterpillars), sawfly larvae, orthopterans (which include grasshoppers, locusts, crickets, and katydids), beetles and their grubs, flies, phasmids (otherwise known as walking sticks), flies, cockroaches, and termites.

Weevils deserve a special mention here, as they exact a particularly fascinating type of chewing damage. Weevils are known for their elongated, snout-like rostrums. The tip of a weevil rostrum has cutting mandibles just like many other insects, but the appendage's elongated shape allows it to operate just like a drill. Using their natural drills, weevils can bore into plant tissues very easily. Weevils drill to create crevices for egg laying, an act that shelters and protects their tiny offspring. Drilling also gives the larvae a jump start by bypassing several layers of plant surface and physical defenses such as the epidermis, trichomes, bark, and seed coats. Weevil eggs hatch into insatiable little grubs. While many weevils feed on plant

Weevils are some of the most common seed predators. Adults use long, drill-like mouthparts to bore into hard-shelled seeds such as acorns. They lay eggs inside the tiny holes. Seen here are acorn weevil grubs, the larval stage of weevils. Once their parents have bypassed the seed's coat, the deposited larvae have free rein to completely devour the plant's seeds.

foliage and stem tissues, what they're best known for is granivory (also called seed predation), the targeted herbivory and destruction of a plant's seeds. Granivory damages a plant's chances for reproduction. Plants with hard seed coats, such as oaks, grains, cotton, palms, and nuts, are hardly a match for weevils and their miniature drills. Once through the seed coat, a weevil larva grows and consumes the majority of the seed embryo and endosperm. Granaries, pantries, and seed stores have long been plagued by weevil species such as the granary weevil, rice weevil, and bean weevil. As we'll learn, however, many plants can mount appropriate counterattacks should a granivore bypass the physical defenses of a seed. The material within the seed, for example, might be lying in wait like a minefield, full of toxins, soaps, tannins, and other chemical defenses.

Vertebrate herbivores, predominantly mammals, also exert chewing damage upon plant parts. Grazers predominantly feed at ground level, clipping vegetation almost to the soil. They target herbivorous vegetation and its leaves and stems. Browsers tend to focus on more woody plant life and glean leaves, bark buds, twigs, and green stems. The specialized types of teeth mammalian herbivores have help them chew, grind, and cut the resilient foliage of plants. Herbivore teeth furthermore grow continuously or replace themselves during the animal's lifetime, keeping pace with the intense wear and abrasion induced by plant defenses. The front teeth, called incisors, have narrow, linear slicing surfaces to help cut through plant tissue. Rodents have chisel-shaped incisors, which expertly cut through plant tissue and seed coats. For this reason, rodents rank among plants' top seed predators right along with weevils and beetles. Herbivorous mammals' wide back teeth, called molars, act as flattened grinding areas for plant tissue. Large grazing mammals such as cattle, horses, and bison have thick, hardened, and well-adapted molars to suit lifestyles of chewing. A group of grazing mammals called ruminants even have specialized stomach chambers called rumens. Rumens use microbial fermentation to break down plant food and efficiently extract the nutrients within.

Mollusks create damage primarily by scraping. These animals eat plants with a sharp, scraping, tongue-like mouthpart called a radula. Radulae are chitinous ribbons lined with thousands of minuscule, hardened, tooth-like projections that cut into vegetal matter. The jaw of a mollusk slices off large pieces of food, while the radula processes smaller pieces. Snails and slugs are examples of animals that feed on plant tissue using a scraping technique.

Leaf Mining and Boring

The larvae of some insects subsist within and eat the leaf or fruit tissue of plants. Leaf-mining insects are a diverse and disparate group that includes species of moths, sawflies, flies, and beetles. Leaf miner larvae live, eat, and pupate inside the softer tissues of a plant's leaf, often avoiding cutting through cellulose-rich veins or the upper and lower epidermal layers. Cellulose is an

Leaf miner larvae eat distinctive squiggly tunnels of damage inside leaves. By feasting underneath this muscadine leaf's epidermis, baby leaf miners evade many common predators as well as plant surface defenses.

insoluble compound and the most abundant organic polymer on Earth. It is used as a structural material in the cell walls of plants and is a major component of vascular tissue, wood, bark, cotton fiber, and seedcoats. We use cellulose to make paper and a wide variety of other materials. What results from this feeding pattern is a protective tunnel that shelters the developing larvae from predators, plant surface defenses (such as trichomes and tough epidermal cells), and chemical defenses found in certain tissue layers of the leaf. As it eats its way through the plant, each species of leaf miner leaves a distinctive tunnel pattern—to us, these just look like unusual scribbles drawn all over a leaf. Fascinatingly, however, what we are seeing are the early parts of a unique herbivore's life cycle. After leaf-miner larvae finish feeding, they pupate and cut their way free from the leaf, hatching into winged adults ready to create the next generation.

Boring and Tunneling

Boring and tunneling occur when herbivores eat or tunnel their way through the wood and cambium of the trunk, branches, and roots. Some of the most prolific boring and tunneling insects are termites, carpenter ants, and some moth caterpillars and beetles, including bark beetles, wood-boring beetles, longhorn beetles, ambrosia beetles, and powderpost beetles.

Termites, a type of tunneling insect, are amazingly successful eusocial insects. Eusociality is the highest form of social organization found in the animal kingdom and occurs predominantly in insects. Many eusocial insects, such as honeybees, operate for the collective good, forming what is known as a superorganism, a group of related animals working for their communal benefit. Termites live in large colonies where brood care is cooperative, generations overlap, and labor is divided into groups called castes that provide specialized reproductive or nonreproductive roles. The diet of a termite colony depends on the species and mostly involves feeding on dead plant material and the cellulose and lignin within it. Because of mutualistic digestive microorganisms in their guts, termites are some of the only creatures capable of digesting the cellulose and lignin that plants make in their wood and bark. Saw-like, serrated mandibles help termites cut into this tough tissue. Specific sources of this plant matter include wood, bark, leaf litter, soil, and the dung of herbivores. Some species of termites also cultivate large crops of fungus to help with the breakdown of this material. Termites use various living arrangements and techniques to accomplish this feeding, ranging from aboveground structures to subterranean constructions. However, the end goal of these decomposers remains the same: to consume cellulose and lignin, difficult-to-digest substances that very few organisms can successfully feed upon.

The life cycle of boring beetles begins as eggs laid by the winged adults on dead or living trees. After hatching, the massive-jawed larvae bore into the tree's tissues. Borer jaws are sharp, durable, and powerful enough to cut into even the tough heartwood of a tree. Borers have a fascinating ecology all their own. Some behaviors include carving special pupation chambers, engaging in varying degrees of presociality (which means living in colonies), collecting fungal spores, and farming colonies of fungi for food. From a plant's perspective, some of the most damaging boring activity targets the cambium—the living layer of cells just underneath the bark. The cambium is precious because it is a source of much of the plant's growth. Here, new cells are generated and begin to differentiate. If too much feeding damage occurs or too many borer tunnels intersect, they might girdle the cambium. Girdling, also called ring-barking, results when cells of the cambium and vascular tissues are severed or removed around the entire circumference of a plant's stem, branch, or trunk. Girdling is one of the most severe threats a plant can face because it can partially or completely sever the shoot and root system and prevent it from fulfilling the plant's needs. If water cannot reach the shoot system or food and nutrients cannot be sent to the roots, then the plant's tissues will begin to weaken and die over time. Insects aren't the only creatures to cause girdling. Humans sometimes inadvertently girdle trees by carving messages into the bark. We have also historically intentionally girdled trees as a form of land clearing, cutting a circle through the bark and cambium layer all the way around the tree. It can take anywhere from one to three years for a girdled tree to exhaust its energy reserves and finally die. Deer rubbing their antlers on a tree can also have the same effect. Rabbits, ground-dwelling rodents, and beavers can also scar or completely girdle woody vegetation by feeding on the cambium and bark as well as constructing nests and dams.

Galling

Galling is an unusual type of attack on plants, but only rarely is it substantially harmful or detrimental. Galls are strange and abnormal growths produced on the leaves, twigs, roots, or flowers of plants. They are caused by the feeding and egg laying of a wide variety of insects, such as gall wasps, gall midges, gall flies, aphids, scale insects, psyllids, thrips, weevils, gall moths, and leaf miners, as well as noninsects such as mites, nematodes, and microorganisms. Gall-forming animals stimulate plant cells to form galls by inducing irritation or other stimulation, sometimes even manipulating the plant's own hormones. This is similar to the reason oysters produce pearls after irritation by sand grains or other debris. Gall growth can take on a variety of unusual shapes and colors, each a diagnostic trait of the particular attacker that caused the gall's formation. A plant part with a gall might appear to have a tumor, wart, lump, ball, or knob tacked onto it.

Galls that form from egg laying typically manipulate the plant into protecting the developing pest's offspring. For example, gall wasps chemically manipulate plant hormones, causing the plant cells around the wasp's egg to swell and develop thick, tough, and protective walls. The product of this growth is a characteristic gall at the center of which is a minute gall wasp larva.

Galls may also occur from feeding. This phenomenon is often stimulated by the attacker in an attempt to increase the safety, nutrient richness, or efficiency of its feeding site. Root-knot nematodes, for example, are parasitic worms that feed on plant roots and create galls in the process. The saliva of root-knot nematodes contains proteins that manipulate root cells near the feeding site to do three things: begin growing rapidly, dissolve their own cell walls, and massively enrich the feeding site with nutrients and other food. This causes the formation of a gall-like giant cell—essentially an enormous single swelling from which the nematodes can easily steal water and nutrients from the roots. Root-knot nematodes get their name from the fact that these feeding galls make root systems look like they've been tied in a bunch of knots.

Parasites

The topic of nematodes brings us to another type of attack on plants: parasitism. Not all nematodes are plant parasites, but some are devastating parasites that affect plants. Plant-parasitic nematodes (PPNs) are microscopically small, soil-dwelling roundworms that feed on and damage plant roots. The crop damage and losses that PPNs collectively exact across the world are valued at more than $125 billion annually. Root-knot, false root-knot, lesion, sting, needle, lance, foliar, wood, seed gall, spiral, ring, reniform, cyst, and burrowing nematodes are just some of the nasty PPNs out there in the world. Nematodes are so widespread, abundant, and diverse that there are even predatory species that kill and eat other nematodes. Gardeners sometimes employ the valuable services of these beneficial nematodes to control their parasitic brethren. The best option in the fight against PPNs in agriculture is the breeding and selection of crop varieties that bear genetic resistances to nematode attacks.

The life cycles of nematodes are as fascinating as they are complex. PPNs have evolved a specialized feeding device called a stylet, essentially a sharp, hypodermic needle–like mouthpart that pierces into and extracts fluid from plant cells. All nematodes begin from eggs, laid by their parents in a variety of ways that facilitate infection of the host. It doesn't take

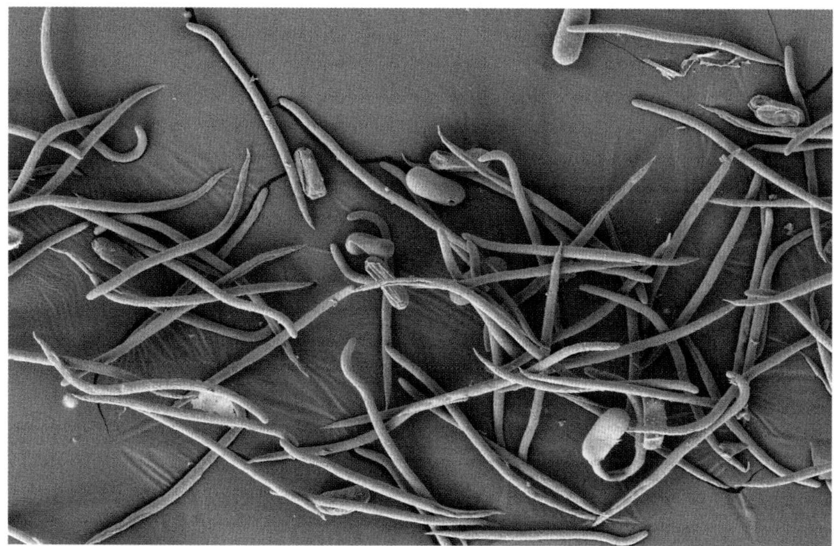

A swarm of soybean cyst nematode juveniles and eggs are seen here under scanning electron microscopy. Nematodes are microscopic, soil-dwelling worms that parasitize plant roots. The worm-like juveniles in the photo are motile and find a feeding site on soybean roots, piercing them with a sharp, needle-like mouthpart called a stylet. Male nematodes will remain worm-like and mobile in order to find females. Once they've bitten into the plant, female nematodes permanently attach to their underground feeding site for the rest of their adult life cycle. The females grow and swell so rapidly that they burst outside of the root at maturity. The cyst that gives cyst nematodes their namesake is actually a hardened sac of eggs formed from the female's own body when she dies. Each cyst contains hundreds of eggs, which the female carefully laid and protected inside herself.

much effort for nematodes to spread. Their eggs and various life stages may be carried or may swim through waterlogged soils or flowing or splashing water. Agricultural equipment, soil shipments, harvested crop roots, tire treads, or even muddy boots are also potential sources of inoculum. PPN eggs hatch into juvenile nematodes, whose job is to migrate into and initiate feeding on plant roots. To this end, juveniles are motile and worm-like, spreading just as easily and through the same mechanisms as the eggs. PPNs are remarkably good at detecting and moving toward potential hosts, identifying a plant's species and other traits by the chemicals exuded by roots. The host preference of PPNs varies considerably, from highly specialized parasites (such as the soybean cyst nematode) to generalist attackers with a wide host range (such as the root-knot nematode). For optimal survival and orientation, even underground, nematodes can additionally sense temperature, humidity, electricity, a plant's water stress, and mechanical forces.

Once juvenile nematodes find their host, the next step is invasion. Anchoring themselves to nearby soil molecules, root nematode juveniles puncture root cells with their stylets and initiate feeding. Some species feed outside the roots, while others invade and migrate through the root tissue as they feed. Aerial nematodes instead prefer feeding sites at the bulbs, stems, and roots, so they specialize in finding these tissues. Nematode feeding can damage and kill plant cells outright, but some species instead choose to inject effectors in their saliva and keep the host cells alive. Effectors are chemicals that trick plant cells in order to stimulate the growth, enrichment, and enlargement of the feeding site. Nematodes such as cyst nematodes and root-knot nematodes use salivary effectors to create food-rich feeding environments. The effectors manipulate the plant to dissolve its own cell walls, thereby growing the feeding site, as well as to deliver extra starch, sugar, and amino acids into the feeding cells.

Nematode feeding is quite damaging not only to the root system of the plant but also to the host's ability to absorb water and nutrients from the soil. As nematode parasitism increases, a plant's roots might be stunted, distorted, and enlarged with galls. The constant nicks, tunnels, and wounding from the nematodes also introduce pathways for infection by pathogens and other attackers. Altogether, a PPN-infected plant might lose vigor, become stunted and weak, develop an abnormal color or appearance, be more prone to water and nutrient stress, or die.

Microorganisms

Some of the smallest organisms can be plants' biggest threats. Infectious microorganisms such as fungi, bacteria, viruses, phytoplasmas, and water molds can cause diseases in plants. The scientific study of these plant pathogens is called plant pathology. The discipline also entails pathogen identification, causation, epidemiology, plant disease resistance, genetics,

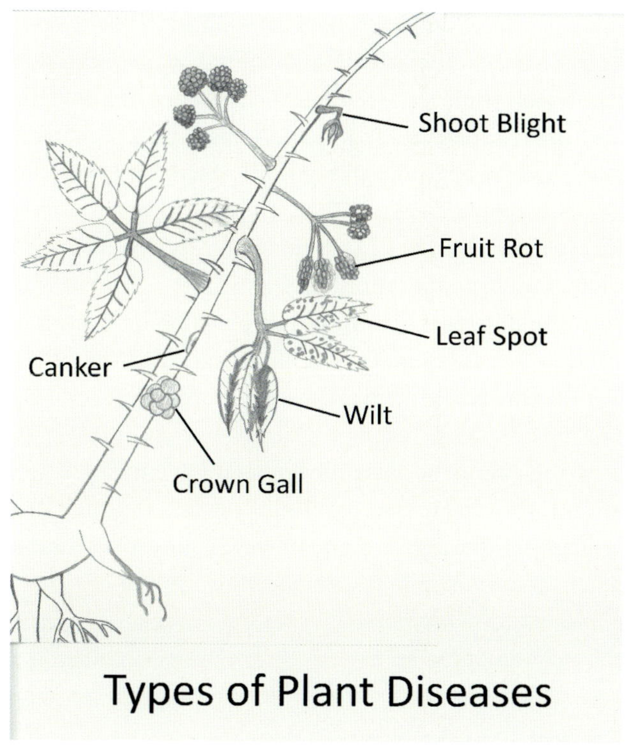

This diagram shows different types of plant diseases affecting various parts of a blackberry plant (genus *Rubus*).

economics, and management. Let's step into the complex hidden world of plant diseases.

Bacteria

Bacteria are a large and widespread group of unicellular microorganisms. They were among the first types of life to evolve on Earth and are omnipresent in almost all conceivable habitats. Bacteria are just micrometers long and shaped like spheres, rods, and spirals. Their ability to multiply quickly and proficiently is legendary. As a result of their small size, bacteria can pass through a variety of openings in plants such as the stomata, lenticels, intercellular spaces, cell pores, wounds, and roots. *Rhizobium* and *Frankia* bacteria are beneficial, having evolved a symbiotic relationship with certain plants' root systems.

Pathogenic bacteria, on the other hand, parasitize and kill plant cells. They manufacture enzymes that degrade plants' cell walls, turning them into the mushy, smelly, rotten slime that signifies the presence of a rot. The digestive enzymes and other molecules that bacteria pass into plant cells eventually become too much to bear, and the cells die. Bacterial infections often lead to necrosis, causing plant cells to turn dry and brown and then die. Depending on their pattern of spread and mode of infection, bacterial infections can cause an array of symptoms in plants. Rots create a foul slime that results from the digestive dissolution of plant tissue. Bacterial spots are ringed necrotic areas surrounding the infection site. Lesions are large necrotic regions on leaves. Cankers are necrotic areas on stem tissue. Wilts occur when an infection severs the plant's water supply from the roots. Typically, wilting occurs because the bacteria have invaded and choked off the xylem in the vascular system. Blights occur as sudden yellowing, browning, spotting, withering, or dying of leaves, flowers, or stems.

Notable bacterial invaders of plants include the genera *Erwinia*, *Ralstonea*, *Xanthomonas*, and *Pseudomonas*.

Fungi

The fungal kingdom is all around us. The soil beneath our feet, the air we breathe, and the surfaces and substances in our world all show signs of microscopic and omnipresent fungal life. Fungi are some of plants' most helpful allies, teaming up in symbiotic relationships, decomposing organic matter, and improving soil structure. Beneficial fungal species help plants grow. They might assist with the absorption of water from the soil, shield roots from diseases, transfer messages and resources between different plants, or render soil nutrients more available to absorption by plants. In return, plants send these fungi resources they can't manufacture on their own, such as carbon in the form of sugar. A different suite of fungi, the pathogenic fungi, are among the deadliest and quickest-spreading diseases plants can face. Plant pathogenic fungi are widespread and diverse but have several general similarities. The main structure of a fungus is called a mycelium. The mycelium is a collective, branching mat of tiny, filamentous, and root-like hyphae. The cell walls of fungi, like the scales of fish and the exoskeletons of arthropods (the group that includes insects, arachnids, and even crustaceans), are made of chitin, a tough, pliable, fibrous polysaccharide that serves a structural and protective function in these organisms. Fungi do not photosynthesize or utilize sunlight for energy (as do autotrophs such as plants). Instead, fungi are heterotrophs, meaning they acquire their carbon through the consumption of plants and animals. Fungi accomplish this process via absorption, during which they excrete enzymes into the environment to break down food. Plant pathogenic fungi glean the food they need by infecting and colonizing the living parts of plants and then stealing and digesting them. Digestive enzymes and other attack strategies assist pathogenic fungi in their efforts. Fungi are immobile organisms but are capable of growing toward different sources of food. Their airborne or waterborne spores also help them spread and reproduce.

Fungi are excellent survivors. They reproduce with a sexually produced structure called a spore. To house and disperse their spores, fungi have a wide variety of spore-bearing structures. You're familiar with these

Black rot, brown rot, and bitter rot are fungal diseases that infect pears and other fruits in the rose family such as apples and peaches. The diseases begin when the fungi's airborne spores infect the leaves, stems, fruit, or flowers of the trees, moving steadily into the developing fruit. The reproductive chances of the plant—its seeds—are decimated. Each disease creates dark, shriveled, fuzzy "mummy" fruits, seen here. Fruit mummies are loaded with millions of spores, which initiate further infections. Mummy fruits falling on the ground stimulate further reproduction of the fungi and even more spore production. You might say that infected plants will always bear the curse of the mummy!

structures if you've ever seen food get moldy and fuzzy when it goes bad. Mushrooms, a ubiquitous sign of fungi, are one of the most well known spore-bearing structures. The mushroom is often just the tip of the iceberg for a fungus. As the reproductive fruiting bodies of the organism, mushrooms produce and allow spores to be dispersed in a variety of ways. Fungi also invest in backup plans for their own survival, often in the form of asexual resting or dormancy structures. These structures might have thick walls or other protections that enable the fungus to survive long periods of harsh or otherwise unfavorable conditions. Fungi have specialized, just as plants have, to persist despite drought, cold, heat, tilling, temporary shortages of hosts, and almost any imaginable circumstance. Fungal spores and propagules can spread over long distances by the movement of air or water, by animal movement, by splashing, or through the soil. Some spores have even been adapted to specifically travel through plant cells.

For fungi that invade through a plant's foliage, such as rusts, the onslaught begins when a spore lands on a leaf. The spore is signaled to germinate by the presence of water or the tissues of its host plant. For example, the very protective waxes from a plant's cuticle can induce spore germination of some pathogenic fungi. A germ tube, which grows from the germinating spore, eventually differentiates into the root-like hyphae of the fungus. From the moment of germination forward, spores and germ tubes secrete a gluey mucilage that adheres them to the leaf surface. The hyphae and spores of fungi are much tinier than mature fungal cells and as such are expertly capable of fitting through small openings in the plant such as cell pits, vascular system filters, intercellular spaces, and other minute nooks and crannies.

Fungi have further tools beyond their hyphae and spores. For example, fungi that specialize in infecting the shoot system, such as rusts and blasts, use a specialized invasion cell called an appressorium. A tiny, needle-like infection peg grows from the appressorium and punctures its way into the plant's cells using brute force. Appressoria and infection pegs invade with such high pressures that they have even been able to punch through sheets of plastic.

Fungi infect and affect plants in numerous ways. Some generalized attack strategies and symptoms are highlighted here. Rusts are generally colonizers of leaves, stems, and inflorescences. The rust-colored, dust-like spores that

What you're seeing here is a massive infection of a daylily leaf by a fungal pathogen called daylily rust (*Puccinia hemerocallidis*). Rust fungi specialize in infecting plant shoot systems, primarily leaves and stems. Their bright orange, dust-like "summer spores" are a type of spore spread via the air for long distances in the favorable wet weather of summertime. A second spore type, a tougher "resting" spore, is black in color and produced in the fall to ensure winter survival. To puncture through and infect the tough leaf epidermal cells, rusts use a pressurized, bolt gun–like organ called an appressorium and infection peg.

impart their name can drift for long distances. Mildews cause powdery-white spots to form on leaf and stem surfaces. Rots use enzymes to digest plant tissue and cause dark lesions or completely dissolve roots and stem tissue, turning them into slimy masses. Cankers often occur on woody plants, manifesting as dead areas on stem tissue that appear sunken, distended, split, or discolored. Spots are fungi that attack leaves, typically causing necrotic or discolored spots to appear. Wilts attack the plant's vascular system, causing damage that affects the plant's ability to conduct water and nutrients. For this reason, the wilts cause plant leaves to lose turgor pressure and wilt. Plants face a relentless assault from all of these trajectories. Only one point of failure is required for a fungal invasion to take hold.

Water Molds

The water molds (known scientifically as oomycetes) are fungus-like microorganisms that often manifest in plants as root rots and other belowground diseases. Water molds have specialized to infect plants in wet soils, floods, and other stagnant and moist conditions. Water molds have swimming spores called zoospores that are propelled by filamentous appendages called flagella.

Plant pathogenic water molds can cause devastating diseases such as seedling blights, damping-off, root rots, foliar blights, and downy mildews. Downy mildews, *Pythium*, and *Phytophthora* are some of the most aggressive and destructive water molds for plants.

Viruses

Viruses also infect plant cells. Since viruses are unable to replicate without a host, they use plant cells for this purpose. Viruses are spread by passing sap from one plant to another. Any animal or activity that creates a wound in multiple plants risks transmitting viral particles. These

This tomato plant is infected with a plant virus called tomato yellow leaf curl. It is passed between tomato plants by the bites of piercing-sucking feeders called whiteflies. Note the pigmented cellular "inflammation" surrounding the infection sites across the leaves. Plant pigments are signs of stress and possess powerful disease-fighting abilities.

infection pathways are called vectors and can include hands, cutting tools, and equipment such as shears, tillers, and mowers. Insects and nematodes are also excellent viral vectors. Once a virus has gained entry through a wound, it replicates and attempts to invade the rest of the plant's cells. In response to an infection, plants actively try and halt the virus's movement by shutting down pathways such as the plasmodesmata, or transfer pores between cells. Viruses usually weaken plants rather than causing death, maintaining the host in a faint, distorted, and drained state of living. Plant viruses can induce mottling, distortion, stunting, discoloration, and dwarfing in a plant. The stress makes the host prone to dying of secondary infections by other microorganisms, pest damage, or environmental pressures. Twenty percent of plant viruses are also transmissible in the host's seeds. Without specialized biotechnology such as tissue culture, which can occasionally generate sterile meristem cells from infected plants, plant viruses are untreatable and cause lifelong infection.

Competition

If it weren't enough that plants constantly struggle with herbivores and pathogens, they also have to compete against other plants. Climate, geology, soil properties, hydrology, and the availability of light are all factors that pit plants against one another. In deserts, roots compete for mere drops of precipitation. In rainforests, thickets of leaves and vines clamber over one another toward the light-rich upper canopy. Nutrient-poor soils have driven the evolution of carnivorous plants that are capable of killing insects in order to extract precious nutrients from the decaying bodies. Virtually any environmental variable has induced heavy selection pressures on the ways plants compete with one another. Plant survival systems have evolved because resilience is demanded at all times. The plants here with us today have proven the most successful competitors in trying conditions such as cold, drought, fire, limiting and toxic soils, windstorms, and flooding. The winners stand and the losers fall. To keep standing involves a fascinating hidden world of adaptations and competitive advancements.

Not only is plant competition a struggle to succeed in Earth's demanding conditions, but plants also face challenges from other plants' "secret weapons." One example is allelopathy, when plants produce chemicals that slow or kill the growth of other plants and their offspring. The caffeine in tea and coffee plants slows the root growth of nearby plants. Black walnuts produce juglone, an allelopathic chemical exuded by their roots and seed hulls to interfere with the photosynthesis of nearby competitors. The successful spread of invasive species such as spotted knapweed, nutsedge, and she-oaks occurs partly because of the powerful allelopathic effects they exert upon competing flora.

A second secret weapon is plant parasitism, among the most unique and fascinating of plant adaptations. Earlier, we discussed how plants rely on photosynthesis to produce sugars. However, we mentioned that there are almost always exceptions to every rule. Parasitic plants steal nutrients, water, and sometimes sugar from other plants. Some species, called hemiparasites, make just a small portion of their own food through photosynthesis and steal the remainder. Others, called holoparasites, must glean all of their food from the plants they parasitize. Parasites have specialized roots called haustoria that have been modified to penetrate into host plant tissues for these resources. A species of parasitic plant might specialize in invading the leaves, stems, or roots of its host. The sensory perception of parasitic plants is also superbly enhanced, allowing them to successfully locate and navigate to their hosts. Some parasites can detect the distinct airborne chemicals emitted by their hosts, while others detect the specific blend of chemical exudates from their hosts' roots. Parasitic plants have plant identification skills that rival the best botanists—after all, the faster and more correctly a parasite infects the correct host, the better its chances of survival and reproduction.

Parasitic plants such as the dodder (genus *Cuscuta*) and love vine (genus *Cassytha*) are vines that wrap around, pierce into, and drain their hosts in fields across the world. Mistletoes, on the other hand, are arboreal parasites that germinate from sticky, bird-deposited seeds on branches. Mistletoe haustoria then bore their way into the stems of their hosts and live as epiphytes. Subterranean parasitism is another strategy, employed when the roots of one plant tap into and siphon nearby hosts. Examples of species with vampiric roots include the giant-flowered *Rafflesia*, witchweed, broomrape, Indian paintbrush, American bluehearts, and beechdrops.

Some of the most extreme parasitic plants, such as dodder, so expertly drain sustenance from their hosts that these species completely lack chlorophyll and photosynthetic leaves. Over time, they have lost the ability to make their own food from sunlight, specializing in botanical thievery instead. Their strange, otherworldly pigmentation is thus a sign of their obligately parasitic lifestyle. All of these unusual and sneaky little parasites are still plants, however. In a strange twist of evolution, they evolved from the photosynthetic plants we are more familiar with. In fact, plant parasitism has evolved independently at least thirteen times within the plant kingdom. In each of these lineages of plants, their peculiar and novel way of life is the result of millions of years of evolutionary changes that helped them survive the intense competition with other plants.

Plants Fight Back

Now that we are a little more familiar with plants and the ecological pressures they face, we might begin to wonder how in the world they survive this constant assault. As we relax in our safe, comfortable homes, a world of swarming organisms descends on the plant life outside our windows. Beyond our yard, the plants we depend on for food, spices, fiber, building materials, natural beauty, and chemicals face threats to their existence every day. Thankfully, however, plants have powerful lines of defense that are constantly evolving and improving. Earth's flora is well armed both inside and out with sturdy tissues; sensitive perception and communication abilities; chemicals that are toxic, protective, and repellent; and a dazzling array of physical structures and internal systems to thwart attacks. Over the next chapters, we will begin to open the hidden world of plants. Plants may never seem the same again as the amazing things they are capable of become apparent. All the more fascinating is how their incredible activities take place alongside and within our daily lives.

Part One

A Journey through Plants' Physical Defense Mechanisms

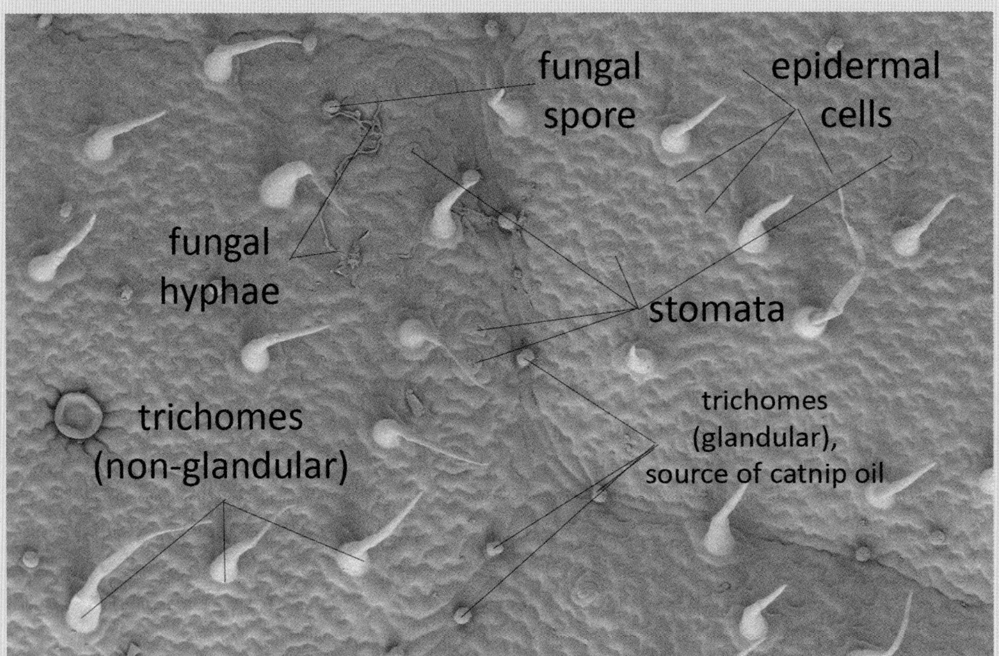

Welcome to the surfaces of plants, among nature's most complex and arduous obstacle courses. A gauntlet of spines (not seen here), trichomes, and jigsaw puzzle–shaped epidermal cells confront all incoming attackers. A fungal spore is here captured in the act of landing on this leaf and germinating. The root-like hyphae it has made will slither their way toward any opening in the leaf, such as the stomata nearby. If the plant's surface can't fight this, a pathogenic invasion has begun.

CHAPTER 3

The Surface of the Plant

A Deadly Obstacle Course

Plants aren't always what they seem at first glance. The closer we get, the more treacherous their hidden world becomes. Plant surfaces are strange, almost otherworldly microscopic landscapes brimming with defenses. Waiting for attackers are unrelenting obstacle courses of hazards: painfully sharp spikes, surveillance systems, scents that mark bugs as targets for predators, Velcro-like traps, poisons, glues, and some of the slickest nonstick surfaces ever invented in all of nature. This chapter invites you on a journey to explore the rarely explored world of survival systems covering the surfaces of plants. Imagine that, together, we'll shrink to microscopic size to visit these systems up close and personal. Like an insect or a fungal spore, we'll close in on our target: a juicy, nutritious leaf. Little do we suspect that our voyage will take us through a formidable gauntlet of botanical defenses. Welcome to the surfaces of plants, some of nature's most intricate and arduous obstacle courses.

Yikes! Plants and Their Spikes

As we make our way to the surface of the plant, some of the first structures we'll pass are spines. To ward off herbivores, many plant surfaces come complete with built-in stabbing devices. If you've ever encountered a cactus, you're well aware of the antisocial behavior of nature's botanical porcupines. Rich in resources, plants have a strong reason to protect themselves.

Plant species around the world have evolved an assortment of piercing structures on their leaves, stems, buds, or roots. Each of these armaments is tough, rigid, and hardened. Cells at the tips and edges of spines are

Rattan palms such as this *Calamus roti* are protected with fearsome prickles. To ward off herbivores, many plant surfaces come complete with built-in stabbing devices. These needle-pointed structures double as grappling hooks, gripping onto taller plants as the vining rattans climb up to the rainforest canopy.

strengthened with extra deposits of lignin, silica, and calcium. Spikes also usually lack standard components such as chlorophyll and vascular tissue.

Let's run through some of the types of plant spikes. The spine is modified from leaf tissues such as the leaves themselves (as in the cactus family), leaf stems (called petioles, present in the ocotillo plant), and stipules (present in acacias and the euphorbia family). Black locust, honeylocust, barberry, and many palms are also spiny. Thorns are modified branches or stems. In eleagnus, bougainvillea, *Citrus* trees, and Osage orange (a noncitrus member of the fig family), thorns are quite simple and unbranched. Branched thorns can be found on the Callery pear, firethorn, or hawthorn.

Prickles are formed from the tissues of the cortex, bark, or epidermis. Prickled plants include devil's walking stick (*Aralia spinosa*), prickly ash (*Zanthoxylum americanum*), roses, brambles, horse nettle, Oregon grape, hollies, cycads, naranjilla (*Solanum quitoense*), and thistles. A spinose apical process is the botanical term for a spiny extension of the leaf tip, found readily on yuccas, agaves, and the snake plant (genus *Sansevieria*). Spinose teeth, often found in monocots, turn the edges of plant leaves into sawblades. They're the reason you can get some nasty cuts from grasses such as sawgrass (genus *Cladium*) and pampas grass (*Cortaderia selloana*), screw pines (genus *Pandanus*), and bromeliads (family Bromeliaceae).

A plant that bears spines has developed a strange and painful way to communicate its intention to survive with you. These vegetable urchins are sending out the message that they do not want to be eaten. They want to make it crystal clear to hungry animals that these plants are not on the menu. In fact, a plant that has survived previous damage is induced to ramp up its production of spine quantities, densities, and toughness. Surrounded in a spiny phalanx, plants are fully prepared to defend their leaves, stems, buds, seeds, and roots. Spines are engineered to cause maximal damage and pain to herbivores. Stout and needle-shaped, spines easily pierce the tissues of the skin, muscles, face, mouthparts, and gastrointestinal system. Spines impose a severe threat to herbivores' ability to eat and digest, as well as to their senses and mobility. The blinding pain response of the animal nervous system is a rapid indication to avoid plant defenses like these. The majority of plants also color-code their spiny defenses, visually distinguishing them as dangerous by utilizing conspicuous pigmentation.

If we were to approach spines on a microscopic scale, we'd see a hidden world of adaptations that make them much more sinister than we could ever imagine. Let's take the tiny, highly evolved spines called glochids. Glochids are specialized spines that stud the surfaces of prickly pear cacti (genus *Opuntia*) in small, pincushion-like pads. These irritating, bristly spines operate like harpoons because they are covered with microscopic barbs. Once they've punctured your skin cells, glochids are there to stay. The barbs resist being pulled out, and, in fact, their shape helps to drive the spine unidirectionally deeper. A spiral arrangement to the barbs allows the glochid to operate like a drill inside your flesh. As you move and

Grass leaves are literally called blades because of their spinose teeth, seen here on a pampas grass (*Cortaderia selloana*) plant with the help of electron microscopy. Spinose teeth slash, abrade, and shred the mouthparts and digestive systems of herbivores. They're also responsible for the tiny "grass cuts" grasses leave in your skin.

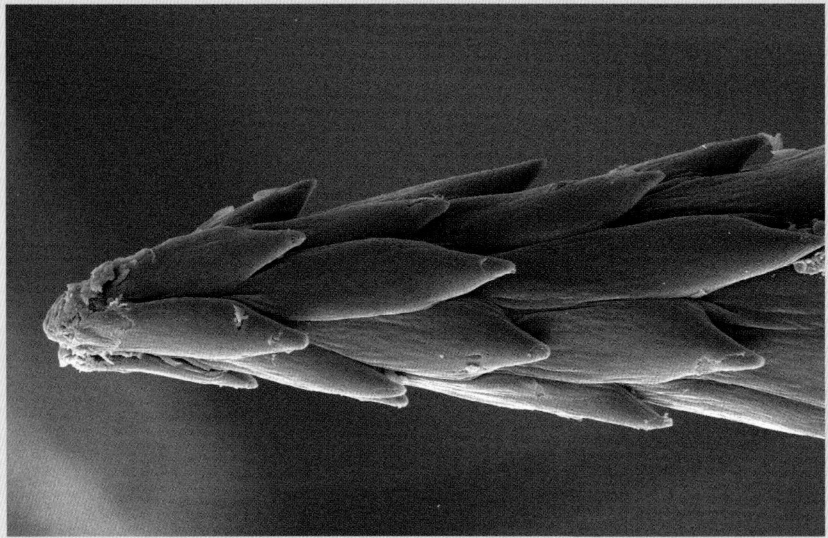

If we were to approach spines on a microscopic scale, we'd see a hidden world of adaptations that make them much more sinister than we could ever imagine. Seen here are glochids, the tiny and easily dislodged harpoon-like spines on prickly pear cacti. Their tiny barbs ensure they remain in your skin, while their drill-like shape drives them farther into your tissues with every movement. It'd be a pretty rough day should these little demons wind up in an herbivore's mouth and throat.

flex your skin, these small movements, no matter how minor, power the glochid's drilling process. Glochids are so fragilely attached to the main plant that these little spines stay inside you as a parting gift. They help animals like you remember your cactus encounter for a long time. Woe unto any animal hapless enough to admit the glochids entry into the mouth and throat!

Spines and thorn fragments that break off in the skin and tissues of herbivores are excellent inoculation devices for diseases and infections. Plant-vectored infections can be painful and irritating, but some of the pathogens inserted can even be lethal. In a phenomenon called bioprotection, plant spines have been found to harbor colonies of pathogenic microbes and facilitate their insertion into herbivores. Plant surfaces are among the most teeming terrestrial habitats for microbes. Many of these microbes are harmless, but sometimes an animal gets a mouthful of truly nefarious pathogens. Hitching a ride on plant spines allows all microbes—both pathogenic and not—an effective entry method into animals' bloodstreams and other ideal sites of infection. By causing harmful diseases and infections to herbivores, pathogenic microbes bioprotect the plants that host them.

In 2006 plant scientists sifted through numerous medical case reports involving injuries from plant spines and thorns. The findings were fearsome, showing a pattern of inflammations, deep and persistent infections, and diseases that could be traced back to contact with spiny plants. Numerous tetanus infections from the bacterium *Clostridium tetani* were traced back to gardening activities, frequently tied to run-ins with the thorns on rose bushes. The deadly gas gangrene bacterium (*C. perfringens*), whose toxins generate gas inside the body, was found living and infecting victims from the thorns of hawthorns and the spines of the date palm (*Phoenix dactylifera*). Date farmers around the world began laboriously slicing the spines off their date palms to avoid the deluge of workers being sent to hospitals for gangrene infections. Anthrax (*Bacillus anthracis*) was also found on these species' spines. Other spine-dwelling pathogens capable of causing severe infections include *Staphylococcus aureus*, *S. epidermidis*, *Streptococcus hemolyticus*, *Nocardia pyarthrosis*, *Pantoea agglomerans*, *Enterococcus faecalis*, *E. faecium*, and *Shigella*. Since these surfaces are scientifically understudied areas, it's quite possible that each spiny plant hosts its own custom blend of pathogenic microbes. Plant surfaces are thriving, evolving bacterial ecosystems, and the result is the hidden world of bioprotection. Be mindful of plants and their microbes while you're gardening. Let's now remember that every rose has not only its thorns (technically, prickles) but also its tetanus!

The function of sinister puncturing devices isn't always to fight herbivory, however. Spines are spectacular temperature and water regulation devices as well. Spines on desert plants are often specialized to not only deflect heat away with their light and reflective coloration but also operate like heatsinks to channel heat away from the plant's stem. A thick coating of white spines is both sunscreen and air conditioning for the plant's surface. Drying winds are also stifled by thick spines, minimizing water loss from the stomata. Deserts are also prone to extreme fluctuations in night temperatures, so fluffy spines also insulate a plant from cold and frost. Spray a spiny plant with a few mists from a spray bottle, and you'll also reveal another hidden purpose: water collection. Spines are superb condensation surface areas for the collection of fog and light precipitation. Some prime examples of spiny plants that feature advanced climate-control systems include the feather cactus (*Mammillaria plumosa*), old man of the Andes (*Oreocereus celsianus*), and silver torch (*Cleistocactus strausii*).

Thorns and prickles (modified stem tissue and epidermal tissue, respectively) do jobs similar to those we have seen with spines. The location and the cellular origin of these devices just happen to differ. Some examples of thorns are present on citrus trees, hawthorns, Osage oranges, honeylocusts, and bougainvilleas. Prickles are found on plants such as hollies, devil's walking stick, roses, and blackberries.

Provided an animal makes it past a plant's spikes, a fortress of other perilous surface defenses await . . .

Treacherous Trichomes

If we were shrunk down, landing on a plant's surface would feel like crash-landing on another planet. The sentinels of this hidden alien world are strange, microscopic projections that dangle over the plant's surfaces. Stalked and tentacular, these are the trichomes, a diverse array of specialized leaf hairs that originate from the epidermis. If you've ever felt a

Stalked and tentacular, the sentinels of plants' surface defense systems are trichomes, a diverse array of specialized leaf hairs. Seen here are two different types of glandular trichomes on the rachis (leaf stem) of a sword fern (*Macrothelypteris torresiana*). Glandular trichomes secrete all manner of antiherbivory chemicals such as toxins, glues, repellents, essential oils, and VOCs. Any contact with a plant must contend with its diminutive but expansive tangle of trichomes.

plant leaf that seems rough, fuzzy, hairy, or gloopy, you've been brushing and mashing hundreds of tiny trichomes with your fingers. To have any chance of eating, moving around, or reproducing on or within a plant, attackers must successfully contend with the dreaded trichome fields, which are among plants' most deadly defenses. Importantly, some trichomes also double as protective shields, blocking severe abiotic threats such as extreme temperatures and moisture loss.

The vast majority of plant species everywhere on Earth have evolved their own unique assemblage of different trichome shapes, sizes, and functions.

Over three hundred different types of trichomes have been discovered by plant scientists. Each structure is classified by how many cells it is comprised of as well as whether it has a predominantly mechanical or secretory function. Trichomes can be unicellular or multicellular. They can secrete chemicals (glandular), or they can structurally protect the plant from feeding, egg laying, or environmental extremes (nonglandular). Glandular trichomes can secrete all sorts of chemicals, such as oils, toxins, glues, and volatile organic compounds (VOCs). VOCs are naturally occurring chemicals made by living things that vaporize at room temperature. They drift easily through the air and serve a wide variety of purposes for plants. Trichomes are seemingly simple structures with a multitude of functions.

Trichome fields are plants' active, living surveillance systems. Trichomes are touch-sensitive trip wires, jutting and branching in all directions. Insects walking across a plant's surface are liable to brush against or break

A plant's trichome fields are active, living surveillance systems. Trichomes are touch-sensitive trip wires, jutting and branching in all directions. Landing and moving around on plants this hairy are tricky tasks. Seen here is a filiform nonglandular trichome thicket guarding the stem of a soybean plant.

Filiform or hair-shaped nonglandular trichomes coat this fig leaf's succulent veins. Fig leaves feel fuzzy because of trichomes like these. They are harmless to us, but their breakage allows plants to sense and prepare defenses against pests. Dense trichomes also slow down the feeding, growth, and reproduction of pests. Subtly, this exposes them to the elements and predators longer as well.

dozens of trichomes as they walk or feed. The plant learns the attacker's physical location as well as its approximate size. The breakage of these trip wires begins processes within the plant that induce the rapid production of defenses such as stronger cell walls, toxins, and greater trichome densities. Common shapes of trichomes that provide mechanical surveillance include filiform trichomes, which look like elongated, tapering hairs; branched trichomes, which look like tiny trees, consisting of single hairs replete with branches in all directions; and stellate trichomes, which are star-shaped, with spreading branches that radiate from one origin. Adding a huge amount of surface area to the plant's exterior, trichomes are bound to make contact with any attackers.

Other trichomes are hooked, barbed, and pointed. We might call them "trap-chomes" because they operate on the plant like herbivore booby traps. Much like many spines, trichomes feature tips and edges hardened with deposits of silica and calcium. These are sinister weapons designed to pin, harpoon, grapple, and impale. Some even inject toxins or deposit harmful bacteria as they pierce an herbivore's flesh. Chewed and broken, the hardened bits of these trichomes can pierce, scrape, jam, or slice into the mouthparts and gastrointestinal systems of herbivores. Even without toxins, pests can be severely wounded or simply pinned by the plant's trichomes until they desiccate, starve, or are discovered by predators. Grappling trichomes are specially designed to hook into the hooks that give insect feet their grip or onto the numerous bristle or hair-like structures on insect bodies (called setae). There are three types of grappling trichomes: hooked, malpighian, and barbed. Hooked (or uncinate) trichomes can be found on garden beans (*Phaseolus*), passionflowers (genus *Passiflora*), and bedstraw plants (*Galium aparine*). Malpighian trichomes are T-shaped, with two broad outstretched arms. Hops (*Humulus lupulus*) are an example of a plant that exploits this type of trichome, which helps this vining plant grip and climb up its competitors. Ferociously barbed trichomes can be found on stickleaf plants (genus *Mentzelia*), hound's-tongue (genus *Cynoglossum*), electric shock plant (*Blumenbachia insignis*), and plants in the genera *Caiophora*, *Nasa*, and *Loasa*. Altogether, grappling trichomes set some incredibly nasty traps for pest insects.

You might think of stinging organisms as predominantly insects such as ants, bees, and wasps. Did you know, however, that some plants have evolved the ability to sting herbivores with their trichomes? The stinging nettle (*Urtica dioica*) is an infamous plant that lives up to its name. The leaves and stems of this species bear sharp pointed glandular trichomes. Brittle, hollow, and tipped with silica, the hairs function exactly like hypodermic needles. As these spear-like trichomes pierce the skin, they break and inject a concoction of incendiary chemicals found in the swollen, saclike base of each trichome. The first components of this mixture are neurotransmitters such as histamine, serotonin, and acetylcholine. This combination creates an intense sensation of pain and signals the immune system into a severe inflammatory response. An accompaniment of formic, oxalic, and tartaric acids intensifies and lengthens the pain symptoms.

If you thought the stinging nettles were wicked, meet their big, bad cousins: the stinging trees from Australia (genus *Dendrocnide*). Stinging trees belong to the nettle family right along with stinging nettles, but the painful punch they pack is far more perfidious. Trichomes protruding from a stinging tree are similarly silica-tipped, hollow, and needle-shaped. Even slight contact injects unique plant-manufactured neurotoxins called gympietides (after the common name for an infamous stinging tree species known as the gympie-gympie). Broken pieces of these hairs and the toxins they contain are easily flung into the air, forming a microscopic venomous dust that can cause considerable respiratory irritation. This sting is one of the most painful plant defense–related experiences an animal can encounter. Pain can last anywhere from several hours to two days—and it might sporadically reoccur for months, especially if the surrounding nerves are touched, wetted, or exposed to heat and cold. Gympietides cause permanent alterations in nerve cells. It takes merely a second to infringe on a stinging tree but a lifetime to atone for it. Stinging trees have even been known to kill larger animals such as horses and dogs whose only fault was to walk too close to these formidable nasties.

Some plant species take a more amiable but still antisocial approach with their trichomes. Barrier trichomes are usually nonglandular trichomes that work by physically blockading and slowing down insect activity. Trichomes involved in this function are so precisely sized that they repel specific attackers. When a phalanx-like row of leaf hairs juts out from a plant surface, they can prevent insect feet from finding a grip. Bugs might be simply unable to land or move along the plant's surface. At the very least, the route the pest takes across the foliage becomes sluggish, inconvenient, and inefficient. This result correlates directly to slower feeding and growth as well as greater exposure to the elements and predation. Trichomes can even cause pests to drop off the plant completely. Flightless insects such as mites, larvae, caterpillars, and aphid nymphs can fall from their host plant to the ground, leaving them vulnerable to starvation, desiccation, and predation.

When an insect tries to feed, its mouthparts can swipe or stab at the plant all they like, but they'll have difficulty reaching juicy plant tissue because of the interference of the trichomes. Many female insects often use an elongated, tube-like appendage called an ovipositor to lay their eggs. If, however, a plant grows trichomes that are intentionally just a little too shaggy, insect ovipositors might not be long enough to reach and glue eggs to the plant's surface. A complex of trichomes may seem like a simple adaptation, but it truly causes insect feeding and reproduction to suffer. You'll commonly find trichome lengths on plants that are just long enough to slightly outpace the feeding strategies and ovipositor lengths of common herbivores such as sawflies, leaf miners, gall wasps, whiteflies, thrips, butterflies, and moths. Plants utilizing this type of strategy might have leaves that feel hairy or fuzzy because of thick trichome growth. The quantity, shape, and density of trichomes a plant makes are a delicate balancing game. If there are not enough trichomes, plants risk allowing herbivores to run rampant over their tissues. If trichomes grow too dense, however, small insects and their eggs might actually be provided cover from predators and sweeping forces such as wind and rain.

Trichomes enable plants to turn their surfaces into deadly weapons other than spears and claws. What if a trichome—no, *thousands* of trichomes—started pumping out poison on plant leaves? These hairs can do more than be a formidable barrier; they can also be laced with chemicals that can be injected or consumed. These glandular trichomes, which lie in wait on leaves like gloopy booby traps, can provide a surprising impact if broken or eaten. These chemical minefields quickly spew glue and poison onto anything nearby. The leaves of the tobacco plant, for example, are festooned with nicotine-bearing glandular trichomes. Each looks like a tiny hair tipped with a juicy glob of poison. Each touch of a tobacco trichome injects a shot of the neurotoxin nicotine, which quickly causes paralysis

The leaves of the tobacco plant (*Nicotiana tabacum*) course with nicotine-bearing glandular trichomes. Each touch of a tobacco trichome injects a shot of the neurotoxin nicotine, which quickly causes paralysis and death. Larger herbivores might need to consume a fair amount of vegetation to become sick from this poison, but tiny insects need only walk past a few millimeters of trichomes before accumulating a fatal dose. This unfortunate sweat bee is actually a pollinator—a beneficial insect rather than an attacker—but now suffers the fate of having gotten too close.

Okra, a plant in the hibiscus family, produces hordes of sticky glandular trichomes on its pods. Okra trichomes protect the plant's seeds from insect herbivores. Studies have found that these dense trichome fields are fantastic at fighting spider mite infestations.

and death. Larger herbivores might need to consume a fair amount of vegetation to become sick from this poison, but tiny insects need only walk past a few millimeters of trichomes before accumulating a fatal dose. Common plants around us such as petunias and potatoes have trichomes that spew glue onto bugs such as aphids. This "plant cement" comes with dissolved, instant-hardening compounds that react with air, drying the cement as fast as super glue. Sundews, a group of carnivorous plants featured throughout this book, have some of the stickiest glandular trichomes on the planet. These mucilaginous wonders employ trichomes that capture, kill, and break down insects in order to survive in nutrient-poor soils.

Other types of glandular trichomes might burst when touched, give herbivores a unique scent when consumed, exude repellent tastes, or secrete specific oils or volatile compounds. Bugs moving across a plant's surface are smeared and painted with conspicuous, strongly scented chemicals by the trichome fields. These chemicals are attractants, all part of plants' silent alarm system. But attractants for what? Plants actually use their trichomes to coat or infuse their attackers with volatile scents that mark them as targets for predation. Numerous predators fly around plants all the time, hunting for these telltale chemical traces to find prey. No matter the level of an herbivore's camouflage or speed of its escape, its predators will more easily find it and consume it, providing efficient pest control.

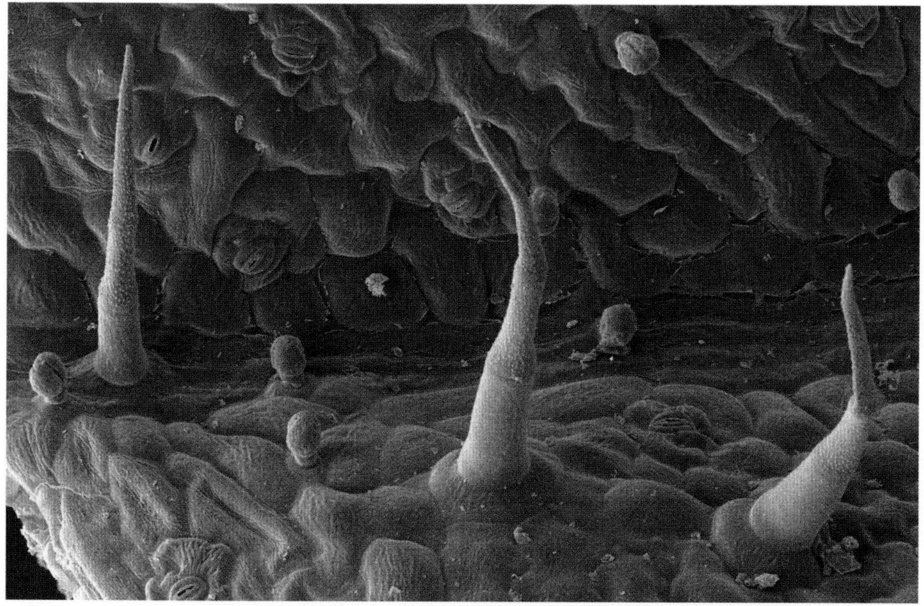

Seen here under a scanning electron microscope is a catnip (*Nepeta cataria*) leaf. Hair-like (filiform) nonglandular trichomes are what render catnip leaves delightfully fuzzy. The small, button-shaped trichomes are glandular trichomes, the source of the strongly scented oils that drive cats crazy. Other surface features seen in this photo are flakes of the waxy cuticle covering the jigsaw puzzle–shaped epidermal cells. A few eye-shaped stomata breathe in the foreground and background, helping the plant regulate its photosynthesis. Would you have ever guessed that this strange hidden world is what plants look like up close?

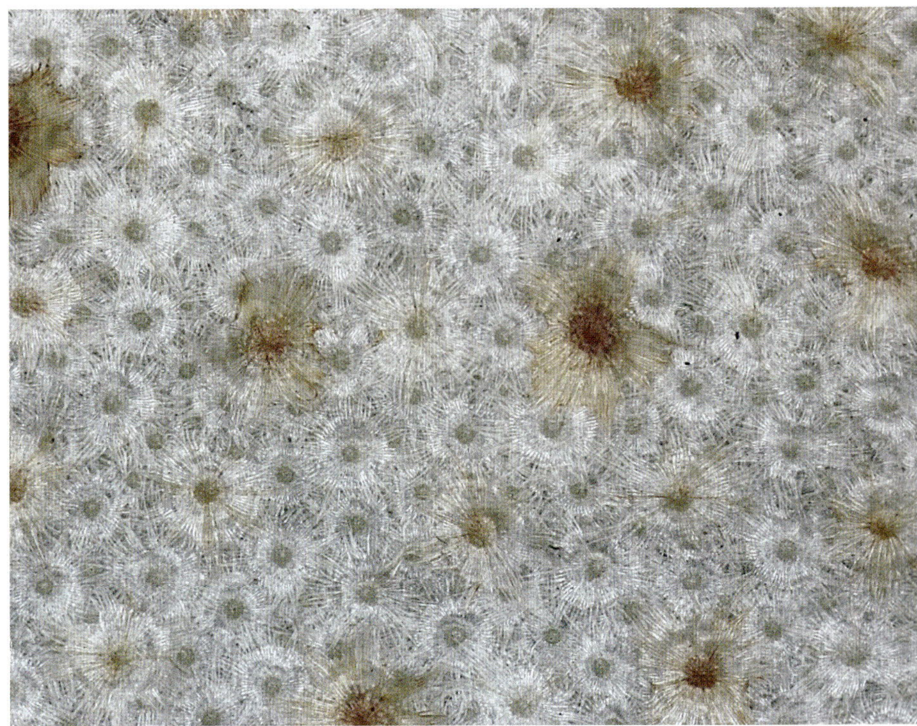

Russian olive (*Eleagnus angustifolia*) is a highly drought-tolerant plant. A major reason for this is the shield-shaped trichomes seen here. Like feathery umbrellas, shield-shaped trichomes protect plant surfaces from the elements. They reflect heat, UV radiation, and high light intensities, and they block harsh winds and frost. The stomata on the leaf surface can respire deeply and safely in a cool, moist, humid microclimate created all over the leaf surface beneath the trichome fields. Desert plants and epiphytes, both of which have lifestyles at high risk of desiccation, have commonly evolved these shielding trichomes.

Glandular trichomes also serve a variety of other plant survival needs. Trichomes are often responsible for secreting noxiously scented or disagreeable-tasting repellents. Strongly antimicrobial secretions such as essential oils might coat and protect foliage from disease. While repellent to pests and diseases, the chemicals that glandular trichomes exude are partly responsible for the powerful smells and flavors of many of our food, herb, and spice crops. Plants such as basil, mint, oregano, rosemary, lavender, thyme, and sage get many of their characteristic and delectable properties as a side effect of their evolutionary efforts to drive away attackers and germs. The glandular trichomes that defend them are the same ones that provide culinary and olfactory sensations. Further, volatiles emitted from trichomes can even play a role in plant communication. Contrary to popular belief, plants do have ways of communicating both with other plants and with other organisms. We will come back to this surprising use of trichomes later in this book.

Trichomes excel during war between plants and other organisms, but some play a much more peaceful role. These trichomes assume a shield-like or scale-like shape. In general, both trichome shapes fulfill two functions: they protect plant tissues from the elements, and they enhance surfaces' ability to absorb water and nutrients. Shield trichomes usually have circular, wide-spreading shapes that provide superb coverage of the majority of the plant's surface. Their appearance is much like an umbrella, with a plate-like or disc-like growth topping a short, sturdy stalk. Scale-like trichomes are wide, papery, overlapping hairs that cover a plant's surface.

One of the best places to find a mix of shield and scale trichomes is on epiphytes, plants that grow anchored to nonsoil surfaces such as tree branches and trunks, rock faces, and even artificial surfaces such as buildings and power lines. Epiphytes are an incredibly diverse bunch. There are more than twenty-seven thousand species of epiphytic plants hailing from numerous families within the plant kingdom. Some of the largest epiphyte groups include the orchid family, ant plants, ferns such as resurrection fern and staghorn ferns, mosses, clubmosses, and bromeliads. The air plants, a group within the bromeliad family (the same group that includes the pineapple), have evolved some of the world's most masterful scale-like and shield-like trichomes. You might be familiar with these wispy plants if you've ever seen glass orb terrariums in any garden centers. The trichomes found on air plants are effective enough at absorption that they can meet the plant's daily needs using only the leaves. This is quite a change from all the land plants that came before. An air plant's trichomes filter and absorb water and nutrients in both liquid and gaseous forms from the air, fog, dust, droppings, debris, and other components of the surrounding environment. As a result, air plants have successfully abandoned the soil in which traditional plants grow. They've taken their entire life cycles to the air, attaching themselves among tree canopies and to other tall supports and surfaces. While air plants do possess roots, they're short-lived and used strictly for anchorage rather than water or nutrient absorption. Air plants have such highly evolved leaves and trichomes that these organs not only harvest sunlight but also capture resources for which most plants use their roots. The way air plant trichomes do this is by acting as wicks. The trichomes of air plants are made up of nonliving, hollow tissue at maturity that is anchored to living cells at the base of each trichome. Water

Seen here is the majestic Spanish moss, *Tillandsia usneoides*, in four separate photos of increasing magnification. It's actually not a moss but rather an air plant in the pineapple/bromeliad family. It is one of the southeastern United States' characteristic sights, and it features highly advanced scale-like and shield-shaped trichomes. Plants in the air plant genus (*Tillandsia*) have a mysterious livelihood as epiphytes, growing on nonsoil surfaces. These mysterious plants are called air plants because two different types of trichomes—shield-like and scale-like—allow their leaves to absorb all of their daily need for water, gases, and nutrients. They so expertly filter water and nutrients from their surroundings that their roots are absent, completely dead, or strictly for attachment. The low-maintenance lifestyle afforded by their trichomes makes air plants popular houseplants.

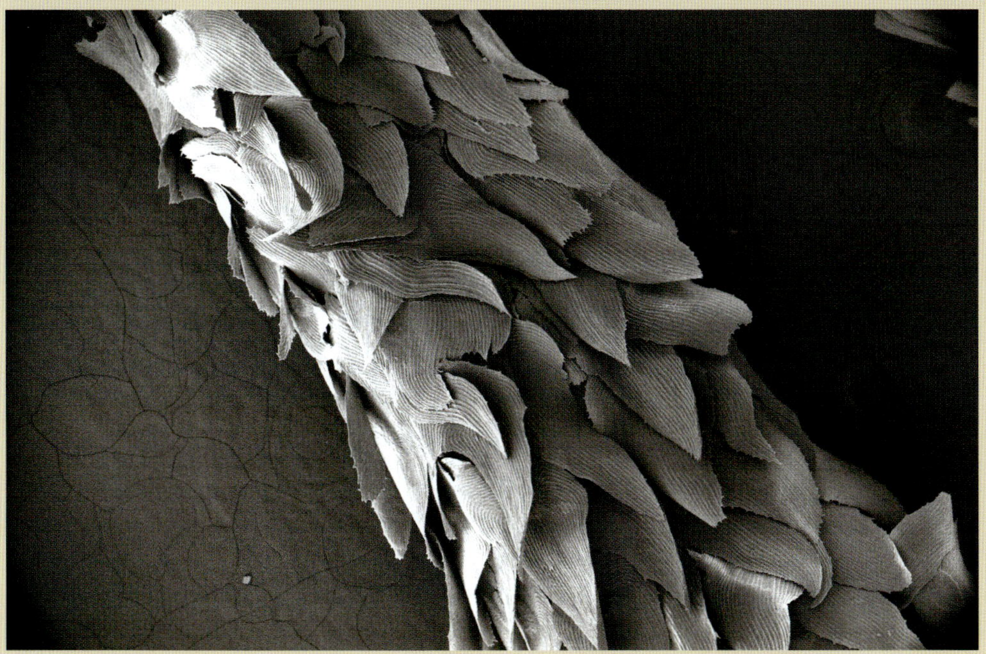

and nutrients in liquid form are easily captured and wicked through this dry, papery tissue into the living leaf cells. These overlapping trichomes not only provide air plants with the resources usually pulled in by the roots but also serve the more traditional function of shielding the leaf from excessive light, heat, UV radiation, wind, evaporation, and moisture loss. Air plants' trichomes establish a humid, stable microclimate across the plant's entire surface area. The great popularity of air plants, in large part, is thanks to their trichomes. Protective trichomes render air plants mysteriously low maintenance and capable of providing many of their own resource needs. Thick trichome growth also renders air plants visually attractive, lending them a characteristic silvery appearance and pettable fuzziness. The next time you see air plants thriving in glass terrariums or hanging baskets, look close and appreciate the hidden little hairs that do so much work for these impressive plants.

Trichomes may be tiny, but, as we've seen, they play a huge role in plant survival. Plants are so well armed with these vicious obstacles that it's a wonder anything can successfully feed on them at all! On the plant surfaces all around us, insects are being repulsed, surveilled, stabbed, hooked, trapped, and stopped dead by overwhelming hordes of trichomes. Plants' trichome fields are amazing and intricate microscopic war zones. Only the strongest attacks may pass further, finally landing a blow on the surface of the plant: the epidermis.

Plants' Final Barrier: The Epidermis and Cuticle

After navigating spines, thorns, prickles, and trichomes, we've successfully arrived upon the surface of the plant: the epidermis. The epidermis is a stalwart layer of cells that covers the surfaces of all plant leaves, stems, flowers, and roots. Under a microscope, a plant's epidermal cells look like an intricate jigsaw puzzle. These jigsaw piece–shaped cells are called pavement cells. The unique shape of pavement cells helps them interlock, lending incredible strength and plasticity to the epidermal layer. Pavement cells must hold up to both external and internal pressures. They must be sturdy enough to resist the cutting, piercing, and tearing exerted by hungry herbivores as well as entry or colonization attempts by pathogens. Plant cells operate with intensely high internal fluid pressures, so pavement cells help

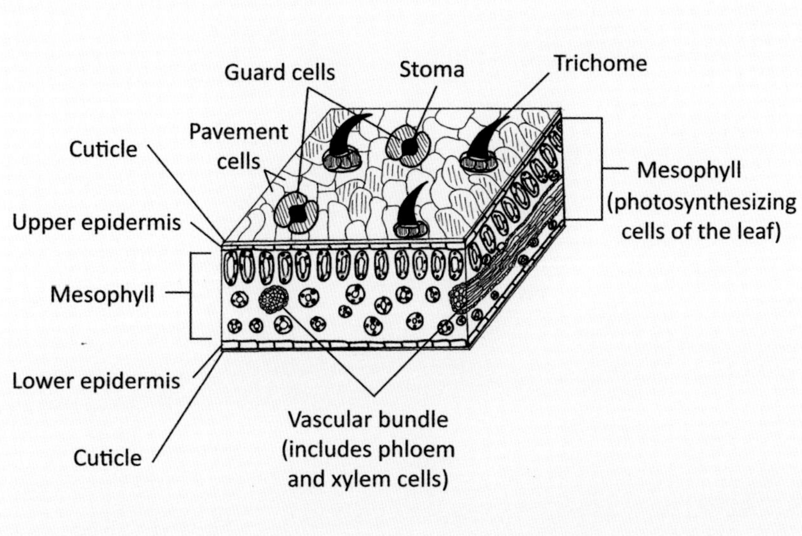

to keep internal cells in place as well as contain potential bursts and leaks. Mixed in among these pavement cells are the guard cells that surround the stomata as well as the support cells that surround the trichomes. These three types of cells make up the epidermis of the plant and regulate the movement of elements into and out of the plant.

The epidermis is the primary boundary between a plant and its surrounding environment. One of the most important purposes of the epidermis is to restrict the plant's precious moisture from being lost. All abiotic and biotic damage, infection, and stresses to the plant must also pass through or best this formidable barrier. Epidermal cells also secrete and absorb chemicals and other substances. The epidermis even manufactures its own protective coating, called the cuticle. Plant cuticles are pretty *slick* adaptations—literally! They're among the most critical of survival adaptations to a plant cell. The cuticle is a lifeless, film-like layer outside the epidermis. This crucial covering is waterproof and waxy. The hardworking epidermal cells build the cuticle, forming a shell of protection at the plant's surface. Depending on plant species, numerous botanical waxes

Palm leaves, such as those of this Bismarck palm (*Bismarckia nobilis*), have incredibly tough and waxy epidermises and cuticles. Their silver-blue hue is caused by a heavy wax buildup that acts as sunscreen against heat and drought.

can impregnate the cuticle or be exuded protectively outside it. Little water passes through the cuticle in either direction without being regulated by the plant.

Were it not for the cuticle, plants would be completely unable to grow on dry land. Their internal water would simply seep out and evaporate, causing them to dry out and die. If plants were unable to contain and pressurize the water they have collected, they would cease to even have structure. The botanical world as we know it—all of the leaves, branches, and flowers we have ever seen—would simply leak until they disintegrated into a desiccated, lifeless mass of dust. Cuticles evolved almost 450 million years ago when plant life began colonizing dry land, first in humble bryophytes such as mosses and liverworts and later in more advanced plants such as ferns, gymnosperms, and angiosperms. Cuticles are just one of the vital components of all the thriving plant life around us.

By enclosing themselves in a protective coating, plants engage in numerous critical survival functions. The cuticle makes it possible for plants to grow other structures that regulate what flows into and out of their tissues. For example, cuticles prevent external fluids from passing straight into the plant cell. Let's say a plant was growing near the saline spray of the ocean. Salt spray on tissues lacking a cuticle would be immediately desiccating and fatal. Plant roots couldn't function either, as excesses of salts, metals, and a wide variety of other potentially toxic substances in the soil would diffuse straight into the plant's cells. Viruses, bacteria, fungal hyphae, fungal spores, and the toxins and digestive fluids they produce would also have no physical barrier in their way. Insects, nematodes, and other animal herbivores would also have an easy feast.

Too much waterproofing can also be fatal for plants. For example, a cuticle that entirely surrounded a plant with no openings would cause problems with overheating as well as exchanging necessities with the outside environment such as gases, water, communications, and nutrients. Regulated exchange of these resources with the outside environment is crucial to plant survival. Regulatory pores in the epidermis are called stomata, and they're in turn encircled and regulated by a pair of guard cells. Plants respire in their own special way using their stomata and guard cells. Guard cells open and close the stomata throughout each day, tightly controlling the water and gas exchange of the leaf. In stems, pores called lenticels perform similar functions. As discussed in chapter 1, these pores both allow water to evaporate and pull water up through the plant. This process is critical for photosynthesis, but too much water loss can also cause issues. Thus, the cuticle assists in keeping transpiration where it belongs: at the regulated pores.

Some cuticles are of such advanced design that they function as self-cleaning surfaces. The cuticle makes possible a phenomenon called the lotus effect, which we cover in greater detail in our chapter on flood survival. Essentially, a cuticle can be so waxy and repellent that water, debris, insect eggs, and microorganisms have difficulty clinging to the plant's surfaces. The nanotechnological properties of the world's superslippery plant cuticles are incredibly advanced, so much so that engineers are creating all sorts of self-cleaning paints, parts, and surfaces based on this principle.

A plant's surface is its foremost survival barrier against every horrible thing the outside world can muster. Plants might appear to be simple,

The slick, waxy cuticle on this collard green (*Brassica oleracea*) leaf forms a self-cleaning surface.

Wax is one cool adaptation! You're looking at extrusions of wax on the outside surface of a silver dollar gum (*Eucalyptus cinerea*) leaf. Under this silly string–like layer of cuticular wax lies the epidermal cells. While the cuticle itself is waxy, extra wax layers can often be found on plants adapted to dry climates, in this case, Australia. A thick wax layer on leaves acts like sunscreen. It reflects heat, reduces water loss, and keeps the leaf cells cool.

passive organisms, slowly growing as the seasons pass by. But they are much more dynamic and exciting than this. A plant forms a life-or-death obstacle course for its foes. To achieve a single bite of a plant, a small bug might have to dodge all manner of botanical booby traps: slippery surfaces, impenetrable thickets of wiry trichomes, stabbing devices, deadly imprisonment traps, spews of glue, or injections of poison. The insect also has to be on high alert for its most feared natural predators. Nature is a harsh and ongoing evolutionary crucible where only the toughest and most committed organisms are permitted to continue their saga of survival. Hopefully, this chapter has shown you how the plant surface is one of the many ways plants have prepared themselves to succeed on our planet's ecological battleground.

CHAPTER 4

Minefield

The Crazy Defenses inside Cells

Thus far, we've focused on the external armor of plants. This chapter continues our journey into plant survival. Now let's peer inside the hidden world of plant cells and the multiple lines of defense coursing through them. Plants are built like fortresses both inside and out. Their outer surfaces are merely the first line of defense against incoming pathogens and herbivores. Chewing, digesting, or infecting a plant is an attack on the plant's very life, which it will fight mightily to preserve. The potent defenses plants pack internally have decisive effects on battles with the hostile organisms of Earth. Hiding discreetly in all plant tissues floats a minefield of specialized superweapons such as crazy cells, virulent poisons, hypersensitive suicide cells, and secretory cells that "goop" attackers to death. Step with us into the toxic, explosive, and viscid world of plant cellular defenses.

Crazy Cells: How Plants Set the Table with Needles and Bombs

Plants feed their attackers all sorts of nasty things, such as needles, crystals, and payloads of poison. To set a disturbingly hazardous table for pests and diseases, plants employ defensive cells called crazy cells or idioblasts to deliver poison to the hungry herbivores. Crazy cells are isolated, "weirdo" cells interspersed throughout all types of plant tissue. They look and function differently from their neighbors, often being larger, having thicker walls, or lacking chlorophyll along with the ability to harvest sunlight. One of the major reasons idioblasts look so strange is because plants have literally turned them into bombs. Each crazy cell is packed with an arsenal of toxins, puncturing crystals, goop, and other harmful weaponry. Inside a crazy cell,

you might find tough, sharp, or abrasive minerals such as calcium oxalate crystals or silica phytoliths (literally, "plant stones"). Oil, latex, gum, resin, tannins, antioxidants, and toxins are also common constituents, depending on the plant species. Many of these defenses are toxic to the plant itself, so idioblasts serve as well-protected containers that release the flood of their contents only when the cell is fatally wounded. Once broken, the cell releases a variety of chemicals that can have a range of negative impacts. One such chemical, the crystalline-shaped calcium oxalate, might just make you develop a fear of needles. A crystal is a structure whose atoms are highly ordered in the way that they form chemical bonds with each other. From this simple principle, the minerals inside a crystal can create all sorts of multifaceted shapes. Diamonds, salt, and the quartz in sand are all great examples of crystals. What plants like about calcium oxalate is that this mineral forms bonds that result in sharp points and angular, spiny edges. Using calcium oxalate, plants turn their idioblasts into devilish little sea urchins!

The Arum family, represented by this tree philodendron (*Philodendron bipinnatifidum*), relies on needle-shaped calcium oxalate crystals called raphides as a defense. The needles are housed in special defensive cells called crazy cells or idioblasts. Even without observable defenses such as spines or thorns, this plant is incredibly well-protected, and its massive leaves remain uneaten.

You're looking at a cross section of the leaf stem of an American white water lily (*Nymphaea odorata*). The spiny structures seen here hold painful calcium oxalate crystals.

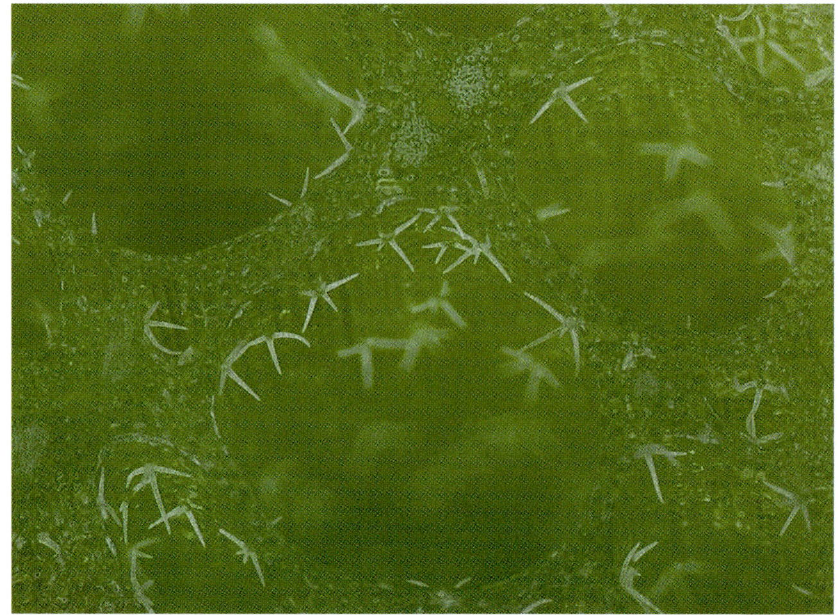

The use of calcium oxalate is inherent to the survival systems of more than 215 plant families. These crystals are often massed inside idioblasts, the location of preference for the angiosperm group. Another location is within each cell wall, a strategy preferred by gymnosperms. Crystal defenses are widely used in plants. Algae, mosses, conifers, and around three-quarters of flowering plants utilize these crystals in almost every type of organ and tissue. Calcium oxalate has been found to make up anywhere from 1 to 90 percent of the dry mass of plant cells.

Plants regulate the levels and forms of crystals they contain based on environmental needs. A plant constantly sensing herbivory attacks, for example, might be induced to maintain a large stockpile. In defense mode, plants precipitate (crystallize) their calcium oxalate out of liquid cellular solution and into solid form. This is how plants form their spiny crystals. Plants often stockpile calcium oxalate because it acts doubly as a calcium reserve. Should calcium ever be found in short supply, plants can dissolve any of these solid crystals into cellular solution and use them as a nutrient for growth. This makes the calcium supply mobile and capable of being reallocated around the plant in times of nutritional stress. Though there are many defenses that can be found within idioblasts, calcium oxalate is a common and effective one.

Ingesting calcium oxalate crystals exerts quite a sinister effect on herbivores. Calcium oxalate is designed to cause intense pain and inflammation of the mouthparts, throat, and gastrointestinal and urinary systems. These effects are almost instantaneous upon feeding, can last for several hours, and can create life-threatening damage. An animal crunching through idioblast-rich plant tissue releases thousands of tiny needles and other sharp, abrasive crystals. Soft, nerve ending–laden tissues in the mouth, gums, and tongue are no match for the slicing and piercing abilities of these structures. Should the herbivore decide to swallow its meal, the inflammatory swelling response of the throat and breathing passages can be severe and might even induce asphyxiation. Chewing insects such as caterpillars, beetles, and grasshoppers are deeply affected by calcium oxalate, too, experiencing debilitating effects such as reduced growth rates and deterred feeding.

Rest assured, however, that animals, including humans, have adapted a variety of deactivation mechanisms for calcium oxalate–based plant defenses. Our body discards calcium oxalate waste from our diet in our kidneys. However, if deposited in excess, calcium oxalate can precipitate into the painful crystals we know as kidney stones. Boiling and cooking plant food helps destroy calcium oxalate, so our food preparation behaviors have helped save our lives from plant defenses. Mammals' gut microorganisms also help them deactivate calcium oxalate. Certain insects have developed their own work-arounds. The piercing-sucking strategy of insects such as aphids is such a precise feeding technique that they are often able to bypass this defense, allowing them to feed on cells without rupturing any idioblasts.

Plants' idioblasts are effectively crystal factories. A plant's genetic code tightly controls the formation and shape of calcium oxalate crystals. Sometimes these tendencies can be so distinct that plant identification can be possible strictly via the telltale traits of its calcium oxalate crystal makeup. The shape, distribution, and quantity of crystals often indicate their function for the plant's survival.

These painful crystals can assume many different shapes, all of which are tightly regulated and tailor-made for each purpose a plant desires. Let's

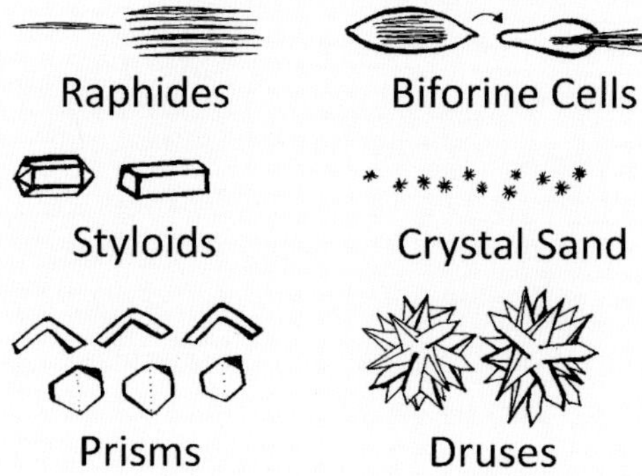

The shape of calcium oxalate crystals determines their function. Crystals such as raphides and druses can be sharp enough to cause severe damage. Crystal sand creates abrasiveness. Prism-shaped crystals reflect intense heat and light.

take a closer look at some of them. Perhaps the most infamous crystal shape is a long, thin, sharp needle called the raphide. Raphides are usually packed inside idioblasts in bundles. One crazy cell may release hundreds of needles should it be ruptured by feeding damage such as chewing, tearing, or grinding. An especially insidious type of raphide-laden crazy cell is called the biforine cell. These football-shaped cells are packed full of raphide bundles, and they are also pressurized. If either end of the football is crushed and broken, its entire load of needles spurts immediately into the mouth of the attacker. Raphides and biforine cells are part of the defensive strategy of plants such as the arum family. Philodendrons, calla lilies, dumb cane, Jack-in-the-pulpit, and elephant ears are rendered quite noxious to consumption as a result. Pokeweed, spinach, beets, spiderworts, palms, agaves, and rhubarb are members of other plant families that also employ raphides. Styloid crystals, shaped like columns or rods, are similar in form and function to raphides but have less formidably sharpened ends. Styloids are employed by asparagus, agaves, hyacinths, onions, garlic, irises, and aloes, to name a few.

Some calcium oxalate crystals are shaped like prisms and serve as sunlight modifiers. The primary function of prism-shaped crystals is to help plants survive extremes of light intensity. In the high light and heat of dry climates, prismatic calcium oxalate crystals are reflective, helping cool and protect the plant's cells. Oleander and the ice plant family (which contains groups such as the living rocks) feature prismatic crystals. In low lighting, such as a rainforest floor, prisms help evenly distribute light within a leaf. Prisms spread light to starving chloroplasts and capitalize on holes in the canopy that might create small patches of intense sunlight. The begonia family and the radiator plants (genus *Peperomia*, black pepper family) feature common houseplants utilizing this type of crystal.

Calcium oxalate might also take the form of crystal sand. Small, angular crystals that look and act like grains of sand can be abrasive to the mouthparts and gastrointestinal tracts of herbivores. Sugar beets, hops, and the yam family (Dioscoreaceae) are examples of plants featuring this type of defense. The druse is the most complex shape of calcium oxalate that plants can concoct. A druse is a multifaceted, spiny crystalline ball that is usually made up of an aggregation of smaller crystals. Druses are typically shaped like stars or the end of a medieval mace, and they are capable of causing extreme damage in any direction. Common plant genera featuring druses include roses, almonds, cherries, peaches, plums, apricots, onions, leeks, garlic, shallots, grapes, muscadines, mulberries, and beans.

Last Meal: Plant Poisons

Don't be allured and disarmed by their static green leafiness and beautiful flowers. Plants are living chemistry labs, manufacturing some of the most lethal poisons on Earth. While plants sometimes try out camouflage or repellents to confuse or inconvenience pests, the intent of poisons is entirely different. Poisons are concocted as last meals. Their sole purpose is to kill, maim, or permanently disable an attacker. Plant life evolved and has existed on our planet long before animal life. Essentially, plants hold an advantage through millions of years of evolution. This head start in the chemistry lab has allowed them to prepare some stunningly powerful and nasty defenses.

Chemical defenses are quite costly for plants to produce. Valuable nutrients such as nitrogen, sulfur, and carbon must be diverted from growth and development processes and forged into weaponry. For this reason,

You are looking at one of the most lethal plants on Earth, the castor bean (*Ricinus communis*). Native to Africa, the castor bean protects its tissues and seeds with an incredibly toxic protein called ricin. The spiny red seed pods in the photo forcefully explode when ripe, hurling the seed far from the parent. Woe betide anything that chews a castor bean. Ricin kills animal cells within hours by preventing the manufacture of protein. The amount of ricin in a handful of shiny, patterned castor beans is enough to kill a human being. Ricin has been involved in the darker side of human history, including suicides, warfare, biochemical weapons, assassinations, and terrorism. Ironically, the high castor oil content of the seeds is entirely harmless and renders one of the most useful plant-derived lubricants on the planet.

chemical defenses cannot afford to fail. They remain among the most lethal counterattacks within plant survival systems.

Let's discuss some of the terrible effects plant poisons are capable of creating. Poisons can target any type of threat in any way. Both generalist and specific toxins have been discovered that operate on mammals, mollusks, insects, fish, reptiles, nematodes, bacteria, viruses, and fungi. Some poisons have evolved to be produced only by an individual species or family, while others may be produced generally among hundreds or thousands of plant species. Poisons can kill cells outright, deprive them of vital resources, or interfere with critical cell processes. There are different types of poisons with a variety of modes of action. For example, neurotoxins can disrupt the normal operation of the nervous system, creating the paralysis of muscle groups such as the heart, lungs, and gastrointestinal system. Phytohormones, on the other hand, can hijack animal endocrine systems, shutting down growth, reproduction, and development. Some plants emit allelopathic chemicals, which wage war with neighboring plants. These chemicals have the power to kill competing plants or impair their growth and germination. Juglone in the black walnut tree is one example of a plant-produced chemical (phytochemical) that causes allelopathic effects in competing vegetation. Even more interestingly, some chemicals can serve multiple functions concurrently. For example, the caffeine in coffee plants has neurotoxic, allelopathic, and pollination-enhancing properties.

One of the largest groups of plant defense compounds is called alkaloids. These chemicals contain at least one nitrogen atom. More than three thousand alkaloids are known to science, and more than four thousand species of plants are known to produce at least one alkaloid. The effects of alkaloids are as myriad as their numbers, and the strength of their toxicity also depends on variables such as the dosage ingested. Alkaloids often taste incredibly bitter. We might want to thank our taste buds for protecting us from alkaloids. Our taste buds are part of our body's advance-detection system to warn us of the presence and potential danger of various chemicals in our food. Alkaloids aren't always lethal. On the lower end of the toxicity spectrum, alkaloids exist that induce severe pain, nervous system manipulation, vomiting, repellency, sleepiness, appetite suppression, and

Toxins called glycosides are literally candy-coated. These poisons include an attached sugar group as a chemical bait, disguising the poison as a sweet energy source craved by animal cells. When an animal cell unwraps the glycoside by breaking off the attached sugar, the plant's toxin is activated. Cardiac glycosides, found in the purple passionflower (*Passiflora incarnata*) shown here, immediately cause heart damage and even cardiac arrest.

debilitating hallucinations. However, more dangerous alkaloids do exist and can impair growth and development or cause paralysis, convulsions, and even death.

Members of another group of phytochemicals, the glycosides, are literally candy-coated. These poisons include an attached sugar group as a chemical bait, disguising the poison as a deceptively delectable "Trojan horse." Animal cells crave the energy that sugar provides and consume the poisonous molecules without a second thought. Unwittingly, the sugary glycosides are rapidly distributed to muscles and other energy-demanding animal cells. When an animal cell unwraps the glycoside by breaking off the attached sugar, the plant's toxin becomes active. Cardiac glycosides, found in the dogbane and passionflower families, immediately cause heart damage and even cardiac arrest. Cardiac glycosides impact anything with a heart. Insects are included in the list of animals with a heart structure, so we might expect plants that contain this powerful and quite fatal chemical to be pest-free. For the most part, these plants are surprisingly clear of feeding herbivores. However, some specialist insects have gained adaptations to handle this troublesome chemical. Enter monarch butterflies—the iconic great migrator. The eastern populations of monarchs are especially impressive, with the final generation migrating from the northern United States and southern Canada all the way down to central Mexico on fragile, paper-thin wings. Their striking orange color is a warning, though. These little insects have found a way to eat from milkweeds, a plant family that laces its latex sap with cardiac glycosides. Instead of passing the cardiac glycosides through their system and experiencing the damage that they would ultimately cause, monarchs sequester these chemicals. This means they store the chemicals in deposits within their bodies. This process not only allows the crafty caterpillar to feed on a plant that no one else dares to touch, thereby reducing competition pressures, but also gives the caterpillar and future adult the ultimate defense. Monarchs make themselves poisonous with all of the sequestered cardiac glycosides brimming within them. Monarchs warn the world of this danger with their bright orange pattern, which gives the clear message to birds and other predators to stay away. Another type of glycoside poison, the cyanogenic glycosides, is found in the euphorbia family and stone fruits of the rose family. These glycosides generate cell-killing cyanide, whose damage can begin occurring in as little as seven seconds. Cyanide poisons animals by interfering with their ability to process oxygen. Cherry pits, wild almonds, and cassava, the source of tapioca, are examples of plants with cyanogenic properties.

The topic of plant chemistry is a deep one. Much has already been written about poisonous plants, and many more chemicals likely await discovery in the botanical world. A sampling of toxins, many of which feature in the chapters of this book, are listed in table 1. If you're reading this book, you have managed to avoid deadly plant poisons throughout your lifetime. For our early ancestors, avoidance was no easy feat. Perhaps, though, you've brushed up against the powerful effects of some of these classes of chemicals, as humans have found many ways to enjoy or utilize plant components that originally evolved to deter.

Table 1 Examples of Plant Poisons

Toxin	*Chemical type*	*Example plant producers*	*Effect*	*Human use*
Caffeine	Alkaloid	Coffee, tea, chocolate, kola nut	Addictive, nervous system stimulant, paralytic, allelopathic	Stimulant
Cocaine	Alkaloid	Coca plant	Stimulant, addictive, harmful neurological effects	Recreational drug
Coniine	Alkaloid	Poison hemlock, North American trumpet pitcher plants	Disruption of nervous system, respiratory paralysis, death	Poison
Cyanide generators	Cyanogenic glycosides	Euphorbia (tapioca), rose (stone fruit pits), legume, and passionflower families, sorghum	Generates hydrogen cyanide when eaten, inhibits cells from uptaking oxygen and producing energy, immediately shuts down cellular respiration and kills cells	Industrial processes such as mining, chemical weapons
Cycasin	Glucoside	Cycads such as *Cycas revoluta* and *Zamila pumila*	Neurotoxic and carcinogenic	None
Heart toxins	Cardiac glycosides	Dogbane family (milkweeds, oleander, desert rose), foxglove, kalanchoe, lily-of-the-valley	Alters function of heart muscle cells, can result in cardiac arrest	Heart drugs
Juglone	Organic compound	Walnuts (especially black walnut), hickories, pecan	Strongly allelopathic, insecticidal, fish poison	Clothing dye, ink, cosmetics, hair dye, herbicide, fish poison
Laxatives	Anthraquinones	Aloe, senna	Operates gut muscles, laxative, emetic, causes electrolyte imbalances, dehydration	Herbal medicine
Mescaline	Amine	Cactus family (peyote, San Pedro cactus, Peruvian torch cactus)	Hallucinogen	Psychoactive drug
Morphine, codeine, papaverine	Alkaloids	Poppy family	Nervous system depressant, paralytic, appetite reduction	Painkillers, cough suppressants, treatment of muscle spasms, opium

Toxin	*Chemical type*	*Example plant producers*	*Effect*	*Human use*
Mustard oil bombs	Glucosinolates	Mustard family	Repellent, foul odor, bitter and spicy taste, antimicrobial, generates toxic organosulfurs	Food, spices
Nicotine	Alkaloid	Tobacco	Addictive, nervous system stimulant, paralytic, appetite reduction	Stimulant, pesticide
Phytohormones	Herbivore hormone mimic	Yam family	Reduced herbivore fertility and reproduction	Birth control drugs
Quinine	Alkaloid	Cinchona tree	Toxic doses can cause vision, gastrointestinal, and hearing problems, highly antimicrobial	Antimalarial drugs
Ricin	Lectin (carbohydrate-binding protein)	Castor bean	Extremely lethal, interrupts cell metabolism and protein synthesis	Poison
Sharp crystals	Calcium oxalate	Arum family, ice plant family, rhubarb, pokeweed	Long-lasting pain, cell damage to mouthparts, GI tract, kidney stones	Ceramic glazes
Soap	Saponins	Soapwort, quinoa, legumes, soapberry, yucca	Dissolves and removes pathogens, erodes red blood cells	Soap, fish poison
Solanine, Solanidine	Glycoalkaloids (alkaloids with attached sugar)	Tomato family (tomato, potato, eggplant, deadly nightshade)	Disrupts cell membranes and neurotransmitters	
Strychnine	Alkaloid	Strychnine tree	Muscular convulsions and death via asphyxiation	Pesticide, tranquilizer dart guns
THC, CBD	Cannabinoids	Cannabis	Psychoactive	Psychoactive drug
Tropane alkaloids (atropine, hyoscyamine, scopolamine)	Alkaloids	Nightshade family (deadly nightshade, angel's trumpet, devil's trumpet, henbane, mandrake)	Disrupts neurotransmitters, hallucinogenic and deliriant	Poison
Urushiol	Caustic oil	Sumac family (poison Ivy, mango)	Severe irritant, inflammation	None

Scorched Leaf Tactics: Cell Suicide and the Hypersensitive Response

Plants excel at feeding attackers an extremely treacherous diet using the toxins described here. However, some plants find it equally effective to deliberately starve pests and diseases to death. Mammals, insects, bacteria, and fungi can experience debilitating nutritional deficiencies as they munch, pierce, dissolve, or bite into various plant tissues. Plants have many ways to render their nutrients inert, difficult to acquire, or restructured in a way to induce malnourishment and impair growth. We cover many of these strategies in other places in this book. They include the extreme dilution rates inherent within plant sap, the blockade of the vascular system in response to disease, antinutrients (nutrient-binding chemicals such as tannins), and the high carbon ratio inside the cellulose and lignin comprising cell walls, wood, bark, and seed coats. Stay tuned to learn more about these options. In this section, though, we'll talk about what might be the most extreme and effective starvation defense of them all: intentional cell suicide.

Called the hypersensitive response, cell suicide is a plant's last line of cellular defense. It works just like scorched earth tactics, when human armies deliberately destroy resources that might be captured by their enemies. Common scorched earth tactics include razing crops and food supplies, poisoning water sources, transferring resources such as fuel and raw materials away from the front lines, and destroying valuable equipment and munitions. The signature of the hypersensitive response in plants is the visible, preprogrammed suicide of entire waves of cells in the path of an attack. Additional "counterattacks" from the dying plant cells include the intentional removal of nutrients, the deposition of blockages (using tough materials such as lignin, cellulose, callose, and tyloses), and a massive release of toxins. (A tylosis in woody plants is a bladder-like swelling that closes off xylem vascular tissue in response to injury, embolism, or disease. The tylosis originates in parenchyma cells, which line the xylem vessels of vascular tissue. The swelling distends through pores between parenchyma and vessel cells into the vessels, causing a deliberate blockage that halts the spread of problems and the leakage of plant fluids. The term "tylosis" can

Early leaf spot (*Cercospora arachidicola*), a fungal disease of peanuts, causes a hypersensitive response. This particular infection causes multicolored rings to develop, which are war zones between the plant and the fungus. In the darkened areas, the invasion of the fungus has induced the plant's cells to suicide in an attempt to deprive the pathogen of resources. In the yellow and orange areas, plant cells are on the retreat. Preparing for suicide, these cells are shipping their nutrients away from the pathogen and leaving behind powerful defenses such as antioxidant flavonoid and carotenoid pigments. These "leaf wars" are visible to the naked eye all around us.

refer to the physiological process of this occlusion happening or to the actual swelling itself.) Let's look at some hypersensitive tactics in action.

Attacking bacteria, fungi, viruses, water molds, nematodes, and insects can all generate a severe defensive response in plants. Since a plant's cells are all interconnected, an unchecked pathogen or invasion can spread at an alarming rate. Plant surveillance systems are ever watchful for chemical signs of the presence of any of these organisms. These chemical signatures are called elicitors, and they might include the obligate components of attackers' cells, their specific digestive enzymes and saliva, and damaged plant cell components. The hypersensitive response begins to loom as the entire plant coordinates how it will fight for its life. The most common part of the hypersensitive response is the mass sacrifice of cells surrounding an invasion site. In a biological process called apoptosis, plant cells deliberately use triggering molecules to initiate a programmed "self-destruct sequence" and die. Trigger molecules for cell death include salicylic acid, plant hormones, and proteins. A plant's self-destructive directives to suicide cells might prompt the termination of vital life processes such as energy production, the degradation of the cell's own DNA, the collapse of cell structures and barriers, or a flooding of chemicals so noxious that they kill the plant cell itself as well as any invaders.

As the cascade of cell death ends, left behind in their withered husks is a minefield of nasty antimicrobial defenses. Even though the tissue is completely dead, the suicide cell barrier remains one of the most powerful barriers a plant can muster. It has become a wasteland, a dumping ground for the plant's perilous chemical soup of destructive and obstructive substances. For convenience, scientists collectively refer to the numerous chemicals in a plant's defensive concoction as phytoalexins, which make targeted counterattacks against disease cells, either killing them outright or at least disrupting their growth and reproduction. Phytoalexin effects include poisoning, puncturing, degrading, digesting, neutralizing, or disrupting microbes. Some phytoalexins featured throughout this chapter and in the rest of the book include lignin, phenolics, sulfur, mucilage, gum, resin, and VOCs. Plants also release their own digestive enzymes, such as peroxidase, chitinase, and oxalate oxidase, hoping to rip apart bacterial and fungal cells. Dying plant cells generate hazardous reactive oxygen species such as superoxide and hydrogen peroxide. Also called free radicals, reactive oxygen species are violently reactive chemicals that, if left unchecked, cause major damage to microbial cells and their internal processes.

Physiological changes also occur in collapsing cells. The vascular channels lie plugged with callose and tyloses. Cell walls begin to shrink and close in. Formerly living tissues have become inert, dry, and barren. It's not looking good for invading pathogens.

As total war rages across the plant tissue in the hidden world, visual signs of these struggles begin to appear all around us. Perishing plant cells are inflamed into colorful patterns surrounding the disease. As

You're looking at a plant-versus-microbe war zone raging across a banana leaf. As rot pathogens (brown) attempt to infect healthy new cells (green), the plant signals programmed cell suicides in the yellow boundary layer. This layer is rich in defensive flavonoid pigments. Altogether, these sacrificial cells provide a united front against microbial attack, releasing a barrage of defensive chemicals called phytoalexins as well as depriving pathogens of resources such as food and water.

chlorophyll breaks down and retreats, the remaining pigments flare. Pigments are powerful antioxidants that nullify common chemical attacks from microbes. The naked eye can easily spot the hypersensitive response in action on plant leaves. Blights, lesions, cankers, spots, splotches, and other patterns on leaves are telltale signs of the natural battle for survival that plants face every day. Between the brown or black infection sites and the healthy green tissue, you'll note defensive pigments flaring into radiant colors and patterns. You might see disease-induced lesions ringed by purple, orange, yellow, and brown. Each of these colors is the visual manifestation of powerful defensive pigments, which we'll learn more about later. When a disease hits, these vivid barriers often indicate severe plant stress. It is here, on these intricately colorful borders, that waves of suicide cells prepare for the last and most important task of their lives. They are ticking down to detonation: 3 . . . 2 . . . 1 . . . boom! Keep in mind, cell suicide is extremely costly. However, when an organism lacks an immune system, cell suicide can be the best bet, especially regarding disease. Walling off an infected area can keep a pathogen from spreading. Sometimes, plants can go overboard with this response when faced with a novel outbreak.

The Scoop on Goop: Resin, Latex, Gum, and Mucilage

Plant cells are like tiny balloons filled with water and poisonous goop. It might captivate you to learn how plants have prepared even their fluids for battle. Once a cell has been popped by an attacker, plants release a deluge of different noxious, gooey, sticky, viscous, and lethal fluids. "Goop" is our unofficial term to refer to all defensive liquids that exude from damaged plants. While the term "sap" is often used for this purpose, it is slightly misleading. Botanically, sap refers to the fluid flowing within a plant's vascular system, usually a mix of water and solutes. Some fluid defenses such as resin, latex, and mucilage are not exclusively housed in the vascular system; instead, they are stored in specialized cells called organelles, which operate like organs by carrying out specific tasks, or in ducts. Therefore, they do not qualify as sap in a strict botanical sense. Because natural gums, however, often flow through a plant's vessels, they can frequently be a component of sap. Altogether, plant goop is a handy catch-all term that we'll use to compare and contrast these oozy substances. Let's flow into the first of these: resin.

RESIN

One of the stickiest, goopiest things a plant can make is resin. Brush too close to a wounded pine tree, and you'll undoubtedly be smeared with this defensive compound. Resin is a viscous fluid or solid mixture of organic compounds. It is flammable, highly odorous, sticky, waterproof, brown or amber in color, and semitranslucent. Resin is secreted by plants inside special channels called resin ducts. Any breakage in these ducts causes a leakage of resin.

Resin has several jobs that it must perform very quickly in order to assure a plant's survival. The first is to fight all types of attackers simultaneously. Should a disease infect its way into a resin duct, antimicrobial compounds will be hand-delivered by the streaming resin. Resin hardens and solidifies on contact with air. The whitish product that results is called rosin, which acts like a liquid bandage to seal, disinfect, and waterproof the wound. Sap leaks, water loss, and future herbivore and pathogen invasions are stymied by these effects. Boring insects are common targets of resin as well. Any insect with a mouthful of resin soon finds its mandibles cemented together. The hardening rosin binds bug legs and mouthparts in almost superglue-like fashion.

A resin leak from this loblolly pine (*Pinus taeda*) is protecting an open wound in the tree. Resin's powerful chemistry glues up insects, fights microorganisms, and seals and waterproofs wounds.

Interestingly, humans have made use of this material in a variety of ways. Musicians who play stringed instruments rub a cake of rosin over the hair of their instrument's bow; the hair can then grab the instrument's strings, causing the vibrations that produce sound. Gymnasts and baseball players rub powdered rosin on their hands to improve their grip. The strong volatile chemicals that give resin its characteristic smell are not only a warning repellent but also an advertisement to predators that prey has been ensnared or marked by plant defenses.

Resin is found in a variety of plants. Most conifers such as pines, spruces, cypresses, cedars, junipers, firs, kauris, and hemlocks employ resin. Conifer needles have secretory channels called resin ducts running lengthwise through them. Greater numbers of ducts also course through the stems of resin-bearing plants. Resins from angiosperms include frankincense, dragon's blood, copal, dammar gum, elemi, hashish, galbanum, gum guaiacum, labdanum, sandarax, styrax, and spinifex.

The southeastern United States has long been a hot spot for an entire resin-based industry: the production of naval stores, which are products derived from pine resin. Originally, throughout the age of wooden sailing ships, naval stores referred to resin products that were used to waterproof and rotproof ship timbers. Before synthetic caulks and sealants, a length of cordage saturated with resin provided the best flexible, watertight, and insoluble solution for sealing between the planks of a ship. Naval stores included resin-saturated cordage, turpentine (distilled pine resin), rosin, pitch, and tar. Numerous industrial resin distillation operations thrived along the Coastal Plain, heating longleaf resin and creating purified by-products such as turpentine. Nowadays, though many of the Southeast's original longleaf pine forests are gone, the management of extensive plantations of various pine species, including not only longleaf but also loblolly and slash pine, still fuels a global market for resin. Resin can be found as an additive in soap, paint, varnish, shoe polish, lubricants, linoleum, and roofing materials.

One of the main reasons this industry was situated along the southeastern Coastal Plain was the historic predominance of the longleaf pine. This tree enabled the southeastern United States to become the world's largest producer and exporter of naval stores from the eighteenth to the twentieth century. The Coastal Plain's intense lightning-induced fire regime and xeric sandy soils helped evolve the strongly resin-based defenses that allowed the longleaf to proliferate. Because of these stresses in its ecology and life history, this species has some of the heaviest resin production among conifer species. Copious resin is a survival strategy that bestows resistance to fire, insects, disease, and rot upon these long-lived trees. Longleaf pines are North America's largest and longest-lived pine species, but their straight and rot-resistant trunks were being felled for resin and timber at an alarming rate throughout the eighteenth, nineteenth, and early twentieth centuries. Faster-growing species began replacing longleaf as lands throughout the South were cleared for pine plantations. Because of that clearance and the loss of cyclic fire patterns, longleaf pines have unfortunately vanished from much of their historic range.

Charles Herty (1867–1938), a chemist at the University of Georgia, revolutionized the turpentine industry by devising a sustainable pine-tapping technique. Previously, entire pine forests had been clear-cut for timber,

A flood of resin from a longleaf pine (*Pinus palustris*) has air-dried, creating this whitish deposit of rosin. Longleaf pine is one of the most prolific resin-bearing plants in North America. Federally endangered red-cockaded woodpeckers nest only in cavities in this tree so they can take advantage of the plant's resin defenses. The woodpeckers peck fresh wounds around their entrance holes so that the pine's constant flow of resin keeps pestiferous insects and predators away. The resin in the longleaf pine also fueled a pine-tapping industry in the southeastern United States. Naval stores, produced from the tree's resin, included valuable products such as pitch, tar, and turpentine. Pine tappers were often called "tar heels" for the dirt and resin glued to their feet during their work.

their resin a simple by-product. The deaths of these expanses of mature pines made it hard to achieve consistent resin production. Herty's system was different. It allowed mature pines to survive, albeit drained of their precious resin. Importantly, however, the Herty system allowed plantations to be maintained in a living state and minimized the need for replanting and decades of regrowth. This improvement undoubtedly saved the already fragmented longleaf pine species from the perils of complete overharvesting and endangerment. The new tapping process employed a plain clay pot called a Herty pot. Herty pots are still found buried across piney woods in the Southeast today. Shaped just like small clay flowerpots, Herty pots are often mistaken as such, but these small pieces of pottery have a much richer meaning. They are historical relics of a noteworthy bygone age and tradition. Herty pots were nailed to the base of the longleaf to collect its resin. Above each pot, a series of sheet metal gutters were placed to channel resin flow. Finally, distinctive, shallow, V-shaped notches called catfaces were cut across the tree. The depth of each notch had to be precise; too deep, and pine tappers risked severing the living cambium layer within the tree and killing it in a type of damage called girdling. Girdling occurs when this critical cambium layer, located just beneath a tree's bark, is completely severed around the tree's entire circumference. Girdling is fatal because water, nutrients, and food can no longer pass between a tree's roots and its shoots. By keeping the cuts shallow, the tappers could induce the plant to employ its defenses without dying. Upon detecting the damage, the pines initiated their favorite self-defense strategy: a mass flood of resin. Catfacing allowed the longleafs to give up their precious resin while enabling the forest to survive.

Resin flow can also be induced in plants by a variety of other types of wounds. A coniferous tree that has died abruptly and immediately after being toppled by a storm or felled by a saw will impregnate its stump with a huge concentration of resin. This happens because the upwelling of hydraulic pressure from the roots, upon failing to reach the tree's now nonexistent crown, deposits all of the plant's xylem contents, including defenses such as resin, into the stump tissue. The extremely resin-rich heartwood and tissues left behind are called fatwood, which is highly fragrant, rot-resistant, and hardened. Fatwood, also known as fat lighter or lighterwood, has been renowned for its fire-starting capabilities for much of human history and prehistory. A sliver of fatwood is like a natural match or piece of kindling. Because of resin's flammability and waterproofing, fatwood has remarkable properties to light easily, burn hot, and remain dry.

You might also have run into resin before if you've ever encountered a piece of amber. Amber is simply plant resin that has been fossilized under the heat and pressures of the earth. As plants throughout time defended themselves with goop, they left a piece of history behind. Each piece of amber is a time capsule of these microbattles. Locked within a chunk of amber are a variety of treasures called inclusions. Inclusions may include microscopic air bubbles from eons past, prehistoric plant and animal life, and microbes. Ancient insects, worms, arachnids, frogs, crustaceans, marine life, pollen, spores, wood, flowers, fruit, hair, and feathers have all been recovered from amber deposits. Humans have appreciated the natural exquisiteness, glowing orange color, translucence, and idiosyncratic inclusions of amber since the Stone Age. Amber inclusions were even the plot foundation of *Jurassic Park*, a science-fiction story. In this 1990s book and movie adaptation, scientists cloned living dinosaurs from DNA they recovered from mosquito blood trapped inside a remarkable piece of amber.

Amber is simply plant resin that has been fossilized under the heat and pressures of the earth. As plants throughout time defended themselves with goop, they left a piece of history behind. Each piece of amber is a time capsule of these microbattles. Locked within a chunk of amber are a variety of treasures called inclusions. Inclusions may include microscopic air bubbles from eons past, prehistoric plant and animal life, and microbes.

LATEX

One of the most noxious and potent fluid defenses of plants is a concoction called latex. Latex is a complex emulsion, meaning a mixture of two or more liquids. Additional chemicals and particles can be dissolved within the latex, too. Latex is a caustic, smelly, sticky, multifaceted weapon against herbivores. A single drop of latex is packed with everything a plant can produce as a counterattack to herbivory. For example, the latex of the opium poppy plant contains more than forty toxic constituents. Latex is usually an opaque, milky white fluid, but, depending on plant species, it can be clear, yellow, or red. Latex is as widespread as it is effective at helping plants survive. More than twenty thousand plant species in forty families, including almost 10 percent of all angiosperms, produce latex as part of their survival systems.

Latex is produced by cells called lactifers, which can be arranged in elongated tubes called vessels or branching networks. Both patterns extend complete protection through the tissues of the entire plant. Lactifers are often highly pressurized, ready to spurt if cut or disrupted in any way. This force is often enough to coat insect attackers in latex spatter. Unfortunately for herbivores, when they place their mouthparts on a plant to feed, they're also placing their eyes and other important sensory organs in proximity to the plant's explosive lactifers. For example, the latex of many euphorbia family members such as the pencil tree (*Euphorbia tirucalli*) is quite potent, able to cause pain and severe damage to mammal eyes and skin. Ingeniously, some caterpillars and other insects have learned special feeding behaviors that avoid latex defenses. They surgically snip the latex vessels in the leaf stem before feeding, allow the latex to bleed away, and then continue their meal in safety.

Sensory damage isn't the only consequence of interfering with latex. Since latex is sticky and waterproof, it becomes incredibly difficult for an herbivore to wash off. Caustic skin burns, inflammation, and destruction of cells throughout the gastrointestinal system are also common effects. Each of the constituents of latex plays a role in suppressing threats to the plant. Like resin, latex hardens on contact with air, becoming fatally gluey should it spray over an insect's legs, wings, and mouthparts or coat a pathogenic cell. Latex is deliberately blended for fast coating and coagulation over a plant's wound, using incorporated proteins, polymers, resins, and gums.

Any wound on a fig tree releases a spurt of latex, which is a complex emulsion of dozens of different chemical defenses. Altogether, this milky fluid seals, waterproofs, and disinfects the wound. Toxins, glues, coagulants, repellents, and predator attractants are potent components of latex that enable it to fight animal attackers as well.

Oils and other waterproofing agents dry and coat the damaged area, preventing water loss and the entry of disease. Toxins such as alkaloids go to work poisoning and eradicating insects, while antimicrobial compounds and tannins disinfect and purify the wound. On top of everything, a suite of off-gassing VOCs takes to the air, marking the site of the attack and the glooped attacker for an air strike by predatory insects.

Most of us interact with latex every day. Rubber is one of the most useful products on Earth, and it's derived from plant latex. By itself, latex is perishable and sensitive to temperatures. It's brittle when cold and sticky when hot. Rubber, on the other hand, is strong, durable, temperature stable, and more elastic. Rubber is created when latex is processed by heating, vulcanized with chemicals such as sulfur, and molded into shape. It is a major constituent in tires, balls and other sports equipment, balloons, gloves, erasers, seals and gaskets, bandages, rubber bands, shoe soles, clothing, toys, masks, and contraception. Allergies to these items, which are the human immune system's inflammatory reaction to foreign substances, are quite common because of the potent chemicals within natural rubber latex. These allergic reactions are a reminder that latex's purpose is defensive—plants produce this fluid not for our benefit but to enhance their own survival.

Seen here is a rubber tree (*Hevea brasiliensis*) being tapped for latex. When incisions are made into the tree's defensive cells, streams of latex gush onto the attacker. Rubber tapping is a highly skilled endeavor—too imprecise a cut, and the latex yield slows; too deep a cut, and the tree can be girdled and killed.

Natural rubber is derived from the latex of a euphorbia family member called the rubber tree (*Hevea brasiliensis*). Native to South America, this species has been introduced in plantations throughout the worldwide tropics. Southeast Asia, a world away from the rubber tree's natural range, is currently the major world center of rubber production. The Panama rubber tree, in the fig family, is also a species traditionally cropped for latex throughout Central and South America.

Tapping latex from rubber trees is traditional knowledge that indigenous South American peoples have honed for generations. The rubber-tapping industry persists throughout Brazil, Bolivia, and Peru, but its roots predate recorded history. Thankfully, indigenous South Americans developed sustainable techniques to harvest rubber without killing the tree itself. As in pine tapping, a pattern of shallow notches is incised into the trunk, inducing the rubber tree to gush latex as a defense against this attack. Gutters and collection cups channel and collect the valuable latex. The trees remain alive, and careful rotation between trees is used to maintain their health.

Indigenous Americans invented quite a number of uses for rubber, which they created by mixing the rubber tree's latex with sap from other plants such as morning glories. Dried latex by itself was simply too brittle to be useful. The contributions from these additional plant species rendered rubber as we know it: elastic, stretchy, bouncy, and durable. Archaeologists analyzing ancient Mesoamerican rubberized balls determined that this technology was in use at least by 1600 BCE, more than three thousand years prior to the invention of modern vulcanized rubber by Charles Goodyear in 1839. Ancient peoples employed rubber as chewing gum, formed it into balls for ancient Mesoamerican offerings and ritual ball games, and produced rubberized figurines and other artifacts. Spanish conquistadors also noticed indigenous people creating rubberized, waterproof hats, clothing, and shoes by dipping these items into latex.

Clearly, latex is a formidable defense technique with some unusual properties. This useful material shows up in many groups of plants, including those from the euphorbia family, poppy family, mulberry family, dogbane family (which includes milkweeds), and sunflower family. We hope you can now appreciate this threateningly sticky substance that plants employ.

NATURAL GUMS

Chew on this: gum is a plant defense! That's right: some of the first chewing gums ever discovered came from plants trying to survive. Natural gum is a term for another type of goopy exudation that occurs when plants are wounded. Gums are composed of polysaccharides, the same group that includes cellulose, sugar, and starch. The long chains of sugars making up these molecules love to absorb water. Upon doing so, they readily form gels. Sticky and extremely viscous, gums flow out of wounds to act as patches to wounds and deterrents to attack.

While they're great at gumming up bugs, natural gums are generally safe and inert for human digestive systems. For this reason, a wide variety of natural gums have found invaluable service in many industries, especially food processing and production. Most natural gums are so unique

and complex that they're irreplaceable. Quite simply, synthetic substitutes have not yet been created with as great a degree of functionality and safety. Natural gums' myriad uses include:

- stabilizing agents, which create thick and rich consistencies in syrups and sauces
- emulsifying agents, which prevent products from separating by helping oil and water mix
- gelling agents
- encapsulating agents, which coat and protect food
- clarifying agents, which cause suspended solids to aggregate (useful for mining, wastewater processing, and papermaking)
- swelling agents
- crystal inhibitors, which prevent the formation of ice in frozen foods
- foam stabilizers
- adhesives and binding agents

It might be a little mesmerizing that gums can do all of these things. Perhaps the thing they do best, though, is something you're already acquainted with: they're chewy. Before recorded history, the very first chewing gums were sourced from trees in the natural world. Chicle is a natural gum that comes from chicle trees, a species of the *Manilkara* genus in Mesoamerica. The word "chicle" is actually derived from the Nahuatl word for the gum, *tzictli*, which translates to "sticky stuff." The Aztecs and Maya chewed chicle gum for thousands of years. Learning this, in 1871 American inventor Thomas Adams used chicle as a gum base for his new confection—you might know it better as Chiclets. While resin, rubber, and synthetic sources of gum base dominate the chewing gum market nowadays, the industry's origin was the natural gum inside the humble chicle tree. Chicle, still used today in small-batch specialty gums, is tapped in the same way as resin and latex, with crisscrossing incisions made across the tree. The polysaccharides in the natural gum render it quite unique and pleasing to the taste.

Your pantry has most likely been hiding plant defenses. Just take a look at the ingredient labels of some of the things you eat, drink, and use daily. You're bound to encounter a few common natural gums concealed as food additives in some of your favorite things. Table 2 lists some natural gums.

Table 2 Examples of Natural Gums

Natural gum	*Plant source*
Gum arabic	Acacia trees
Gum ghatti	*Anogeissus* trees
Gum tragacanth	*Astragalus* shrubs
Karaya gum	*Sterculia* trees
Guar gum	Guar beans
Locust bean gum	Carob tree seeds
Beta-glucan	Oat or barley bran
Chicle	*Manilkara* trees
Kino	Eucalyptus trees
Konjac gum / glucomannan	*Amorphophallus konjac*
Tara gum	Tara tree

MUCILAGE

Mucilage is a very unusual part of plant survival systems. Mucilage is transparent, clear, thick, gluey, extremely water absorbent, and sticky. It is made up of sugary polysaccharides and proteins. Plants produce mucilage from a variety of tissues, including seed coats, secretory cells, glands, mucilage ducts, and trichomes. Cacti, aloes, monkeyflowers, flax, licorice, jute, okra, psyllium, slippery elm, basil, chia, and carnivorous plants are just a few examples of prolific mucilage-producing plants. Even sweet, delicious marshmallows owe their origins to mucilage. Long before these light, fluffy goo-balls made it into your s'mores or crispy rice treats, marshmallows were first made using the heavy mucilage produced in the root of the marshmallow plant (*Althaea officinalis*). What makes plants produce all this slime in the first place?

Mucilage fulfills a strange and wide-ranging array of plant survival functions. First, it helps roots grow safely. A constantly flowing coating of mucilaginous gel lubricates and protects root tips as they glide between abrasive rocks and soil particles. Root mucilage even acts as a shield against toxic heavy metal–laden soils. An unusual but potentially useful example is Olotón corn, a variety of corn from the Mixe people of Oaxaca, Mexico. This highly studied variant of corn produces copious amounts of mucilage from its root tips, which in turn attract nitrogen-fixing bacteria. Scientists have found that Olotón's goopy roots are capable of absorbing nitrogen from the air to fertilize the plant. This unique ability may someday prove revolutionary: it could be bred into commercial corn varieties and reduce fertilizer inputs.

Second, mucilage is also produced in deciduous plants' buds and other aboveground tissues to aid in cold and freeze tolerance. Plants can easily lace mucilage with antifreeze proteins and polysaccharides, which lower cellular water's freezing point. Additionally, the water absorbency of mucilage gives plants the power to regulate their cells' moisture content. A slight coating of mucilage outside a cell drains internal water that might expand during a freeze. Otherwise, a frozen cell might explode and die. Frozen, solid water is unusable, so even well-hydrated plants face desiccation should they be frozen. Enter mucilage, whose antifreeze properties maintain cellular water in a salubrious liquid form.

A third task is water collection and preservation. In dry habitats, internal mucilage helps succulent plants such as cacti and aloes desperately absorb and retain any moisture their tissues are able to collect. Mucilage is also crucial to water collection in the upper canopies of some of the tallest trees on the planet. In redwoods, dawn redwoods, eucalyptus, and poplars, copious quantities of mucilage increase the taller the tree grows. The reason is mucilage's superb water absorbency. Hundreds of feet into the forest canopy, mucilage actually becomes more adept at supplying a giant tree's water needs than the plant's own roots and vascular system. At these tremendous heights, the roots and vascular systems of giant trees have reached insurmountable physiological limits. Tall trees encounter great trouble pulling water (which is quite heavy) hundreds of feet above the ground against gravity. Pumping increasingly high levels of mucilage within their upper foliage allows trees to efficiently wring moisture and humidity from the air and pull it right into the plant's cells. Plants use foliar mucilage to extract atmospheric moisture in two ways. The first is simply coating their leaves in thin films of mucilage, a strategy employed by the ulmo tree of the Andes (*Eucryphia cordifolia*) and the kingwood tree of the Amazon (*Astronium fraxinifolium*). A second and more widespread mechanism is the epistomatal mucilage plug, which is simply a plug of mucilage surrounding the plant's stomata. The common fig (*Ficus carica*), eucalyptus trees, and some oaks are just a few of the plants known to employ this widespread strategy. Both strategies offer superb water collection services for plants in need.

The fourth role of mucilage is as a protectant for seeds. When hydrated, such as by precipitation or saliva, the mucilage swells into an odd, gelatinous coating that surrounds and protects the seed from microorganisms, chewing damage, and the ravenous digestive systems of animals. The mucilage stimulates germination and the development of nascent seedlings by retaining extra water and dissolved nutrients around the tiny propagules. Basil and chia, both in the mint family, have seeds equipped with coatings of mucilage. Once damp, a group of these seeds looks and feels like a batch of frog eggs. Billowing within some commercial kombucha blends are harmless, jelly-like masses of chia seeds. As you gulp down your next shot of kombucha, wish these tiny seeds the best of luck in their survival journey through your omnivorous body.

You're looking at the octopus-like leaf of a Cape sundew plant (*Drosera capensis*). These carnivorous plants have plentiful trichomes that drool with a sticky mucilage. Sundews and other sticky-leaved plants use mucilage to capture and digest live insect prey.

The fifth clever use for mucilage can be found in certain carnivorous plants that are perhaps the reigning champions of goopiness. Carnivorous plants live in extreme environments around the world. The lack or difficulty of access to key soil nutrients has driven these plants to catch and break down other living organisms or their waste in order to create readily available fertilizer. These plants create some of the most advanced mucilage on our planet. They use this viscid slime to attract, ensnare, kill, and digest insect prey and the precious nutrients bugs contain. Not all carnivorous plants utilize this particular technique, however. There exist tube-like catching devices in the pitcher plant world and triggered snap-traps found in the Venus flytrap. But those that do employ mucilage are quite adept at its use. Major genera of mucilaginous carnivorous plants include the sundews (genus *Drosera*), butterworts (genus *Pinguicula*), dewy pines (genus *Drosophyllum*), and rainbow plants (genus *Byblis*). These are some of the most widespread and successful insect-capturing plants nature has to offer. A major reason behind their achievement is mucilage. Thousands of glandular trichomes secrete sticky mucilage across the surface of these plants' flypapery foliage. Each drop of mucilage is an example of some of the most intricate and amazing chemistry plants are capable of creating. These globules emit honey-scented volatile attractants, gaining the interest of prey. In smell and overt appearance, the droplets appear to be a harmless free drink of nectar or water. Dissolved glues within the mucilage prepare to grip the exoskeletons of any insect visitors. Nanofibers—fibers made of organic polymers so small that their diameters are in the nanometer range—are there too, invisible but strong and stretchy like mozzarella cheese. Once a sundew has ensnared any moving parts of a hapless insect, there is no escape. Sundew trichomes sense touch as well as chemistry, especially the nutrient-rich proteins possessed by their victims. The very writhing of live prey among the trichomes triggers the plant's digestion phase. The tapped trichomes send internal electrical signals that slowly induce sundew leaves into motion. The leaf blades and trichomes bend, curl, and wrap around the struggling prey. This process prevents the precious, winged packet of fertilizer from escaping, being stolen by predators, or being swept away by wind and rain. The octopus-like leaves wrap prey into intricate "bug sandwiches" or "bug tacos" as the constriction reaches completion. Collectively, as wave after wave of trichomes deposit their mucilage, prey is pinned and drowned right on the leaf. Surfactants spread the mucilage into every nook and cranny of the insect, not only securing a more binding hold but also allowing the mucilage to creep into the insect's spiracles. Fully coating an insect is lethal, since spiracles are vital pores in the exoskeleton through which insects breathe. Death by oozy suffocation soon takes place. Antimicrobial chemicals within the mucilage disinfect all prey before consumption, preventing rot from withering the valuable nutrients inside the carcass and ensuring that disease cannot spread by contact to the plant itself. Even a ravenous carnivorous plant will pause to disinfect its food before its drooling, bizarre, meticulous, and highly germophobic eating ritual. At this point, digestive enzymes within the mucus begin dissolving the prey's tissues into nutrient-rich fluid. The meal finishes when the

Sundews (genus *Drosera*) are carnivorous plants with advanced, mucilage-producing trichomes that function as killing fields. They use their viscid slime to attract, ensnare, kill, and digest insect prey and the precious nutrients bugs contain.

fertilizing, mucilaginous goo is slurped up eagerly by special absorptive glands all over the plant's leaves. Delicious!

For plants, what's inside really does count. Attackers breaching the outer surface defenses of plants are truly in for a hidden world of harm. Plant cells on the brink of defeat become ruined wastelands, bursting with deadly poison and slime. With every bite of a plant, animals chomp further and further into the minefield.

CHAPTER 5

Heavy-Duty Construction

How Plants Grow Tough

Plants have had to become remarkable structural engineers throughout their pursuit of life on land. Plants' lives are tough indeed, but they have responded by concocting durable molecules for strength, flexibility, and protection. Eternal stresses such as gravity, drought, and wind as well as the constant threat of attack ensure, quite literally, that only the strongest plants survive. Have you ever wondered what plants are made of? What's holding trees up even though they weigh thousands of pounds? Why are the barriers they make against the outside world so ridiculously effective? Even from a plant's first days as a seed, it's covered in a unique armor that plants created all by themselves. From germination onward, plants must incorporate structural support and armor in order to fend off the many challenges the environment can provide. Wood, bark, silica—these basic building blocks are all a part of the hidden world of plant physical resilience.

The Wooden Revolution: Cellulose and Lignin

The plant cell wall, wood, and bark are among plant life's greatest successes. Approximately 600–500 million years ago, a revolutionary, unprecedented event occurred. It was at this point that algae, after three billion years of ocean life, began adapting to growth on land. As discussed in chapter 1, the first true land plants were the bryophytes: diminutive, flat plants often less than an inch tall. You might know them better as mosses, hornworts, and liverworts. These plants began evolving from green algae roughly half a billion years ago. While rich in strong and flexible cellulose, the tiny plants were still quite limited in their ability to conduct water through their

Two botanical molecules, called cellulose and lignin, were responsible for the wooden revolution of our entire planet. Together, they create wood and bark, giving plants formidable structural and defensive powers.

tissues, contain high fluid pressures, and support their own weight. Lignin, a molecule that evolved around 450–420 million years ago in vascular plants, solved these problems. Vascular plants, the next group of plants to evolve on the planet, are named for their internal lignin-reinforced water-conducting vascular systems. They include most of the plants we see around us in our daily lives—taller, stronger, and more drought-resistant and durable than their ancient predecessors. Vascular plant evolutionary landmarks include the clubmosses (~410 mya), ferns (~360 mya), gymnosperms (~390–319 mya), and angiosperms (~125–100 mya).

The evolution of land plants was an unheralded development. As Earth's land masses began to green for the first time in the planet's history, it paved the way for the evolution of land animals. As plants spread across terrestrial systems, further oxygenation of the atmosphere and the deposition of soils through decomposition began to take place. Plants' natural history is our history. The molecules that plants built ages ago shaped Earth in a profound way that led to the evolution of our own species as well as the plant-derived natural resources that sustain us. Different structural

The first true land plants were the bryophytes: diminutive flat plants often less than an inch tall. You might know them better as mosses, hornworts, and liverworts, and they evolved from green algae around 850 million years ago. While rich in strong and flexible cellulose, the tiny plants were still quite limited in their ability to conduct water through their tissues, contain high fluid pressures, and support their own weight.

components, including two botanical molecules called cellulose and lignin, were responsible for the wooden revolution of our entire planet. Together, they create cell walls, wood, and bark, giving plants formidable structural and defensive powers. The utter profusion of cellulose and lignin on Earth is a testament to their splendid success as keystones of plants' survival systems. Let's saw into this topic and find out just what cellulose and lignin are doing for plants.

Botanical Rope: Cellulose

Thanks to plants, cellulose is Earth's most abundant organic molecule. All plant cells are walled off with cellulose, so it's molded into every leaf, trunk, seed, fruit, or root you've ever encountered. Cellulose is stiff, flexible, and strong. Plants use its mechanical support to assist in anchoring themselves to the ground, attaining tree-sized proportions, and hoisting their branches and leaves into the air. Up to half of wood, a quarter of phloem, and 17 percent of bark are comprised of cellulose.

The fibrous properties of cellulose result from its chemical structure. Cellulose is made of repeated units of a smaller molecule, in this case, the sugar glucose. Glucose, a sugar made by plants during photosynthesis, is usually in plentiful supply. Hundreds or even thousands of glucose units may connect to form a single cellulose molecule. Since the sugar units are connected in long, straight chains, cellulose molecules can easily form elongated fibers with high tensile strength.

If you were to look microscopically close at a plant cell wall, the entire structure would resemble the random weaves of a ball of yarn. A plant cell wall contains a labyrinth of crisscrossing, interwoven layers of cellulose fibers. Much like the steel rebar running back and forth through concrete buildings, cellulose fibers in the cell wall run omnidirectionally like woody botanical mesh. It's an impressive system that allows plant cells to expand as they accumulate water or grow, contract as they respond to drought, and absorb the mechanical forces from their own weight or from windstorms. The cellulose in the cell wall resists the breakages caused by the chewing, puncturing, and cutting action of herbivore feeding. Cellulose is also tough to digest—humans are unable to digest it at all, which is why we simply consider it "dietary fiber." Even though cellulose is made of sugar units, our digestive systems have never been able to outsmart the chemical bonds plants place in cellulose. This simple rearrangement of molecular bonds is all it took for plants to turn sugar into armor. Thus, we merely pass cellulose through.

Only a select few organisms have cracked the cellulose barrier in plant cells: bacteria and fungi in the form of plant diseases, parasites, and decomposers. Termites and grazing mammals called ruminants can digest plant matter because they formed symbiotic relationships with the decomposing bacteria that live within the guts of their animal hosts. To break through the cellulose walls around plant cells, fungi and bacteria assault them with cellulose-dissolving enzymes. However, plant cells are not stagnant structures. Plant cell walls are the most critical part of life support, so they're kept under diligent surveillance and maintenance. Each plant cell is constantly being worked on—built up, strengthened, repaired, and wrapped with more cellulose fibers whenever the need arises. If they detect broken pieces of cell walls, plants send even more lethal defenses to the rescue.

Ever worn clothes before? Thank cellulose! Cotton fiber is composed of more than 90 percent cellulose. Cotton fibers are borne on soft, puffy bolls that surround the plant's seeds. While domesticated cotton cultivars retain their bolls for easy harvest, cotton's wild ancestors used these lightweight fibers to help their seeds take flight.

Cellulose threads are so interwoven throughout our lives that it may astonish you to learn how connected you are with this botanical defense material. The ethanol and fossil fuels at your gas station, the veggies you eat, the cotton and rayon fibers in your clothing, and the timber in your home and furniture are all derived from cellulose. Every herb, shrub, and tree around you stands tall because of cellulose. Cellulose was the basis of one of the first plastics—ever heard of cellophane? That's right, much of the packaging in the food and goods you buy at stores is derived from cellulose! The plant fibers have simply been dissolved, reconstituted, and extruded through slits to form thin, transparent sheets that are then waterproofed with lacquer.

If you have a physical copy of this book, it's printed on thin, pressed matrices of cellulose fibers, otherwise known as paper. Paper has been around since the second century BCE. It is believed that the Chinese inventor Cai Lun came up with the idea after observing paper wasps chewing and secreting plant matter into papery nests. The stems, leaves, bark, and wood pulp of various plants have been preferred throughout human history to produce paper. Historically important paper species include the rice paper plant (*Tetrapanax papyriferus*), papyrus (*Cyperus papyrus*), paper mulberry (*Broussonetia papyrifera*), Florida strangler fig (*Ficus aurea*), Jamaican nettletree (*Trema micrantha*), and Texas mulberry (*Morus celtidifolia*). Today, "pulpwood" typically comes from species such as pines, spruces, firs, larches, hemlocks, eucalyptus, aspens, and birches.

The advent of modern technologies such as the printing press and highly automated wood harvesting has facilitated the production of disposable paper, making it an affordable commodity today. Consequently, there has been a surge in paper consumption and subsequent waste generation. The escalating global environmental challenges—ranging from air and water pollution to climate change, overflowing landfills, and widespread clear-cutting—prompted increased governmental regulations. In response to these concerns, the pulp and paper industry is now embracing a sustainability trend. Efforts are under way to minimize clear-cutting, reduce water usage, curb greenhouse gas emissions and fossil fuel consumption, and mitigate the industry's impact on local water and air quality. Despite these strides, environmental apprehensions persist in papermaking due to the use of harsh chemicals, substantial water requirements, contamination risks, and the carbon sequestration lost through deforestation resulting from clear-cutting trees used for wood pulp.

Despite the tremendous rise in demand for paper, the basic papermaking process hasn't changed much over a couple millennia. Cellulose fibers from plant foliage, bark, stems, and wood are first milled: shredded, pulverized, and suspended in water. To form sheets of paper, the wood "pulp" is then drained, pressed, and dried, a process that reconstitutes and interlocks the cellulose fibers into thin sheets such as office paper or into stronger materials such as cardboard and paperboard. Paper—essentially mashed plant defensive tissue—is extraordinarily important to us. We print words and pictures on it, clean with it, employ it in filtration, toss it around decoratively, box and package things in it, and use it as money. How many paper objects are important to your daily life?

Around 1846 a German Swiss chemist named Christian Friedrich Schönbein accidentally spilled a mixture of nitric and sulfuric acids all over his home "laboratory" (i.e., his kitchen table). Schönbein's frantic fumble around the kitchen inadvertently became one of the seminal "eureka" moments of his life. Out of pure chance and convenience, he swiped his wife's cellulose-based cotton apron to dry the spill. Satisfied that his wife would arrive home to find the kitchen table intact, Schönbein hung the apron to dry, whereupon it shortly burst into flames and disintegrated. The cellulose had reacted with the nitrogen in the nitric acid, forming some of the first nitrocellulose. Nitrocellulose, also called guncotton, is highly flammable, and it soon began replacing gunpowder in firearms, mining explosives, torpedoes, artillery shells, and naval mines.

The discovery of nitrocellulose was the basis of even further discoveries, such as a second type of early plastic called collodion. Collodion, a syrupy, highly flammable mixture, is made by dissolving nitrocellulose in ether or alcohol. In 1847 an American surgeon named John Parker Maynard found that collodion could be smeared on the skin as a surgical dressing. When dry, collodion leaves behind a sticky, flexible, and durable coating of nitrocellulose that helps protect and close wounds. In 1851 early photographers found a second use for collodion. It could be thinly coated over metal plates along with other chemicals such as silver nitrate, allowing the plate to be exposed to light and developed into color-reversed images called negatives. A negative results because light-sensitive photographic chemicals are darkened rather than bleached during the photographic exposure and development process. The true-color pictures we know as photographs are positive prints reproduced from negatives by inverting the dark and light colors. This early process, called the wet plate collodion process, made possible the first mass production of photographs. The oldest photographic method, called the daguerreotype, only yielded a single image that could not be easily copied.

A third plastic, called celluloid, is also derived from cellulose. Celluloid is the result of nitrocellulose being mixed with camphor (a plant VOC). Created in 1856, celluloid was the first thermoplastic, which means that it could be mixed with colored dyes, heated, extruded or pumped into

Celluloid, the first moldable plastic, is derived from cellulose. The first rolls of motion picture film were celluloid—thin, unbreakable, and perfect for holding the more than 150,000 individual photographic frames in a feature-length movie. Celluloid rolls scroll past the light of a projector at an average of twenty-four frames per second. Photographic negatives are also produced on celluloid film.

molds, and then cooled into intricately customized forms. Celluloid began replacing expensive and unsustainable animal-derived substances such as horn and ivory in jewelry boxes, utensil handles, billiard balls, and musical instruments. One of the most groundbreaking celluloid products was the first film for motion pictures, which was the invention that created the motion picture industry. In 1889 American inventors Hannibal Goodwin and George Eastman (founder of the Eastman Kodak Company) were working separately on thin, unbreakable, rollable celluloid films for rapid photography. The resulting film rolls could be projected and visualized so rapidly that they created the illusion that photographs taken in rapid succession were "moving." The motion picture film Goodwin and Eastman created was a staple of the movie industry, photography industry, and X-ray photography for almost sixty years until the invention of "safety film" in 1948. Under the hot lamps of a movie projector, the nitrocellulose-based celluloid film rolls had a notorious reputation for spontaneously combusting. Today, you can still find celluloid as a major component of guitar picks, ping pong balls, and musical instruments such as accordions. Just don't keep them near an open flame!

While we're on the subject of catching things on fire, let's talk about how cellulose powers the internal combustion engine. The internal combustion engines in our vehicles and machines have been igniting ethanol-based fuels for almost two centuries. In 1826 the American inventor Samuel Morey discovered a way to power an internal combustion engine using a liquid fuel: ethanol. You may have heard of ethanol before, but did you know that we produce much of it from plants? It's also the intoxicating alcohol in alcoholic beverages. The industrial production of ethanol utilizes great amounts of cellulose. Ethanol is fermented by bacteria operating on vast amounts of glucose molecules. The current best source for those sugars is cellulose. Agricultural waste is one of the most convenient bulk sources of cellulose. Cellulose is present in all plant tissues, but the most concentrated cellulose wastes come from sugarcane, corn, hemp, miscanthus grass, switchgrass, sugar beets, sorghum, sweet potatoes, potatoes, cassava, sunflower, fruit, wheat, cotton, and the next-generation biofuel, algae.

Cellulose is incredibly useful to plants as the armor and support for their cells. Cellulose is amazingly versatile and useful to us, too! We hope you can begin to see parts of the planet-wide wooden revolution in your own life. What do plant defense molecules such as cellulose do for you? The hidden strength of cellulose may be powering your daily commute, helping you get your daily serving of vegetables, or providing you with goods or information through the mail.

Barking Up Trees: Lignin

During the Carboniferous period, plants invented one of the best trump cards in their survival systems: the phenomenally tough, waterproof, and decay-resistant molecule called lignin. Lignin functions like an organic cement, binding, linking, and protecting cellulose fibers and other polysaccharides in plant cell walls. When animals and microbes evolved digestive acids and enzymes that could digest cellulose, plants had a serious problem. Their answer was lignin, essentially a more advanced, complex, and almost invincible molecule. Before we take a look at how lignin works, let's take a step back in time to see how monumentally important the evolution of lignin was for plants.

Around 360 million years ago, Earth became abruptly and tremendously lush with plant life. We know this because of fossil and geological records. This period, lasting from roughly 360 to 300 million years ago, is referred to as the Carboniferous period. The evolution of lignin was largely responsible for rich, diverse plant growth and subsequent deposition of huge masses of Carboniferous plant fossils. Other factors, such as a wetter and warmer climate, created planet-spanning rainforests, wetlands, swamps, and river systems that allowed plant life to proliferate and surge with vitality. Immense piles of carbon-rich woody debris and peat were laid down for millions of years, the likes of which have never again been witnessed in our planet's history. For tens of millions of years, the lignin-protected logs, branches, and leaves of entire forests sank into the mucky soils completely intact. Rot and decomposition scarcely existed at this point in time. The towering tree-sized ferns, giant clubmosses, horsetails, seeds ferns, and progymnosperms that dominated the Carboniferous were rarely toppled by anything but old age. The Carboniferous marked one of the most unchallenged and prevailing victories ever achieved by plants. Clad in lignin, plants found a way to render themselves virtually unassailable to the assaults of most other coexisting organisms on Earth. Major Carboniferous herbivores included the early ancestors of mammals, reptiles, piercing-sucking insects, and chewing insects such as grasshoppers, crickets, and cockroaches. The excessive botanical deposits lignin created during the Carboniferous are several times greater than during any other geological period in history. Today, Carboniferous geological strata can be easily identified by their dark, carbon-rich color. The plants that made them are now immortalized as fossils and coal. Thankfully, these fossils tell us enough about the story of lignin to explain why it made plants so successful.

The word "lignin" might sound unfamiliar, but it's quite plausible that you've already met some extremely high natural sources of lignin, such as tree bark, wood, seeds, and coal. All it takes to encounter the sturdy might of lignin is to give your nearest tree a hug—more than 50 percent of tree bark is composed of lignin molecules. This is where lignin shines, adding both mechanical and defensive support to plants. Lignin is a dominant molecule in some of the most successful protective structures that plants have ever created. Bark protects trees from herbivory, disease, wildfire, and drought. Seeds, which are comprised of a seed coat, the endosperm, and the embryo, rely on the lignified seed coat to keep them intact through trying environmental catastrophes and animal digestive systems. Lignified

This diminutive clubmoss species is a living relic of scale trees, the moss's ancient and gigantic ancestors that dominated Earth during the Carboniferous. The reason they did so is lignin. At its apex, each scale tree bore some of plant life's first branches, which enabled the trees to spread and disperse spores more efficiently. This particular clubmoss is enjoying the humid rainforest of Mount Kinabalu on the island of Borneo, Malaysia.

Scale trees (extinct genus *Lepidodendron*) and other Carboniferous plant life, unlike modern trees, were supported by massive, thick rings of lignin-rich bark and very little internal wood. This photograph shows the fossilized trunk impression of a *Lepidodendron* scale tree. The diamond-and-scale-shaped pattern that gives these trees their name results from thousands of leaf scars, so the trees must have been green and photosynthetic for parts of their life cycles. The majority of the intensely lignified tissues of Carboniferous vegetation sank intact into the swampy soils dominant at the time, fossilizing into coal and other rich black deposits.

vascular tissues became strong and waterproof enough to exploit water and nutrients in the soil and permit tremendous growth rates. The plant vascular system is such a critical and fascinating part of plant survival systems that it features in its own chapter of this book. To physically support the vigorous new forests growing across ancient Earth, lignin bound and laced cellulose fibers into towers of wood—the first trees! The plant world would never be the same.

Imagine you were one of the first herbivores encountering a three-foot-thick layer of bark in ancient land plants. Lignin turned plants into nearly unbreakable vaults of food. The Carboniferous's humble assemblage of rudimentary herbivores could exert only the most meager pressures on lignified plants. These early creatures included primitive mammals and reptiles, land snails, giant millipedes, primitive cockroaches (order Dictyoptera), primitive grasshoppers and crickets (order Protorthoptera), and sap-sucking insects (order Palaeodictyopteroidea). Advanced attacker equipment such as enamel-coated and self-sharpening teeth, metal-plated cutting mandibles, cell wall–degrading enzymes, symbiotic microbial digestion, and bacterial and fungal decomposition were far from even existing. Lignin's revolutionary success caused ancient plants to rely on it as a defense far more than the majority of plants do today. Log fossils of extinct plant life show massively thick rings of bark, a pattern contrasting with plants today. These bark barriers allowed plants to survive herbivory, disease, fire, drought, and the weight of their own tall and upright growth. Extinct trees such as the giant scale trees (order Lepidodendrales), horsetail trees (class Sphenopsida), and progymnosperms (order Cordaitales) had an estimated eight to twenty times more bark than wood. Modern plants rely far less on lignin, since they've since evolved many of the chemical defenses we highlight throughout this book. Most modern plants have actually reversed the lignin trend, instead making almost four times more wood than bark. This is why most plants we're familiar with bear relatively thin layers of bark. Rather than relying too much on lignin, modern plant tissues have much more powerful and advanced surveillance, communication, and chemical defense systems channeling through them at all times. Lignin is still of vital importance, but it is now only one part of plants' much more complex and effective chemical arsenal. Good thing, because despite the process having taken millions of years, fungi and bacteria have finally figured out how to defeat lignin. Their powerful enzymes and acids snap lignin's chemical bonds, generating sugars and carbon dioxide. Today, the only animal capable of utilizing lignin as a food source are termites, because they employ and support these symbiotic microorganisms in their guts.

The chemical bonds of lignin are special indeed. The way lignin molecules are built is the secret that renders them so successful against digestion. Like cellulose, lignin is an organic polymer comprised of a network of smaller units. Instead of sugars, however, lignin is made from recurring units called phenols, which are aromatic organic compounds. This means that lignin is also a type of molecule called a polyphenol (tannins, featured in this book, are also polyphenols). That is where the similarity between lignin and cellulose ends. Lignin perfectly complements the weaknesses of cellulose because it is waterproof, fire-resistant, largely indigestible, and rigid. What lignified plants do is embed each cellulose fiber in their cell walls within a protective shield of lignin. This process is similar to how rebar is joined with concrete to create a superstrong building material, increasing the strength of both products. While cellulose can absorb and conduct water, catch fire, be cut, or be digested easily with gut acidification or enzymes, the lignified layers of the cell wall prevent these processes from occurring. It is lignin that helps keep plants from drying out, burning up, succumbing to diseases, being eaten by herbivores, and falling over from their own weight.

Lignin is revolutionary for three reasons. First, unlike the linear, fibrous chains of sugars in cellulose, lignin's phenols *branch*. The result is that lignin's strength extends in all directions. This web-like network of chemical bonds is highly resistant to cleavage and mechanical damage by herbivores. Chewing, piercing, and boring through plant cells became much more difficult when lignin entered the picture. This network also meant that plants could support branching for the first time. Think about how many branching patterns exist on plants that we might take for granted. Large, heavy branches, moisture- and nutrient-seeking roots, rhizomes, intricate leaves, and vines were made possible thanks in large part to lignin.

Second, lignin is highly resistant to chemical and biological degradation. It may take thousands of years for a lignin molecule to be fully broken down. Lignin is a *massive* molecule, far too large to make it through microbe cell walls via direct absorption. As a result, the only option to digest lignin is to shoot digestive enzymes at it and hope for the best. White rot fungi, the most effective lignin decomposers on the planet, do exactly this. Even so, you will often find logs lying on the forest floor for many years. Additionally, lignin is insoluble and possesses strong aromatic chemical bonds, meaning that it is strongly unreactive and inert against digestive chemicals. All animals, with the exception of termites, simply pass lignin through their guts intact. When a lignin molecule does break down, it releases toxic, antimicrobial products that further reduce decay. Lignin is also virtually nutritionally sterile, starving the needs of most organisms with excessive carbon to nitrogen ratios between 350 to 1 and up to 500 to 1.

Third, lignin is one of the most extraordinarily heterogeneous molecules in nature. Like fingerprints, snowflakes, and the stripes on a zebra, most lignin molecules are exquisitely unique. Even within the same plant or cell, you'll rarely find two that are the same. The reason behind this is that lignin is designed to defy chemical degradation. Lignin molecules bond in a myriad of randomized and assorted ways, which destroys the ability of specific digestive enzymes to work. Digestive enzymes are usually specialists, holding the chemical keys to unlock only specific types of chemical bonds. In lignin, enzymes find a labyrinth—layers upon layers containing dozens of distinct and randomized chemical bonds. An enzyme that unlocks and digests the bonds of one layer finds itself stymied and useless when the next one is unique. Who knew plants had maze-like security systems at their front doors?

One of the most lignified defensive types of cells in plants is called a sclereid (also known as a stone cell). Stone cells are nonliving at maturity. Their contents are filled with lignin to protect the rest of the plant's living cells. Stone cell walls are thick, durable, layered, and highly lignified. You can often find solid, chunky lignin bundles and silica shards inside stone cells. To enhance their grittiness and abrasiveness, stone cells can be forked, branched, angular, or spherical. Apple cores, the grittiness of pear flesh, nutshells, peach pits, mango seeds, coconuts, peanut shells, sunflower seeds, and many other hardened plant structures feature stone cells as part of their survival systems. A solid layer of stone cells helps a plant protect the offspring encapsulated within its seeds.

Fiber cells, in which bundles of cellulose fibers are wrapped in matrices of lignin, are also formidable structures against herbivory. Like stone cells, fiber cells are nonliving and placed throughout plants, often in their bark and vascular tissues. Fiber defenses are heavily carbon-based. This type of defense is quite useful when plants live in habitats with impoverished soils or arid habitats. Carbon is a biologically inexpensive investment in these environments, since plants there are likely to be exposed to high levels of solar stress. High light intensity is conducive to the generation of carbon-based sugars during photosynthesis. Without enough precipitation or soil moisture to dissolve nutrients, plants become limited in their ability to build nutrient-based chemical defenses such as alkaloids. Let's explore how effective this defense can be against herbivory. Imagine taking a chewy bite off the tip of a piece of rope. The unpleasantly stringy, leathery bundles your teeth would be fighting are highly lignified chains of fiber cells. It's unlikely your teeth would win, and even if they did, the lignin would be immune to the efforts of your stomach and digestive tract. Monocots such as grasses, yuccas, agaves, the houseplant snake plant (genus *Sansevieria*), bananas, and New Zealand flax (genus *Phormium*) are all amazingly fibrous plants that feature this type of defense. In the storm chapter, we will discuss the columnar, rope-like trunks of palms—the lignin there helps them weather all manner of storms and herbivores. Most palms, such as the coconut and vining rattan palms, have incredibly strong fiber cells. Lignified fibers feature prominently in dicots such as jute, hemp, pawpaw, leatherwood, breadfruit, oaks, basswood, mulberries, hickories, willows, and elms. Conifers such as spruces and junipers also bear strong fibers. All of these plants have been identified and used as fiber and cordage sources by humans since prehistoric times.

Despite lignin having been one of plants' most monumental achievements, modern industry considers it mostly a waste product from paper production. Lignin is uneconomically tedious to break down because breaking its myriad bonds results in dozens of distinct breakdown products that all must be dealt with separately. Lignin also causes paper to yellow, so more than 98 percent of lignin separated from cellulose during paper production is simply burned in order to recover the energy of processing. The other 2 percent goes into things such as kraft paper; lignin gives the paper its familiar brown color.

Palm trees, like this Dominican palmetto (*Sabal domingensis*), are basically strong, rope-like columns of lignin. This massive specimen is growing at the Coastal Georgia Botanical Gardens in Savannah, Georgia. Because of lignin, palms can whip back and forth to survive hurricane-force winds. The long tough lignin fibers in their leaves and trunks are virtually invincible to herbivory.

Dig around in the soil sometime, and you'll soon undoubtedly turn up some lignin. The black organic humus that darkens and enriches soil is a leftover product from the breakdown of plant tissues. As you might surmise, it is largely created from remaining bits of lignin that nothing was able to decompose or digest. Once lignin is buried, the oxygen supply that bacteria and fungi need for decomposition is cut off, leaving the lignin intact for long spans of time. As lignin sinks down into deeper geological strata over millions of years, it becomes subjected to intense heat and pressure and can condense into coal. Lignin even lends its name to lignite, the least geologically altered form of coal.

There are many other places to find lignin hiding in your world. The next time you uncork a bottle, note that almost a fourth of the cork is made of lignin. Cork is actually the lignified bark of the cork oak (*Quercus suber*). Native to dry, fire-prone habitats in southwestern Europe and northern Africa, the cork oak bears a protective layer of spongy, two-inch-thick corky bark. When you prune a tough tree or shrub in your landscape, admire a stately forest, or live inside the timber frame of your house, you've also met with lignin. You might also experience it in your food. Vanillin, a phenolic compound known for producing the incredible flavor and aroma of vanilla, is also produced artificially from lignin. The distinctive, smoky taste and aroma of barbecue result when lignin's bonds are oxidized by fire, releasing a flurry of aromatic and taste-imparting by-products. Each plant species' unique blend of lignin is therefore partially responsible for the characteristic flavors woods such as apple, mesquite, and hickory transmit to food during smoking and barbecuing.

As we've seen, the physical materials that plants use to build their tissues, wood, and bark make quite a difference in their survival. Lignin and cellulose are valuable to us as well, giving rise to entire industries such as timber, clothing, paper products, movies, and the uniqueness of different foods. After the wooden revolution, plants became strong, durable, flexible, herbivory- and decay-resistant, and less flammable. This is because the amazingly intricate structures hiding in lignin and cellulose imparted significant resistances to plants' attackers. The invincible might of plant defenses remained mostly uncontested for millions of years during the Carboniferous period. Plant survival systems have expanded even more since then, adding further advancements such as indirect defenses and toxic chemistry. However, no matter how many novel and advanced defenses plants add, it is easily observable how influential lignin and cellulose were to their evolutionary history, shape, size, and resilience. Lignin and cellulose make plants what they are: trunks, sticks, roots, leaves, and branches. Thus, the world's plants, armored and reinforced with wood and bark, stand tall through each and every hardship. These molecules are such a ubiquitous and core advancement of plants that survival is impossible without them.

Silica: Why Plants Have Abrasive Personalities

Do sand, rocks, and glass sound tasty to you? Plants think they do. In fact, ten out of ten plants recommend a diet of sandpaper . . . yum! Every land plant on this planet feeds its herbivores silica, the element behind the gritty abrasiveness of many natural materials. Silicon is the second most abundant element on earth, second only to oxygen. Sand, glass, rocks, and numerous other minerals in the ground are silica-based and, therefore, quite accessible to plant life. While long ignored as a beneficial nutrient, silica is absorbed and utilized by plants at rates equal to or exceeding even macronutrients such as nitrogen, phosphorus, and potassium. The purpose of this section is to discover why plants love building themselves into green sandcastles so much. We will see how plants mine silica from the earth, secrete it into their tissues, and toughen themselves against stress and attack.

Silica deposits in plants are called phytoliths, after the Greek words *phyto* (plant) and *lithos* (stone). There are hundreds of intricate types, sizes, and shapes of phytoliths. They can be rigid, gel-like, glass-like, stony, sharp, crystalline, plate-like, or three-dimensional. Spheres, cubes, saw-blades, needles, footballs, crosses, triangles, dumbbells, and hooks are just a few of the distinct shapes phytoliths assume. Each plant species creates multiple kinds of phytoliths that are all completely unique. A scientist looking at one microscopic phytolith can distinguish the plant species it originated in, what kind of cell the phytolith belonged to, and the purpose of the phytolith. Phytoliths are so hard and decay-resistant that they remain long after the death of the plant that made them, surviving even wildfire, erosion, underwater deposition, and digestion by animals. Plants incorporate phytoliths just about everywhere, including their cells walls, organelles, spines, trichomes, vascular tissues, leaves, stems, and roots. Silica in the form of phytoliths is strong, durable, and waterproof. Plants capitalize on these properties by utilizing silica in structures such as cell walls, seed coats, and trichomes, which defend against herbivores and diseases. Silica coats a plant's surface in the cuticle and epidermis, preventing moisture loss. It also encapsulates a storage organelle such as the vacuole, potentially protecting a plant's cell from any deleterious chemicals housed within. Lastly, silica is a constituent of water-transport vascular tissues. It helps them remain waterproof and strong enough to resist the pressurized fluids within them. Silica and cellulose were the first structural materials manufactured by plant life. The ability to enrich living tissue and cell walls with silica and cellulose originated in ancient microscopic sea life such as algae and cyanobacteria, the ancestors of all plant life.

Because phytoliths can lie preserved in the ground for eons, they have become invaluable for taxonomists and archaeologists. Phytoliths have a fascinating hidden world of stories to tell if we only listen. Phytoliths remained hidden to science until the 1830s, when the German botanist

Christian Gottfried Ehrenberg began isolating the tiny particles from soil samples, coining the term "phytolith." One of Ehrenberg's major discoveries included thousands of unique phytoliths in dust collected by Charles Darwin from the sails of HMS *Beagle*. Sediment cores from Antarctica, for example, have contained phytoliths that allowed scientists to determine that Antarctica once was closer to the equator and hosted ancient rainforests. A simple burnt plant residue can leave behind phytoliths that etch into pottery shards, rocks, soil layers, or ancient fire pits and structures. These residues tell us what plants were cooked or stored in ancient cookware or destroyed in wildfires. The bits of silica in the coprolites (fossilized feces) of dinosaurs and other animals tell us which plants made up the animals' diets. This is all because each phytolith has a telltale shape that reveals what kind of plant made it and the type of tissue it originated from.

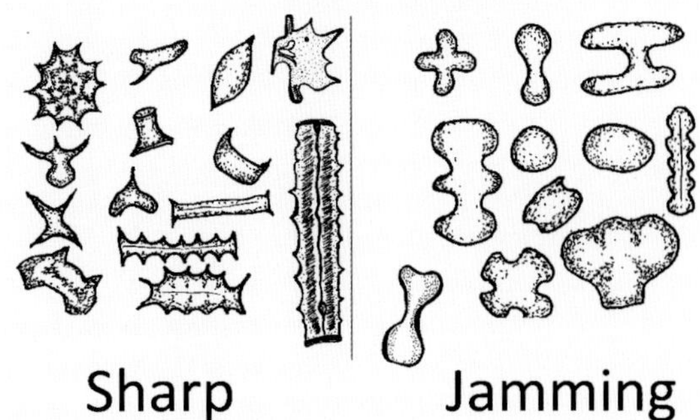

Silica is a botanical multipurpose material. Phytoliths are excellent structural building blocks, so you'll find them bolstering cell walls and vascular tissues to add mechanical support and flexing strength. This strength comes in handy, preventing cell collapse during droughts and bursting during freezes. The brittle, stiff, and sharp properties of silica also feature in the construction of trichomes and puncturing devices such as spines, thorns, and prickles. Silicate barricades in cell walls can slow or halt the mouthparts of pests, the digestive assault of fungi and bacteria, and the oxidizing effects of fire. Phytolith barriers encapsulate poisons and toxins within the vacuole, keeping plant cells safe despite carrying risky payloads. Toxic concentrations of salt, heavy metals, and defensive chemicals are moved and sequestered safely inside silicate barriers. Silica's involvement in these processes is not simply structural; it also occurs on the chemical level. As such, much of silica's role is behind the scenes inside a plant, such as its ability to help manage the production and translocation of nutrients, hormones, and signaling molecules.

To craft biosilicate components, meaning any organic, silica-containing parts of their cells, plants have a fairly straightforward production process.

First, silica is absorbed by plant roots as a naturally occurring liquid in the soil called silicic acid. Traveling with the xylem's upward flow, silica begins depositing itself in tissues such as leaves and stems. As plant leaves evapotranspire, the water carrying the silica evaporates away, leaving the silica to turn into gel-like polymers, salt-like precipitates, or glass-like phytoliths. Silica packs into the epidermis, spines, trichomes, wood, bark, cell walls, intercellular spaces, and membranes of cells. Silica-based defenses are some of the simplest and cheapest for plants to produce because silica is plentifully available in the soil. It requires only to be moved and secreted into tissues. A carbon-based defense such as lignin and nitrogen-based defenses like alkaloids have complex and intensive manufacturing processes that cost cells far more nutrients. Silica is an affordable alternative for plants to utilize in their protection.

The most significant reason plants employ silica is as a defense against herbivores. Gnashing, shredding, chewing, and digestion all become much more difficult when silica is present in plant tissue. Silica's grittiness is not simply unpleasant. It actually turns plant tissue into a grinding stone against the mouthparts of herbivores. This defense is universally effective, affecting minute herbivores such as slugs, termites, and caterpillars all the way up to rodents, cattle, and elephants. Millions of phytoliths scrape across the teeth and mandibles of herbivores as they chew. Each microscopic particle etches the teeth, wearing small channels into their surfaces. Over a lifetime, phytolith action can deform and destabilize teeth completely. These changes are irreversible and intensify over time. As omnivores, humans and their ancestral species have been dealing with this problem for quite some time—fossilized primate teeth often include phytoliths and associated scratch marks. Some phytoliths can be harder on the Mohs scale of mineral hardness than the structural constituents of mammal teeth, which include enamel, dentine, bone, and cementum. Silica can even challenge the hardness of steel. When your lawnmower or power tools go on the attack, it only takes plants a few hours of wear to dull their blades. Controlled feeding experiments on a wide range of herbivores all have documented similarly intense wear, including enamel removal and surface abrasion, caused by phytolith action on mouthparts. This sandpapering effect isn't noticeable right away, but over time it creates substantial feeding inefficiencies, even if the herbivore decides to change

Horsetails (genus *Equisetum*) are spore-bearing plants so rich in silica that bunches of their stems have long been used to clean cookware, giving these abrasive plants a second moniker: scouring rushes. Horsetails are living fossils, the remnants of a much larger order of land plants that originated and dominated Earth during the Carboniferous period more than 350 million years ago. Their extinct relatives, such as the giant genus *Calamites*, were among the first tree-sized land plants, because silica strengthened their stems. The survival of horsetails into the modern era testifies to the power of silica-based defenses.

This horsetail (*Equisetum hyemale*) stem is tough and gritty, loaded with abrasive silica deposits. Herbivores might as well eat cylinders of sandpaper.

diets or molts a fresh set of mandibles. The abrasion from silica-rich diets universally causes long-term developmental effects such as reduced weight and nutritional deficiencies throughout the animal kingdom. Overall, silica helps slow down the assault of attacks on plant parts.

Plants also craft silica-based roadblocks for their attackers. Trichomes, spines, prickles, and other hardened structures slow animals' feeding by forcing them to take smaller bites or inefficient feeding pathways on the leaf. For example, a caterpillar might have to chew awkwardly around silica-rich trichomes, leaf veins, or hardened leaf edges. The slower and less efficiently an herbivore feeds, the slower it grows and damages the plant, giving bitten plants more time to ramp up further defenses such as volatile organic compounds and toxins. Plants bear special silicate structures tailored to different types of feeding. For example, against the rasping damage caused by thrips and the piercing damage from aphids, a leaf might have silica-armored tissues between the leaf surface and vascular bundles. The sharp, silica-rich teeth on the margins of grasses, as seen in the pampas grass photo, can slice open "grass cuts" or "paper cuts" on the skin, mouthparts, or intestines of mammals. Leaf-margin feeders such as caterpillars and grasshoppers are also counterattacked every time they bite into one of these shark-like blades.

Imagine if your next meal was made of marbles and the precious food you desired was compartmentalized inside them. That's very similar to the effect silica has on digestion. The tough, marble-like, silica-rich cells of plants block digestion and hinder the mechanical breakages caused by chewing. Specifically designed "jamming" phytoliths are even shaped like spheres, cubes, hearts, dumbbells, or saddles to wedge in between the cutting edges of teeth. If a plant cell can resist the compression and cutting forces of jaws and the acids and enzymes of digestion, the water, starch, proteins, and nutrients inside never release to benefit the herbivore. Animals can only deal with siliceous plant matter with intense, protracted chewing, which multiplies the wearing effect of even a single mouthful of foliage. This is one reason why the group of large grazing animals called ruminants evolved a rumen, which is a specialized chamber of the stomach where cell walls are slowly broken down by symbiotic bacteria before being passed on to the rest of the stomach. Ruminants include animals

The grass family has some of the most prolific and advanced usage of phytoliths. The reason the leaves of the pampas grass (*Cortaderia selloana*), shown here, can papercut your skin is because of the silica-rich blades on their margins. Parallel trenches of microscopic prickles are poised to slice and abrade the mouth, teeth, and guts of herbivores during chewing and digestion. Peering into the hidden world of grasses transforms them from everyday plants in your lawn into monstrous, terrifying organisms. Eat up, herbivores!

such as cattle, giraffes, deer, sheep, moose, reindeer, elk, bison, antelopes, gazelles, and goats.

But plants get tougher still. Now, imagine if the food-marbles were studded with tiny fragments of broken glass. There are needle-shaped, hook-shaped, branching, cone-shaped, and spiny phytoliths designed to cause cell damage during their passage through animals. Phytoliths scrape not only the mouthparts but also cells along the entire digestive tract. If a digestive cell is injured and in the process of repair, absorbing nutrients becomes much more difficult. Microwounding occurs in addition to the other digestibility problems silica creates. Mammals, which can heal their digestive zones, fare much better in this regard than many insects. Quite a few insects keep their original digestive tracts completely intact through metamorphosis and have a much harder time repairing them.

Ever wondered why bamboo is famous for growing incredibly fast? Part of the reason is because bamboos build their stems, leaves, and defenses out of cheap, plentiful silica. Bamboos, which comprise the Bambusoideae subfamily of vigorous tree-sized grasses, have tissues with 17–23 percent silica. Bamboos are one of the highest natural sources of pure silica.

Silica is a universally employed plant defense, but some botanical families take it to the extreme. The most well armed plant families include horsetails, legumes, cucurbits, asters, grasses, palms, and sedges. Grasses may top the list as the most prolific users of silica defenses, with tissues that can approach 2 to 6 percent silica content. This is several times more than most other flowering plant families. The grasses have such abrasive personalities that they have driven the mouthpart evolution of major herbivore groups throughout history. Grasses evolved in the Cretaceous period, roughly 150 million years ago. Now grass-dominated habitats such as savannahs and prairies range across an estimated 40 percent of Earth's land area. During this colonization period, fossilized grass remains dramatically increased. We know this not only from fossilized tissue but also from the conclusions paleobotanists can infer from the deposition of phytoliths and pollen. This evidence correlates to fossilized coevolutionary adaptations seen in major herbivores.

For example, grazing mammals developed high-crowned teeth (called hypsodonty), which contain much more material to allow wear. Unlike those of humans, the teeth of grazing mammals are several times larger and longer, extending far into the gums and containing much more strengthening material. The size of grazers' teeth in the fossil record matches evolution in the proliferation of grasslands. Hypsodonty, perfect for lifetimes of grinding and mashing, evolved in ungulates such as horses, cattle, deer, camels, zebras, rhinoceroses, oxen, antelopes, bison, and elephants. This wasn't the only solution, however. Other groups of mammals have evolved teeth that grow continuously throughout the animal's lifetime (called hypselodonty). These animals include the rodents (rats, voles, prairie dogs, shrews, mice, beavers, squirrels, chipmunks, chinchillas, guinea pigs, hamsters, porcupines, and capybaras) and lagomorphs (rabbits, hares, and pikas). Ever-growing teeth are just about the only thing that can keep pace with how fast silica wears teeth out.

Insects took a third direction, perfecting mouthparts that cut like garden shears. Insects such as the larval—or caterpillar—stage of lepidopterans (butterflies and moths), termites, coleopterans (beetles, including wood borers), and orthopterans (grasshoppers, crickets, and locusts) have mouthparts called mandibles, which are chewing mouthparts that are excellent at slicing through matter. These cutting devices are among the most advanced biological blades on planet Earth. As grasses have proliferated, insects have consistently grown larger and sharper mandibles over time. Mandibles must remain razor-sharp and durable throughout each stage of the insect's metamorphosis. If a bug's jaws become too dull to open plant cells, it cannot feed, and it will starve to death. Thankfully, mandibles come with a fantastic lifetime warranty; a full replacement awaits each time the insect sheds its exoskeleton, a process otherwise known as molting. The cutting surface of the mandible is called the cuticle. Mandible cuticles are reinforced, multilayered composites of proteins, lipids, metals, and carbohydrates. A polymer such as chitin, the major structural component found in the exoskeleton of insects, is often cross-linked with structural proteins and enriched with metals such as zinc, copper, iron, and manganese. What results is a scalpel-like cutting edge that makes even nanotechnologists envious. While they seem like overkill, such extraordinary biological

The extraordinary slicing mouthparts of herbivorous insects such as this grasshopper are called mandibles. Mandibles have evolved cutting surfaces that must remain sharp and durable throughout the insect's life stages. Some species, such as termites, have even evolved to grow mandibles impregnated with metal ions such as zinc, iron, manganese, and calcium, rendering their mandibles some of nature's most effective cutting and chewing devices.

shears are a necessity to slice repeatedly through fortress-like plant cells.

All around us, steady streams of billions of phytoliths are flowing through the guts of voracious herbivores. Humans are included, too, as phytoliths are undoubtedly awaiting in your next salad or plate of veggies. Sandy grit in your shoes, grass cuts, the crunch of crisp produce, and the strain of push-mowing your lawn are all ways you can come in contact with the hidden power of silica.

Seeds: Tiny Armored Plants

Almost thirty-two thousand years ago in what is now modern-day Siberia, an industrious little Arctic ground squirrel (*Spermophilus parryii*) stocked the subterranean chambers of its nest full of plant seeds for winter, including some of an ancient ancestor of narrow-leaved campion (*Silene stenophylla*). That winter, though, was a cold one—possibly even more severe than the ground squirrel could survive. Some of the seeds inside its burrow went uneaten and remained alive. This event occurred during the early years of an ice age, and the squirrel's burrow slowly became entombed in permafrost, a continuously frozen layer of soil, as the ice age progressed. The nest and its contents would stay hidden, solidly frozen and preserved for tens of thousands of years. Ages passed—hunters and gatherers roamed, dogs were domesticated, rudimentary wooden and stone tools were crafted, pottery was invented, and cave paintings became the first examples of art. Humans progressed through agriculture, writing, civilization, and science before the awe-inspiring survival potential of these humble seeds, encapsulated in this undisturbed squirrel burrow, would be discovered. The seeds finally came to light in 2007, when they were excavated and cultured into live adult plants by Russian scientists. Carbon dating ascertained that the seeds had experienced more than thirty-two thousand years of frigid stasis and natural gamma radiation from the ground. How is it possible for plant life to survive circumstances few other organisms can live through? Their secret lies in one of their most successful inventions: the seed. In the case of the death-defying campion, its tissues were so packed with sugary sucrose that it acted like botanical antifreeze. In this section, we'll crack open the mysteries of seeds, arguably plants' ultimate survival structures.

The power of a seed lies in its tough seed coat, food reserves, dormancy mechanisms, toxins, and suite of dispersal strategies. These adaptations make seeds resistant to drought, fire, cold, flooding, herbivore digestive systems, and disease attacks.

Seeds are primordial plant offspring encased in armored shells, sealed for moisture preservation, and stocked with some food to get them started. Once plants developed the ability to craft hardened cell walls using lignin and cellulose, they quickly began swathing their progeny in these materials. Plants began crafting the first seeds approximately 420–350 million years ago throughout the Devonian period, enabling the successful colonization of new ecosystems. Gymnosperms and angiosperms are plants' modern seed-bearing descendants. Gymnosperms created naked seeds, while angiosperms gained the addition of a fruit, which can be dry or fleshy. Seeds can be carried via wind (such as the parasol of the dandelion), on fur or feathers (such as the cocklebur [genus *Xanthium*] or stick-tights [genus *Desmodium*]), by floating on water (such as the husk of a coconut), or by being eaten (such as the fruit of the grape). Plants make heavy investments in the success of their offspring. The power of a seed lies in its tough seed coat, food reserves, dormancy mechanisms, toxins, and suite of dispersal strategies. These adaptations make seeds resistant to drought, fire, cold, flooding, herbivore digestive systems, and disease attacks.

Some of the earliest seed-producing plants extant today are gymnosperms called the cycads. These *Cycas panzhihuaensis* seeds and their fleshy, fruit-like orange cones have an ancient evolutionary history stretching back to the time of the dinosaurs. Once eaten and dispersed by dinosaurs, the seeds' hard coatings, enormous food stores, and biochemical toxins ensured baby cycads' survival for millions of years.

The Arctic ground squirrel (*Spermophilus parryii*) is an example of a granivore, or seed predator. Inside the comfort of the rodent's burrow are caches of seeds waiting to be gnashed and murdered. Will the seeds' defenses be strong enough to help the seeds survive?

Unfortunately, since seeds are such a highly concentrated food source, they are attractive to a large array of animals called granivores. Granivores are specialized to prey on seeds and originate from all parts of the animal kingdom. Mammals (such as primates and rodents), birds (such as blue jays, turkeys, doves, finches, and parrots), reptiles, fish, crabs, and insects (such as beetles, weevils, and ants) all eat seeds. A seed must be designed to handle the sharp teeth and mechanical crushing forces of all possible granivore jaws. Still more hazards await as the seed passes through digestive systems and their chemical acids, gut bacteria, and abrasive gizzards. Many seeds have evolved to specifically journey through the digestive tract of an animal, though that is not the only mode of distribution. If the seed can emerge from this gauntlet victorious, a rich, moist pile of fertilizer-rich dung awaits it, far away from its competitive originator! To a seed, that's like winning the lottery.

A seed's primary line of defense is the seed coat, its shell of armor. The seed coat is actually comprised of multiple distinct layers of defense. Sometimes, hardened layers of the plant's fruit (called the pericarp) provide additional protection to the seed. Nuts such as acorns, pecans, walnuts, and pistachios are good examples of fruits with sturdy pericarps. Some seed coats, such as those of chia and basil, secrete external chemicals like gloopy mucilage. The mucilage primarily collects and holds water around the seed for germination. It can also glue dirt to the seed as camouflage, creating sand armor. Gloopy seeds can hide like nobody's business. Protection provided by a pericarp, mucilage, or other external defenses is optional depending on plant species, however. The outer layer of the seed coat is called the cuticle. It controls water diffusion into and out of the seed. Preventing water loss from inside the seed keeps the embryo alive. All embryos must stay moist. The cuticle also creates seed dormancy by keeping water from inducing germination prematurely. The most important function of the cuticle is to keep diseases and granivores at bay by repelling attack chemicals such as digestive enzymes before they can penetrate into the seed. Below the cuticle is a layer of strong, dense, waxy, thick, and highly lignified cells called palisade cells. This layer is what actually creates the shape and hardness of a seed coat and further increases its water impermeability. The hard shells of legumes and sandspurs (genus *Cenchrus*) are great examples of robust palisade layers. The palisade layer can render a seed fibrous, hairy, spiny, or quite simply rock-hard. The insane amounts of lignin and wax bound around a seed in this layer enable it to resist mechanical crunching, wildfire, drought, cold, and diseases. Anything that makes it past the palisade cells reaches a third fortified layer of the seed coat, the sclereid layer. As we know from previous chapters, sclereids (stone cells) are gritty, durable cells with spine-shaped chunks of lignin and silica. The hulls of coconuts, nuts (such as acorns and walnuts), and peach pits are robust because of sclereids. Only after grinding through all of these defensive layers can a bacteria, fungus, or digestive tract begin to reach the precious plant embryo. Some plant species have even evolved to require this seed coat erosion process (called scarification) to occur for germination.

Acorns and other nuts are protected by hard, dense, and highly lignified shells. In the case of acorns and many other nuts, this tough layer of armor is technically a part of the fruit called the pericarp. Other seeds, such as legumes, have similarly woody seed coats.

This avocado seed contains baby food—not for humans, mind you, but for the tiny plant embryo inside. The brown seed coat of the avocado is quite thin, but the cream-colored endosperm layer is massive! The humongous stockpile of food allows the seedling to survive in the shade of rainforests as well as sprout through massive piles of disperser dung.

Investing in the Seed Bank

There are more than four hundred thousand seed-producing plant species on Earth. Each species has evolved its own unique strategy to create a substantial, persistent seed bank in the soil. Plants create long-lasting seeds with internal food supplies and dormancy mechanisms. If constructed right, a seed will have the resilience to resist attacks and severe environmental forces, and it will even survive long after the death of the plant that created it. Millions of seeds lie hidden underground around us in stasis. Each one is making a gamble for survival: Is the investment its predecessors made in its seed coat, food stores, chemical defenses, and dormancy programming enough to allow it to survive and produce the next generation of its species?

Seeds, like fortresses, are packed with enough food to survive many years of the sieges of predation and cataclysmic environmental changes. Seeds contain food for the tiny plant, a stockpile of carbohydrates, proteins, and minerals called the endosperm, which provides sustenance to the developing seedling. The endosperm, which results from an intricate process called double fertilization, is an amazing evolutionary feat. Plant pollen actually releases two sperm nuclei. One fertilizes the egg cell, forming the seed's embryo, which will develop into a new plant. A hidden fact to most of us is that the other sperm actually causes a separate, second fertilization inside the embryo sac. The first cell of the endosperm forms from this double fertilization. Thus, a seed's food reserves are an entirely unique organism, sort of like a sibling, that aborts and sacrifices its entire being for the development and nourishment of the embryo.

Little plant seedlings are more precocious than we might imagine. While seeds appear to just lie passively in the dirt, they are brimming with life and activity inside. Along the road of evolutionary advancement, plants began equipping seeds with sensory equipment to record and respond to the surrounding environment. Seeds are fastidious about the day they choose to germinate. All conditions must be optimal, lest the next wildfire, drought, or roving pack of herbivores end the seedling's life as quickly as it began. Once germinated, a seedling is incredibly small and vulnerable, and it cannot reverse its decision. If the seed inadvertently germinates during a volcanic apocalypse, best of luck! The genetically encoded ability of seeds to prevent their own germination is called dormancy. Plants' best chances for survival and procreation occur when they've built a large, persistent

The lotus (*Nelumbo nucifera*) is an aquatic plant that thrives in flooded ecosystems. Its hard, nut-like seeds prevent moisture loss so well that the seed can withstand centuries of drought. It possesses mostly seed coat dormancy. In a famous example, 1,300-year-old lotus seeds were unearthed from a dry Chinese lake bed in 1994. Once scientists placed them in the flooded, fertile soils they were waiting for, the seeds germinated in the same four-day span as a control group of seeds from their modern descendants.

stockpile of dormant seeds in the soil (called a seed bank). Seeds employ physical and chemical methods to distinguish between favorable and unfavorable circumstances for survival. The most important parameters for seeds to measure are temperature, humidity, gravity, moisture, oxygen, and light. Over time, dormancy adaptations allow a plant species to adapt and disperse successfully to specific geographic ranges and climates.

As they craft a batch of seeds, plants closely monitor environmental parameters, too. Each crop of seeds is unique because its dormancy characteristics are meticulously adjusted by its progenitors. A favorable year with abundant light and moisture might favor abundant masses of seeds with thinner seed coats or fewer chemical inhibitors. A tougher year full of drought, fire, or cold leads to fewer seeds with stronger dormancy. There are three general mechanisms that seeds use to become dormant: seed coat dormancy, physiological dormancy, and morphological dormancy.

Seed coat dormancy (also called physical or mechanical dormancy) occurs when a seed coat prevents water or gases from permeating the seed. Natural forces must modify the seed coat in such a way that either of these cues can reach the embryo. Seed coat dormancy can be broken by natural events such as fire, drying, temperature fluctuations, freezing or thawing, melting of waxes, mulching, and light. In a process called scarification, small abrasions or breaks in the seed coat can occur as the shell is eroded by soil particles, affected by weather, or attacked by an animal's digestive system. Mechanical dormancy is a type of seed coat dormancy wherein the seed coat is so strong that it suppresses the growth and expansion of the embryo. Seed coats can also be permeated with chemicals called germination inhibitors that must be leached over time by precipitation before germination can occur.

Physiological dormancy strategies occur internally in seeds. Usually, chemicals called germination inhibitors permeate the seed's tissues. Unless the inhibitors are eliminated, the seed cannot germinate. A common germination inhibitor is the plant hormone called abscisic acid. Abscisic acid's concentration is influenced by light, meaning that both light and darkness can stimulate seeds to germinate depending on a species' preference. Seeds use this hormone to tell whether they've been buried or whether they'll have access to enough light for optimal photosynthesis. Abscisic acid has a special relationship with its antidote, the plant hormone called gibberellic acid, which regulates the growth of leaves, stems, and seed germination by controlling cell division and elongation. Gibberellic acid is a major mechanism allowing seeds to read environmental cues. Seeds begin converting some of their starches to gibberellic acid in response to heat, fire, cold, and water entry. Other germination inhibitors may leach slowly after precipitation events or after elimination by animal digestive systems. When an embryo reaches its preferred concentration of gibberellic acid, it's showtime! These important and amazing chemicals enable each and every seed to be aware of its changing environment.

Morphological dormancy occurs when the embryo itself is immature. A plant such as a magnolia actually releases its seeds before the embryo has completely developed. Even though a fruit may ripen and fall or even be eaten and dispersed, the half-developed embryo is tucked safely inside the seed coat. It continues to grow and mature for a few more weeks or months before germinating. Morphological dormancy is basically like a seed containing a time-delay fuse.

Another way plants go the distance for their seeds is with truly clever dispersal strategies. The quicker and further a seed can flee from its source, the less statistically likely it is that all of the seeds will fall victim to herbivory, granivory, competition with their relatives, environmental catastrophe, or disease. For this reason, we consider dispersal to be an excellent part of plant survival systems. The object of dispersal is not only to reduce intraspecies competition but also to defensively create spatial and temporal shortages in seed crops.

Once ripe and packed full of delicious food, all seeds are essentially fleeing for their lives from granivores. Amazingly creative and advanced mechanical engineering has evolved in seed dispersal. Plant species exist that equip seeds to hitch rides on fur or in digestive systems, fly, swim, catapult, explode, camouflage themselves, and bury themselves. Fascinating and aptly named plants with burr-based dispersal include beggarticks, sock destroyers, devil's claws, burdock, avens, and puncture vines. Winged or tufted appendages for wind dispersal can be found on dandelions, maples, milkweeds, Javan cucumbers, orchids, sycamores, willows, cattails, grasses, tumbleweeds, and many other plants. Some seeds are housed in fruits that act like catapults or explosive bombs, including siliques of the mustard family, bean pods in legumes, oxalis, columbines, jewelweed, geraniums, exploding cucumbers, witchhazel, and euphorbias such as tapiocas and castor beans. The seeds of the common stork's-bill, needle grasses, black needle grass, and needle-and-thread grass have humidity-activated drills. Each seed has a drill-shaped awn that buries it in the soil. The awn flexes hygroscopically—moisture causes the drill to uncoil, while drying tightens the self-drilling seed into the soil another turn. The drilling power and sharpness of these seeds are so powerful that they are able to bore through fur, skin, and even muscle tissue, posing a hazard to livestock. Almost all of these transportation adaptations can be found in angiosperms, since the ovary ripens to become a fruit. Fruits can be fleshy, as found in strawberries, cherries, and oranges, or dry, as found in the parasols of dandelions, the hooks of burrs, and the wings of maple seeds.

The amount of seeds produced in a crop, called seed masting, is also a survival strategy. Plants may decide during certain years to bear "bumper crops" of seeds, followed by years when plants produce severely low or practically no seeds and fruits. The decision is often based on the stresses, temperatures, moisture, and nutrition the plant experiences during the year. Oaks, pecans, hickories, apples, bamboo, almonds, and pistachios are notorious for this seemingly mysterious, finicky, and flip-flopping nature. The reason behind it is that it regulates herbivore, granivore, and disease populations. Without enough food, these organisms either move on to better territory, starve, or perish. If a plant has a boom-and-bust cycle of fruit and seed production as its life history, you can bet its intent is to purge attackers.

Seeds can also be laced with poisons and repellents, many of which are featured in detail throughout this book. Plant offspring are protected by chemicals such as the capsaicin in chili seeds, caffeine in coffee and chocolate, saponins in quinoa and beans, cycasin in cycad seeds, ricin in castor beans, mustard oil bombs in mustards, and cyanogenic glycosides in peach and cherry seeds. More universally employed chemicals are ones that interfere with digestion. Tannins, which imprison nutrients, carbohydrates, and proteins, impede digestion. Other chemicals in seeds inhibit common digestive enzymes such as amylase (used in saliva to digest starch) and proteases (used in the stomach and intestines to digest proteins).

The next time you discover a seed in your world, stop to ponder about its life story. What attackers might it fear the most? How did it flee its source of origin and arrive at its present location? How powerful are its external and internal defenses? How long do you think the seed could survive based on its food stores and dormancy traits? Perhaps you might even consider planting it to watch the next chapter of its story unfold. Each encounter with a seed allows us a chance to tap into the massive, hidden evolutionary story of the tiny propagule now at our feet.

CHAPTER 6

Plumbing Revolution

The Vascular System

Water, minerals, sugars, and defenses are flowing secretly inside the plant life around us. While plants seem like solid organisms, they contain 90–95 percent water. Water regulates their temperatures, supports their cells, transports nutrients, and sustains processes such as photosynthesis. A branching, intricate network of conductive tissue called the vascular system spans most plants' leaves, stems, and roots. Plants' vascular tissues form a busy, dynamic, and interconnected system. If things are going well, all parts of plants are continuously supplied with the water, nutrients, and sugars they require. Both the tallest and the smallest plants are piped with astonishing complexity. When we turn on the faucets in our homes, the water we see has moved through many yards of pipes, flowing clandestinely throughout our homes' floors and walls. Hundreds of millions of years before humans ever came up with the invention of plumbing, plants figured out how to channel and distribute massive amounts of fluid using pipe-shaped cells. The evolution of the vascular system is one of the most critical advancements plants have ever made, enabling the colonization of land and the biodiversification of our planet. Vascular tissues brim not only with valuable nutrient-rich fluids but also with powerful defenses.

Xylem

The vascular system is made up of two primary tissues: xylem and phloem. While phloem primarily moves around sugars, xylem is responsible for conducting the water and nutrients extracted by the root system to the rest of the plant. In general, xylem's flow pattern is upward, even hundreds of feet into the canopies of the tallest trees. If you've ever

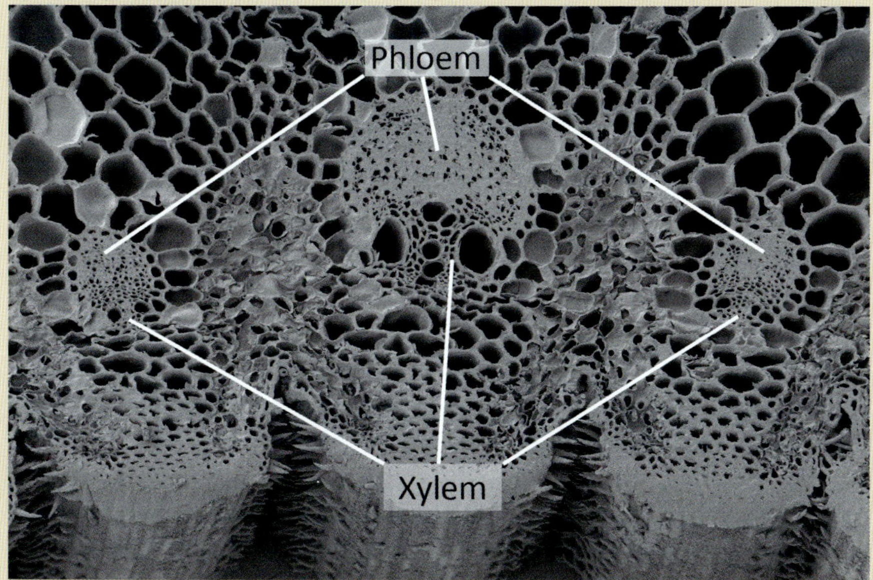

Seen here is a cross section of a pampas grass (*Cortaderia selloana*) leaf. Notice the circular vascular bundles of different sizes across the middle of the photograph. Vascular bundles are tubes of xylem and phloem that run through all parts of plants. The denser cells in the upper sections of each bundle are phloem, closest to where the sun hits the leaves to form sugars. The bottom, coarser section of each bundle is formed of the xylem cells. Each type of vascular tissue has its own suite of defenses. While not featured in this chapter, also apparent are silica-enriched teeth lining the bottom surface of the leaf—yet another defensive layer for shredding herbivores.

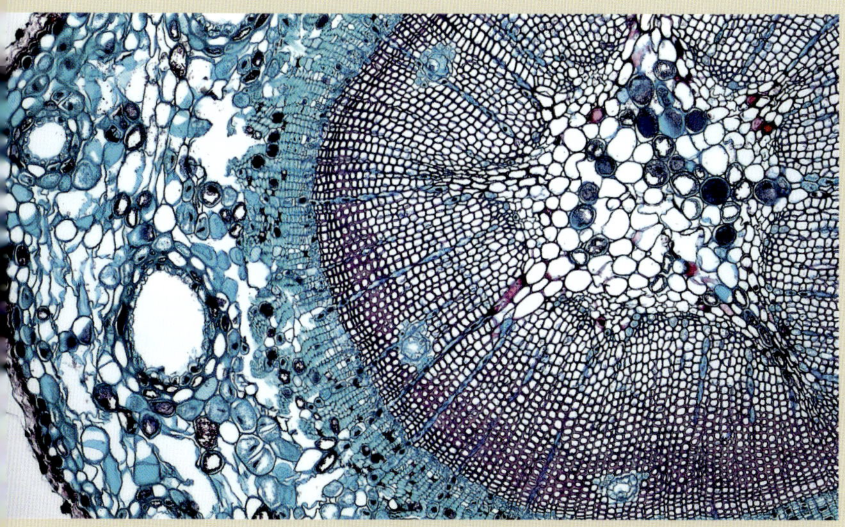

This photo shows a cross section of the stem of a pine tree (genus *Pinus*). The phloem cells occur on the left-hand side of the photo (the exterior of the stem), pierced with large, clear resin ducts. The xylem cells are at the right-hand side of the photo (interior of the stem).

peered closely at a leaf, you've noticed veins. The veins inside a leaf are pipe-like bundles that make up its vascular system, including both xylem and phloem cells embedded in living companion cells called parenchyma. Nonwoody plants are also densely networked with these vessels throughout their leaves, stems, and roots. Let's imagine we were to saw through a tree trunk or stem of a woody plant and look at the resulting disc-like cross section. This disc is quite familiar to those of us who enjoy counting tree rings, and it's also fondly called a "tree cookie" by foresters and dendrologists. The outermost layer of this disc consists of the tree's bark. Just inside the bark is a thin ring of living cells called the cambium, which is the tissue from which xylem and phloem grow and differentiate in stems. Beyond the cambium, the remainder of the cross section is woody tissue—the xylem. The wood, fibers, and hardened core inside woody plants are made of xylem cells. Wood itself consists of old xylem cells that are completely dead but have such a perfectly engineered cellular form that they continue to conduct water and solutes long after their individual lives have ended.

The creation of xylem involves a risky gamble: plants sacrifice healthy cells and introduce vulnerabilities in the process of creating this plumbing system. Xylem is formed when plants designate some of their cells to grow elongated and tube-shaped. Extensive xylem fibers form throughout plants because these tubes connect end to end and develop openings at these connection points. This structure allows plants to transfer fluids more efficiently. Once these cells have finished developing, they are ordered to die. Their contents dissolve, leaving the skeletons of their cell walls behind. Without the inner life support systems of the cell, the transfer openings become vulnerable. However, these hollow tubes are worth the risk because they allow for the efficient flow of water. You've felt the strength of these lignified xylem cells if you've ever held or snapped a piece of wood. Each cell is wrapped in complex, mummy-like layers of strengthening polymers and waterproofing agents. We also know that these materials lend not only mechanical strength to cells but also defensive strength against herbivore and pathogen penetration. Additionally, vascular cells must sustain tremendous fluid pressures as well as the compression forces from the plant's own weight. This is accomplished by the evolution of superstrong lignin, wood, and bark. The earliest land plants—mosses, liverworts, and

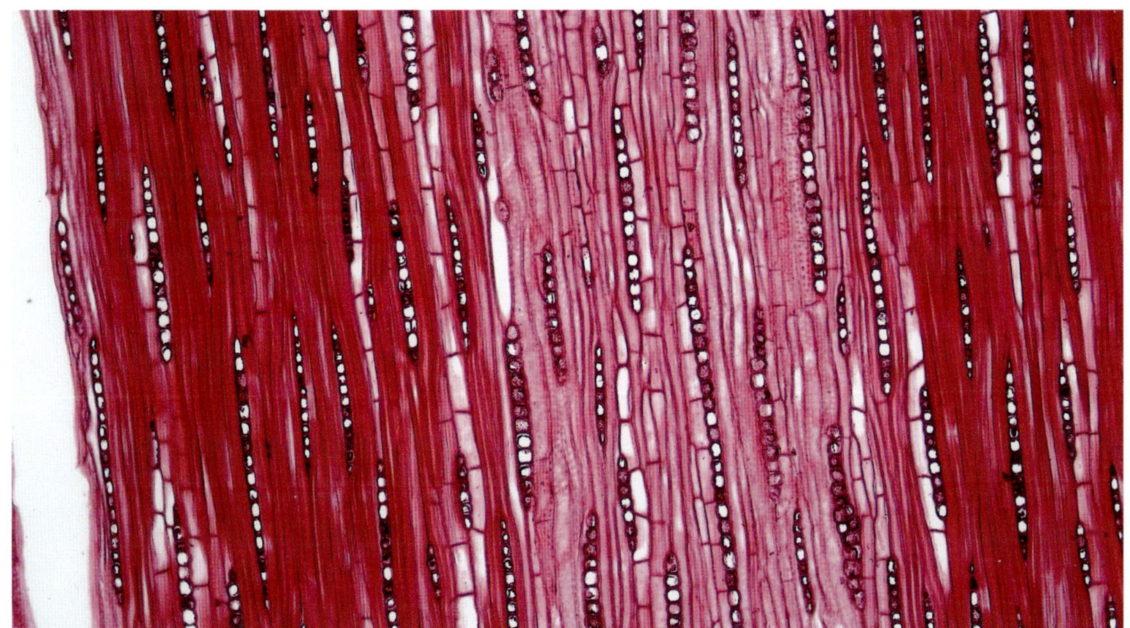

If you had microscopic vision and were facing an oak tree (genus *Quercus*) in real life, this is how the xylem of its vascular system would look. You are face-to-face with the outside surface of a growth ring. Water and nutrients are flowing upward through the different cells you can see: tracheids (pale pink and larger in diameter), vessel elements (dark pink and denser), and ray cells (vertical—they look like chains of beads projecting from the center of the tree toward your face).

hornworts—lacked lignin and are called nonvascular plants. However, in order for future successors to grow upright and tall, plants had to develop the lignin polymer. The first pioneering vascular plants included lycopods, ferns, and horsetails. The vascular system is so successful that it is found in all plant groups.

The sacrificial death of these courageous xylem cells offered massive benefits to plants, but it also came with drawbacks. For starters, the death of a cell also means the deactivation of many innate defense functions. Xylem cells lack many abilities of their living associates, such as signaling, response, communication, and production of defenses. Therefore, vascular strands must be surveilled, guarded, and protected by nearby living cells. Second, as more channels for fluid transfer began to open between xylem cells, the cells' vulnerability to pathogens began to increase. This interconnectivity benefited plant evolution but simultaneously created a need for more complex defenses and screening tools. If a bacterium, fungus, or virus successfully breached a plant's vascular system, it found a network of tubes leading to all remaining healthy parts of the plant.

Thankfully, plants bring a number of powerful defenses to bear on xylem pathogens. Each cell's openings are equipped with their own filtration systems. The openings, called pits and perforation plates, have filters consisting of membranes and valves that monitor and regulate the flow of smaller particles such as water, chemicals, and nutrients. Viruses, bacteria, and fungi are much larger cells and cannot fit through these openings. Despite the passage of more than four hundred million years since the first xylem showed up on the planet, these filters are still virtually impenetrable. To be successful, any microorganism must breach the tremendously reinforced cell walls using brute force or produce smaller propagules, such as fungal spores, which can squeeze through them.

Should microorganisms breach the cell walls and pits, they stumble into the deliberately barren food deserts that are xylem cells. Compared to the sugar-rich and living phloem cells, xylem cells are a nutritional wasteland for microbes. Since xylem cells undergo programmed cell death, the majority of their tasty proteins, sugars, organelles, and nutrients have dissolved and been reallocated elsewhere. Microbes enter expecting a buffet, but they encounter only an empty table. Xylem cells bear enormous quantities of water with just traces of the minerals and nutrients destined for other plant cells. Furthermore, these nutrients are present in forms that are usable by plants but intentionally *not* usable to microbes. For example, while fungi prefer nitrogen in forms such as ammonia and glutamate, xylem nitrogen is usually present as plant-friendly nitrites, nitrates, amino acids, and proteins. To secure enough of what they need, microorganisms secrete enzymes that degrade cell walls and breach surrounding cells. Sugars and nutrients are the by-products of this degradation. These components fuel the microbes and tip the balance of the xylem battleground in their favor. If you've ever seen a plant suffering from rot, the disgusting, slimy mass you encounter is the product of this warfare. Rot is a mix of degraded plant and microbial cells, sugars, digestive enzymes, leaking cell contents, and masses of bacteria or fungi invading and being actively fought by plant defenses.

If microorganisms persist in their attacks, xylem cells enter a defense process called tylosis. When xylem cells have detected a pathogen or drought, the surrounding living cells begin to swell, so much so that their walls and contents balloon through pits and into the xylem cells. These woody, bubble-like projections are called tyloses, and their purpose is to restrict and eventually close off the entire xylem cell and its openings. The resulting xylem cells lose moisture, becoming a desiccated and impregnable physical barrier to the vertical and horizontal spread of pathogens. Tylosis is a critical step in the formation of heartwood, the darker central region of xylem wood. To form heartwood, xylem cells are entirely blocked with tyloses and then become dumping grounds for everything toxic to microbial and insect life: resins, gums, waxes, and antimicrobial chemicals such as VOCs and tannins. Xylem defenses are the reason heartwood is well known for rot resistance. The timber of woods such as longleaf pine, bald cypress, cedar, and redwood can stand up to decay for centuries after the felling of the living tree.

Conquering Xylem: The Vascular Wilt Diseases

A group of pathogens that specialize in invading xylem are called vascular wilt pathogens. They are scary and nefarious organisms responsible for the microscale deaths of our vegetable gardens as well as the macroscale wipeouts of entire crops and plant species. Vascular wilts come from all branches of the evolutionary tree and aren't necessarily genetic relatives. The only thing these bacteria, fungi, and water molds have in common is their preference for the xylem. Some examples of vascular wilt pathogens and their hosts are listed in table 3. They represent some of the worst scourges on the planet, causing enormous ecological, agricultural, and economic catastrophes around the world.

A vascular wilt begins when a disease propagule makes contact with a plant. This contact can happen in a variety of ways. Plant diseases have various means of making their way to the next host. Depending on the disease, the fungus or bacteria might travel by wind or water or transfer itself through the bites of insects such as bark beetles, borers, and other chewing insects. These diseases can make entry through broken xylem cells as an herbivore feeds, through damaged or stressed roots, or even

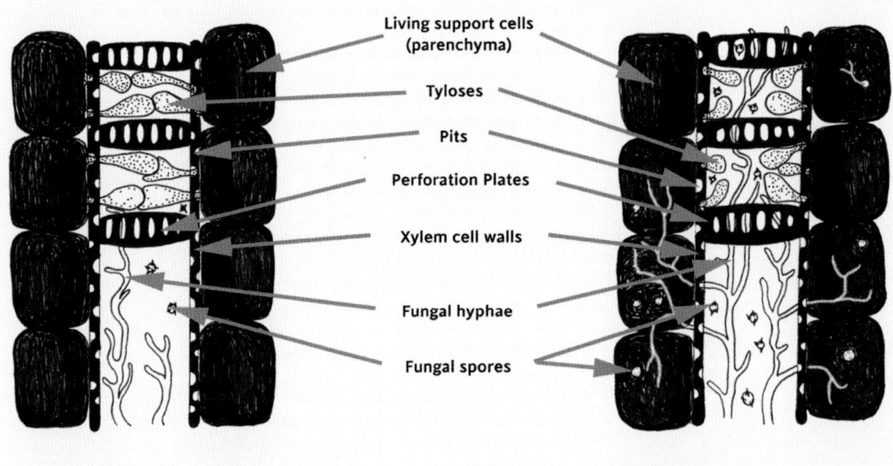

A group of pathogens that specialize in invading xylem are called vascular wilt pathogens. Once inside the xylem, disease cells begin multiplying. New spores and growing bacterial cells begin drifting upward in the xylem's flow, infecting fresh plant cells along the way. To breach xylem filters, microorganism cells squeeze smaller propagules such as spores and hyphae through the openings or simply melt down plant cell walls using specialized digestive enzymes. As a xylem disease's population grows, its parasitic impacts begin taking a fatal toll on the host's nutrient and water supplies.

through foliar gas exchange sites, the stomata. Once inside the xylem, disease cells begin multiplying. New spores and growing bacterial cells begin drifting upward in the xylem's flow, infecting fresh plant cells along the way. These microbes can often squeeze through tiny openings, such as xylem filters, using their spores or hyphae or by employing digestive enzymes to dissolve plant walls. As a xylem disease's population grows, its parasitic impacts begin taking a fatal toll on the host's nutrient and water supplies. As the host's aboveground tissues grow weaker and die from lack of water or nutrients, they become a microbial buffet. Thus, sometimes

Table 3 Vascular Wilt Pathogens: The Xylem Specialists

Common name	*Latin name*	*Type of organism*	*Host plant(s)*
Dutch elm disease	*Ophiostoma*	Fungus	American elm and other North American elm species
Chestnut blight	*Cryphonectria parasitica*	Fungus	American chestnut
Citrus greening	*Liberibacter*	Bacteria	Citrus
Verticillium wilt	*Verticillium*	Fungus	More than four hundred species of plants
Pierce's disease, citrus variegated chlorosis, coffee leaf scorch, olive quick decline, alfalfa dwarf, phony peach disease	*Xylella fastidiosa*	Bacteria	Grapevines, citrus, coffee, olives, oleander, alfalfa, peach
Fire blight	*Erwinia amylovora*	Bacteria	Pears, apples, loquat, crabapples, quinces, hawthorn, cotoneaster, pyracantha, raspberry
Ring spot of potato and bacterial canker/wilt of tomato	*Clavibacter*	Bacteria	Potato, tomato
Bacterial wilt of beans	*Curtobacterium*	Bacteria	Beans
Stewart's wilt of corn	*Pantoea*	Bacteria	Corn
Bacterial wilt of cucurbits	*Erwinia tracheiphila*	Bacteria	Cucumber, muskmelon
Southern bacterial wilt of solanaceous crops and moko disease of banana	*Ralstonia*	Bacteria	Tomato, potato, tobacco, banana
Black rot of crucifers, bacterial blight of rice	*Xanthomonas*	Bacteria	Crucifers, rice
Laurel wilt disease	*Raffaelea lauricola*	Fungus	Redbay, sassafras, avocado, federally endangered pondspice and pondberry
Root rot	*Pythium*	Oomycete	Generalist, large host range
Oak wilt	*Bretziella fagacearum*	Fungus	Oaks
Panama disease	*Fusarium oxysporum*	Fungus	Banana
Pineapple black rot	*Ceratocystis paradoxa*	Fungus	Pineapple, banana, coconut, sugarcane

CODIT, the compartmentalization of decay in trees, is a defense and healing process engaged against internal pathogens. Plant survival systems are fighting diseases to the death in the trunks and branches all around you. The clock starts ticking the instant vascular pathogens begin their onslaught. The botanical victim must successfully detect, respond to, and seal off the pathogen before it can kill the tree by completely severing water and nutrient flow in the xylem.

these microbes successfully defeat a plant's defenses completely. Once this happens and all useful resources have been consumed, xylem diseases prepare to leave the deteriorating remains of the once healthy plant. They enter the propagule production part of their life cycle and migrate toward their next victim.

Together Forever: How Trees Seal Diseases with CODIT

While these terrible microscopic plagues seem insurmountable, plants are fully prepared to respond with a powerful defensive barrage called the compartmentalization of decay in trees (CODIT). CODIT is the neutralization and healing process by which woody plants (not just trees) respond to disease attacks. Plant survival systems are fighting diseases to the death in the trunks and branches all around you. Does photo 73 look familiar? Take a close look at any broken branches, pruning cuts, fallen limbs, or other open wounds on plants in your daily life. Have you ever noticed that the branches on trees heal after a pruning cut? If you've ever spent time in the woods, then no doubt you've come across a fallen tree that looks quite healthy on the outside, but the inside looks quite diseased and dead. What's going on there? We are all unwitting witnesses to the hidden struggle of CODIT to save trees' lives.

Plants engage in CODIT as a costly endeavor to stop an otherwise deadly plant pathogen attack. CODIT is a death knell for plant pathogens and is engaged when plants detect signs of attack. Damage to tissues and different forms of chemical signaling can trigger CODIT to begin. A plant might detect its own broken cell parts or recognize specific chemical attributes possessed by its attackers. For fungi, this "smoking gun" molecule is the polymer chitin, which comprises the cell walls of all fungi and the exoskeletons of insects. For bacteria, certain proteins and digestive enzymes are the telltale sign of their presence. CODIT is extremely costly and results in many irreversible effects to xylem cells. However, it works to blockade, entrap, starve, toxify, suffocate, envelop, and permanently kill a disease in its entirety. If CODIT fails or overreacts, as we see in chestnut blight or Dutch elm disease, the entire plant is at risk of dying. The clock starts ticking the instant vascular pathogens begin their onslaught. The botanical victim must successfully detect, respond to, and seal off the pathogen before it can kill the tree by completely severing water and nutrient flow in the xylem.

CODIT consists of the construction of four intensifyingly hostile barriers against disease. First, xylem cells engage tylosis, quickly sealing off avenues for vertical spread. The second barrier is made up of growth rings—those rings on a tree commonly used to find its age. They actually have a survival purpose, secretly acting as barricades to radial pathogen spread. Growth ring cells, which are inactive cells, have sturdy walls with a high lignin content and form continuous bands in front of and behind pathogens. The third wall, ray cells, are the strongest walls that trees prepare in advance of attacks. They divide stems into pie wedges that block tangential spread. Rays are designed like a labyrinth, with variable walls of almost random lengths, heights, and thicknesses. Rays need to be absolutely impenetrable because they prevent girdling, which occurs when damage from either physical damage or pathogenic invasion kills an entire ring of cambium cells around a tree's trunk or branch circumference. Once this occurs, the entire plant above that girdle becomes unable to translocate resources and dies. Thus, it is essential for plants to prevent girdling from happening. To further ward off the invader, all three of the existing walls also begin filling with antimicrobial chemicals such as tannins, resins, waxes, and VOCs following an attack. The fourth wall, called the barrier zone, nails the woody coffin lid shut on pathogens once and for all. It is the only wall that is produced subsequent to attacks and wounds. Having surrounded the disease on all other sides, the fourth wall grows from new, healthy tissue that encloses the wound area and continues growing. This tissue is highly impregnated with antimicrobial compounds and strong cell walls. Most importantly, the barrier zone creates an air- and watertight seal around a disease—few microorganisms can live in anoxic and desiccated environments.

The amazing survival strategies we've seen in action are just part of plants' repertoire. What happens when attackers target living plant cells in the phloem, the vascular tissue that moves around sugars and other nutrients?

Phloem

While xylem cells are equipped with defenses and resistances, they are ultimately easier to invade than phloem cells. Phloem cells are *living*, teeming with a full arsenal of equipment for surveillance, protection, and chemical deployment. Let's take a microscopic trip into the phloem to see how it works to keep plants alive.

Phloem distributes the sugar-rich food produced by the leaves throughout the rest of the plant. Phloem consists of two types of conductive cells: sieve tube elements and companion cells. Sieve tube elements are elongated conducting cells that form fibrous, pipe-like connections with each other. While still alive, they have lost certain organelles such as the nucleus and vacuole to make more space for fluid movement. Each end of a sieve tube element has a porous connecting wall called a sieve plate, which acts

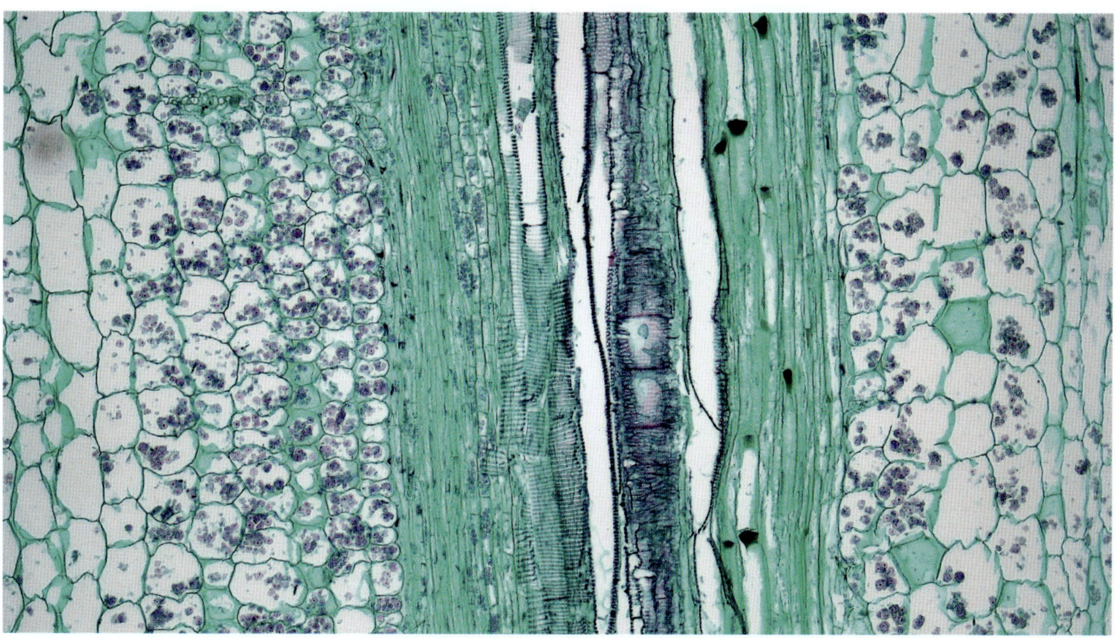

This section of the stem of a squash plant shows what you would see if you were directly facing the stem in real life. Phloem is taking the sugars from the leaves downward, while xylem is taking water and minerals upward. Phloem cells are the open, curvy cells occurring on the left and right sides of the photo, while a few dense xylem fibers occupy the middle of the photo.

as a cellular filter. Each sieve tube element has a buddy called a companion cell. Companion cells stand watch over sieve tube elements. They have superstrong cell walls, which add structural strength to the vascular system and allow the cells to share resources, secrete defensive chemicals, and conduct long-distance defense signaling.

Phloem's innate characteristics render it a difficult target for attack or herbivory. First, the cells are under high fluid pressures. Once a pathogen or insect cracks into a phloem cell, it's like drinking from a fire hose. The pressures contained within a phloem cell can be lethal, more than two to five times that of the attacker's own body fluids. When punctured, phloem cells can spurt their fluid contents forcefully enough to blow minute herbivores such as aphids and spider mites completely off the plant! These pressures are also designed to damage small pests' feeding tubes, eject microbial pathogen cells, and coat attackers in toxic fluids such as resin, latex, mucilage, and natural gums.

Second, phloem possesses lethally high levels of sugar. While sugar may be a large part of what an insect is after, the extremely high levels found in phloem actually act as a natural desiccant to insects, drying them out. When phloem's superconcentrated liquid candy gets into an insect's gut, it begins absorbing moisture and body fluids from the rest of the insect's cells. Sugar's attraction and absorption of moisture exert the same effects as other common desiccants such as salt, charcoal, and silica gel. We have experienced this property whenever we notice old sugar in our pantries becoming a rock-solid block after absorbing humidity from the air. Except for a specially adapted few, insects targeting phloem risk desiccation and dying from sugar poisoning.

Third, phloem's nitrogen content is specifically formulated to starve insects. Phloem contains gigantic quantities of sugars but merely trace levels of amino acids, proteins, and other nitrogen-rich compounds that are intended for the plant's own needs. This isn't ideal for animals. To get enough protein, herbivores must find a way to drink vast quantities of sap and safely process the deadly sugars it contains. Provided animals solve the sugar problem, they then discover that phloem's blend of sap lacks critical amino acids that animals can't synthesize on their own. Twenty different amino acids are necessary for animals to create proteins. On average, phloem purposefully leaves almost half of them in short supply. If an animal comes up short in even one amino acid, it can't create protein, so its growth and development will be impaired.

Fourth, plants deliberately make phloem a dynamic and unreliable food resource. The nutritional blend of a phloem cell changes in response to every environmental condition a plant experiences. Phloem is enriched during the day as sunlight stimulates photosynthesis, but then production backs off at night. Cold, drought, attacks, and virtually any stress can also shut off the tap at a moment's notice.

All we've discussed thus far is simply the survival power inherent to the makeup of phloem. On top of it all, actual defenses are hiding there, too, just waiting to spring into action. Within seconds after damage or the detection of an attacker, phloem tubes engage multiple defenses simultaneously. Any break in a phloem cell, such as during herbivore or pathogen attack, is immediately patched with phloem protein plugs (or P-plugs). P-proteins are filamentous and strong, lying at the ready along the membranes of phloem cells. Phloem's high-pressure leaks jam tangled masses of P-plug fibers into even tiny cuts, sealing further water loss within seconds. P-proteins are also potentially lethal if they clog up an insect's mouthparts and feeding tubes. Imagine how difficult it might be to drink through a straw clogged with cotton balls. Plants attempt the same strategy with P-proteins. Another plug, the carbohydrate polymer called callose, is also generated. Similar to cholesterol clogging arteries, callose begins growing and depositing in the sieve pores and plates of a phloem vessel during herbivory, disease outbreaks, or environmental stressors. Callose can restrict the openings of sieve tubes like filters and slowly dissolve after a stress has passed. Water, chemicals, and microorganisms are unable to pass the callose barrier, so herbivores find their supply of sap choked off. Unlike the tyloses in xylem, both the callose and forisomes of phloem are rapid-response and reversible valves. (Forisomes are protein polymers found only in plants from the bean family. They have the remarkable ability to undergo reversible "muscle-like" contractions that cause deliberate occlusions within sieve elements of phloem tissue. The goal of this process is to provide a barrier to attackers and limit the leakage of plant fluids.)

Showdown: Aphid versus Phloem

On a verdant early spring day, an aphid crawls menacingly over a leaf. It taps the leaf as it goes, feeling and monitoring the resounding vibrations throughout the plant's cells. The aphid, one of the most specialized and amazing phloem feeders in the animal kingdom, is prowling for sap. Once the vibrations indicate a plump vessel, the aphid prepares to stab the plant with its highly developed, needle-like feeding tube (called the stylet). This fraction-of-a-second event begins one of the most fascinating evolutionary battles between plants and herbivores.

Aphids are a member of the insect order Hemiptera, a very special group indeed. Hemipterans, commonly called the true bugs, are the only evolutionary lineage of insects that have evolved to feed predominantly on plant phloem. This is because they have evolved ways to circumvent phloem defenses. Members of this order also include mealybugs, scale, whiteflies, plant hoppers, leafhoppers, and psyllids—all common garden and agricultural pests. One trait that unifies these herbivores is their piercing-sucking feeding strategy, which relies on a sharp, syringe-like mouth.

We'll use the aphid as an example of how this type of insect attacks. An aphid's stylet is protected by a stiff sheath that contains nerve endings that sense where to place the stylet in a leaf or stem vein. You'd expect these needle-like mouthparts to stab directly through the leaf's cells into the phloem. While that may be true of some piercing-sucking feeders, the opposite is true for aphids. Instead, aphid stylets move like a long tongue, tiptoeing around the heavily defended photosynthesizing cells of the leaf's mesophyll layer. Stylets snake a contorted route to the phloem layer below, choosing the safest path to the phloem in order to avoid triggering plant defenses. However, aphids need to know what they're getting into, so they collect information before they begin feeding. As the stylet navigates through the leaf, it pierces mesophyll cells and takes small tastes from them. Chemical sensors in the aphid compile information about the plant's defenses, sugar and nutrient contents, and pH and keep the stylet on track toward the phloem.

An aphid's saliva is its cloaking device. With the right mixture of spit, an aphid can shield itself from poison and deactivate many of the defense signals found within plants. Plant cells become so deceived that they might

An aphid swarm feeds together on a daylily leaf. Aphids are specialized herbivores of plant phloem, possessing one of the most fascinatingly complex feeding strategies of any insect. The dance between herbivory such as this and plant defenses is under way all around us.

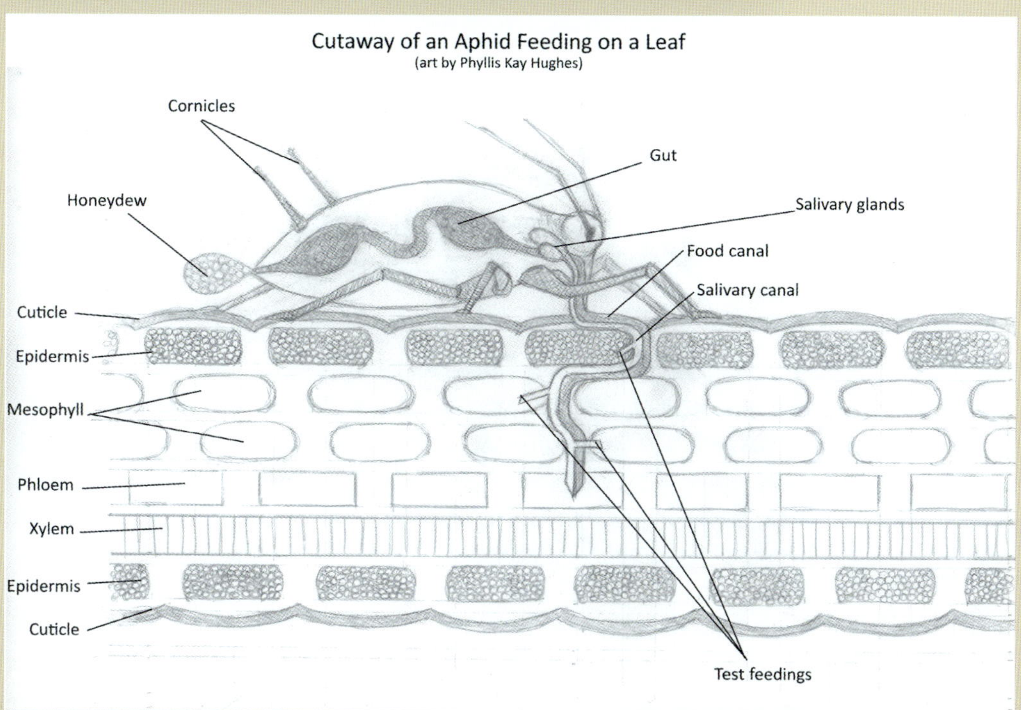

Aphids are specialized phloem feeders. Their sharp mouthparts (called stylets) tiptoe around the heavily defended photosynthesizing cells of the leaf's mesophyll layer. Stylets snake a tortuous route to the phloem layer below, choosing the safest path to the phloem in order to avoid triggering plant defenses.

fail to perceive and respond to an aphid's attack. Inside and along the length of the stylet are two parallel ducts, a small one for saliva and a larger one to extract sap. Aphids stock two different kinds of saliva, a gel-like one and a watery one. The gel saliva is like caulk, rapidly hardening into a water- and chemical-proof protective layer after being extruded from the tip of the stylet. It seals the stylet against the plant's leaf so that the phloem pressure can't eject the aphid from its impending feeding frenzy. Even more gel saliva coats the stylet as it twists through the leaf, forming a continuous, solid, and protective barrier against the plant's internal defense chemicals. If they're lucky, however, plants can sense the elicitors within this saliva or the breakage of trichomes and conclude they are under an aphid attack.

Aphids secrete their watery saliva inside plant cells once the stylet has found the perfect place to begin feeding. Both gel and watery saliva are like utility belts, carrying a huge range of proteins that can execute a specific surgical strike on the plant. Watery saliva strikes at plant cell walls, defense signaling molecules, and the phloem filters. To keep the phloem tap open and access fresh cells, aphids spit their own specialized counter-proteins, which digest cell walls and interfere with the formation of protein plugs and forisomes. Watery saliva also mops up and deactivates a defense signaling molecule such as calcium as well as defensive hormones such as jasmonic acid and salicylic acid. Without messenger molecules, plants become unable to recognize the presence of an attack, much less coordinate a defense response. Aphids usually feed en masse, and their collective secretions cause a defense "blackout" across wide regions of the plant. Provided aphids succeed undetected, they must still confront the pressure, nitrogen, and sugar barriers inherent to phloem sap.

To guzzle from the firehose-like phloem, aphids open their rear ends wide as they feed, venting the deadly pressures and excreting a rich, sticky, and sugary liquid called honeydew. This also helps cut down the concentrated sugar of phloem, rendering aphids safe from desiccation. Furthermore, aphids and their relatives are coated with waxy deposits to escape being glued to death by their own sticky honeydew. Interestingly, an entire web of life thrives on this honeydew as a food source, whether by harvesting it directly from the aphid colonies or by consuming honeydew once it has landed on other surfaces. Notable honeydew feeders include ants, flies, wasps, bees, beetles, butterflies, moths, birds, and flying foxes.

Some animals love honeydew so much that they protect the aphids from harm and carry young aphids to inoculate new plants. This behavior is called tending, and it's performed by hemipteran-farming ants, polybiine wasps, silvanid beetles, and even nectar-loving Madagascar day geckos.

The salivary glands and guts of aphids are teeming with life, too. With the power bestowed by symbiotic microorganisms, aphids can circumvent the amino acid deficiencies within phloem. Starting around 280–160 million years ago, aphids have been internally colonized by endosymbiotic bacteria in the genus *Buchnera*, which live inside aphid cells, turning them into protein factories. As aphids feed, they steal energy, carbon, and amino acids from the phloem and provide it to their gut bacteria. One amino acid, glutamine, is available in virtually limitless quantities in phloem sap. *Buchnera* bacteria convert this excess of glutamine into the amino acids that are lacking in phloem, all of which aphids are unable to produce on their own. With their sap-processing abilities, *Buchnera* help aphids overcome the starvation risk of drinking phloem sap. Each adult aphid can hold over five million *Buchnera* cells. These two organisms are so intimately tied to each other that neither can survive alone.

As it turns out, aphid guts are quite overpopulated. *Buchnera* isn't the only microorganism housed inside insects. Hemipteran insects are some of the most efficient vectors for phloem diseases. Fungi, bacteria, phytoplasmas, and viruses can hitch an easy ride into the phloem via the stylet. They never have to deal with the pressure and cell wall barriers of phloem cells because their host insects perform the work for them. Let's take a look at one example of a phloem-specific disease: the phytoplasmas.

Phytoplasmas

Phytoplasmas are one of the best examples of plant diseases that attack the phloem. They are bacteria that are highly adapted to jumping between life inside the bodies of insects and the phloem of plants. We find these bacteria in the guts of leafhoppers and plant hoppers, though aphids can carry them as well. Phytoplasmic diseases are an immense economic strain on crop plants, causing untreatable and exponentially spreading losses around the world. Symptoms of affected plants can range from a mild yellowing to abnormal growth and even death. These impacts have been

Witches'-broom grows on bamboo (*Dendrocalamus strictus*), a symptom of phytoplasma infection. A leafhopper, having bitten this plant, has injected phytoplasma into the bamboo's phloem. Normal growth is shown in the background, but a phytoplasma infection has corrupted the rest of the growth, turning leaves small and yellow. Further insect feeding will take the phytoplasma up again, where it can colonize new insects and be transmitted to other plants.

observed on over seven hundred different plant species. Phytoplasmas occur worldwide but are especially pervasive in tropical and subtropical areas, where they affect crops such as coconuts, sandalwood, and sugarcane. Some other victims of phytoplasmas are noted in table 4.

Phytoplasmas are quite small, and in order to enhance their own transmissibility, they've lost cell walls, leaving only a thin, amorphous membrane behind. It's a heck of a gamble to have committed oneself that completely to an aqueous existence, but phytoplasmas are massively successful inside the juicy guts of their insect hosts and watery phloem cells.

In fluid form, phytoplasmas can transfer between plant and animal with ease. As stated, they're tiny—usually less than one micrometer. To understand how incredibly tiny that is, a human hair ranges anywhere from 17

Table 4 The Phloem Specialists

Common name	*Taxonomy*	*Type of organism*	*Plant host(s)*
Aphid	Order Hemiptera, family Aphidoidea	Insect	Numerous plants
Scale	Order Hemiptera, superfamily Coccoidea	Insect	Numerous plants
Whitefly	Order Hemiptera, family Aleyrodoidea	Insect	Numerous plants
Mealybug	Order Hemiptera, superfamily Coccoidea	Insect	Numerous plants
Seed bugs, milkweed bugs	Order Hemiptera, family Lygaeidae	Insect	Numerous plants
Leafhopper	Order Hemiptera, family Cicadellidae	Insect	Numerous plants
Planthopper	Order Hemiptera, family Fulgoridae	Insect	Numerous plants
Psyllid, jumping plant lice	Order Hemiptera, family Psyllidae	Insect	Numerous plants
Stink bugs, shield bugs	Order Hemiptera, family Pentatomidae	Insect	Numerous plants
Leaf-footed bugs, squash bugs	Order Hemiptera, family Coreidae	Insect	Numerous plants
Aster yellows	Genus *Phytoplasma*	Bacteria	Three hundred species in thirty-eight families, including aster, carrot, grass, and onion
Elm yellows	Genus *Phytoplasma*	Bacteria	Elms
Grapevine yellows	Genus *Phytoplasma*	Bacteria	Grapes
Sugarcane grassy shoot disease	Genus *Phytoplasma*	Bacteria	Sugarcane
Palm lethal yellowing disease	Genus *Phytoplasma*	Bacteria	Palm family
Peanut witches'-broom	Genus *Phytoplasma*	Bacteria	Peanut
Potato purple top	Genus *Phytoplasma*	Bacteria	Potato
Tomato big bud	Genus *Phytoplasma*	Bacteria	Tomato
Stolbur disease	Genus *Phytoplasma*	Bacteria	Tomato family crops
Clover proliferation	Genus *Phytoplasma*	Bacteria	Clover, bean, potato, alfalfa
Papaya bunchy top disease	Genus *Rickettsia*	Bacteria	Papaya
Citrus stubborn disease	*Spiroplasma citri*	Bacteria	Citrus
Corn stunt disease	*Spiroplasma kunkelii*	Bacteria	Corn

to 180 micrometers. The next-largest plant pathogens, fungal and bacteria cells, seem huge by comparison with phytoplasmas, coming in at around ten micrometers. Phytoplasmas have a second adaptation for virulence: lacking cell walls, they're shapeless enough to pass through phloem filters, a major line of defense for plants. Phytoplasmas are often filamentous but can change their shape in order to better navigate through their host insects and plants. Even more importantly, phytoplasmas are uptaken through the stylet into any leafhopper's guts. After proliferating in the digestive tract, they colonize the insect's hemolymph (the insect equivalent of blood) and eventually the salivary glands. At this point, phytoplasmas are once again weaponized, ready to be inserted into any plant the leafhopper decides to feed on.

A phytoplasma infection is virtually irreversible for plants. Phytoplasmas attack phloem cells, raiding and stealing their sugars and nutrients. As the plant cells succumb, they begin to show characteristic symptoms of phytoplasma infection. Plant hormone and protein production is impacted, creating growth form changes that are diagnostic symptoms induced by phytoplasmas. You may have noticed many of these common signs before without realizing their hidden meaning. Phytoplasmas create a variety of symptoms. The entire plant may become stunted, yellow, and weak and can even die. Infected plants may begin bearing unusually small leaves or overproduce leaves and flowers. Flower parts may abnormally transform into leaves or turn green, while buds may grow incredibly large. Phytoplasmas also induce witches'-broom, a phenomenon wherein the normal growth of plants is deformed into a dense mass of stems and leaves. All of these weak, leafy, and proliferating tissues are great opportunities for further leafhopper attack. Additionally, plant defense production shuts down following this onslaught, enabling burgeoning numbers of leafhoppers to colonize the plant and giving the phytoplasma a transmissivity field day. A single phytoplasma species can be spread by multiple insect species simultaneously. A single leafhopper might also have more than one phytoplasma species riding around in its cells, all ready for transmission on the next puncture. This team spirit is often enough to do plant defenses in. Only under the treatment of the same antibiotics we use on patients in hospitals, such as tetracycline, can a plant be cured, but this is uneconomical. Furthermore, phytoplasmas can simply bide their time, riding in leafhoppers until they can reinfect a host after the drugs have run their course. Ultimately, it is quite a damaging attack.

The plant vascular system is an absolutely amazing contraption, pumping with complexity and intrigue all around us. Sometimes a plant wins, sealing off its diseases with CODIT. On the other hand, attackers like aphids can completely overwhelm a plant's protective elements. When all internal defenses have failed, there is one last trick up plants' leaves. They ask for help from the predators of their attackers. For all the powerful abilities herbivores such as aphids possess, we must remember that they still remain subject to the huge pressure of predation. Plants emit chemical distress signals into the air when they are under severe attack. Let's drift into the next chapter and see how plants raise armies of minions to rescue them from their herbivores.

CHAPTER 7

Plant-Speak

Botanical Signaling and Communication

Our world parallels the hidden world of plant communication. Plants have secret, intricate systems that allow them to "talk" right under our noses. Our leafy friends have a clandestine language all their own that is replete with messages encoded into signaling molecules, hormones, gases, and electricity. Gathering and sharing information about their surrounding environment helps plants survive. Plants can transfer a wealth of information such as warnings and weather reports internally to other plants and even to animals. The air around us swirls with botanical clouds of information, while the ground below us is a living signaling network. This world of plant communication has eluded us until only recently simply because we are not the intended recipients of plants' messages. To decode the clandestine communiqués plants send, we need to understand what kind of information they're sending, how they send it, and why. We hope this chapter helps you decipher some of the peculiar botanical "language" in the world around you. It is quite different from the audible signals we hear from birds, bugs, and other animals. We hope you'll begin to perceive the more subtle plant voices that have always been there right alongside us.

While what plants can do is astounding, it is important to realize that plant communication is much different from animal communication. Plants cannot speak or transmit language in the strictest sense, yet, wordlessly, they have no problem understanding each other. Plant communication is quite a simple process: plants collect information about their world, channel it through a variety of mechanisms for different purposes, receive these signals, and make a calculated conditional response. Common responses include changing the way they grow (tropism), allocating defense priorities (such as ramping up trichome, spine, or poison production), and preparing for severe environmental events such as droughts and freezes.

Plant Detectives

Plants are extremely effective at gleaning information about the world. It might just surprise you to learn how much plants sense and the degree of detective work they perform on their environment. Plant survival systems are always engaged in around-the-clock, multifaceted surveillance operations. Plants watch environmental events such as the weather, monitor their own well-being, and stay vigilant against attackers. Their cells collect and send this information so that the plant can adjust its growth for best survival. The result of this complex process, which involves coordinated intercellular communication, is the many kinds of plant tropisms. A tropism is the biological reaction of a plant to an environmental stimulus and is often the perceivable result of plant cells in tacit communication with each other. Quite simply, plants that are better at detecting, communicating, and reacting to stimuli are also better survivors.

Plants are extremely effective at gleaning information about the world. It might just surprise you to learn how much plants sense and the degree of detective work they perform on their environment.

Plants keep track of environmental events very closely. Plants can judge the direction, photoperiod, and intensity of sunlight, allowing them to optimize their growth and pigmentation accordingly. A plant knows to send its roots belowground and its stems above because of special gravity-sensing cells called statocytes. Statocytes detect gravity with an ingeniously simple measurement tool called statoliths, which are simply dense packets of starch. Because of their density, statoliths are always induced by gravity to settle toward the bottom of statocytes. Plants can thus sense direction at any point in time regardless of their orientation. Roots are also sensitive detectors of nutrient and water sources. Touch sensitivity allows vines and tendrils to coil around other plants, while broken trichomes tip plants off to lurking insect attacks. Temperature readings, taken by light receptors known as phytochromes, allow a plant to prepare in advance against droughts, heat waves, fires, or freezes. The regular mechanical stresses of wind allow plants to strengthen their physical structures against treacherous windstorms. Aquatic plants are equipped with specialized flood detection systems, which can monitor conditions impacted by floods, such as oxygen levels. Chemical sensory perception by plants is also extremely common. Cells take constant chemical readings of their surroundings in order to detect toxic excesses or deadly shortages of gases, minerals, and many other chemicals.

Cells even monitor their own health and well-being. Specialized receptors throughout plant cells detect "out-of-place" components that only occur following severe damage and require immediate reaction. These may include broken cell wall chunks, proteins leaking from damaged organelles, and warning chemicals secreted by other injured cells. These chemicals are a type of defense response trigger called damage-associated molecular patterns (DAMPs). DAMPs can trigger cell wall repair, leak-plugging activities, hypersensitive responses, and the production of numerous nasty defenses such as toxins and disinfectants.

While it is amazing that plants can sense if an attack has occurred, the degree of specificity they record about the nature of the attacker is mind-blowing. The breakage of trichomes and the vibrations made from the walking and chewing of pests are some of a plant's first clues of nefariousness. In their own way, plants can hear what's afoot. This allows plants to estimate the invasion's size and location. Plants have special chemical receptors just for their worst nemeses, such as invading insects, nematodes, and diseases. Plants have prepared receptors whose sole purpose is to scan for bug spit and identify the type of creature that made it. Chemicals that trigger a defense response in plants are called elicitors. There are quite a lot of elicitors, among them the unique chemical signature within insect saliva. Plants have evolved an entire part of their survival system toward the molecular recognition of each attacker's specific chemical traces. While we identify objects and organisms using our senses, a plant's "senses" are different. For example, plants read the event "insect attack" as the breakage of trichomes, the crushing of cells, and the deposition of chemicals that are obligately produced in insect saliva and egg-laying fluid. Even further, the plant can make the distinction between herbivore species, identifying an attacker as a caterpillar, an aphid, thrips, a mite, a beetle, a leaf miner, a borer, a large browser, and more. Plants also become aware when pests lay eggs on or inside plants. The sticky fluid that glues insect eggs to a plant bears a distinctive chemical trace that elicits the plant's defenses. Even before a hatching insect's first day of life, its plant host knows of its existence and is pumping a flurry of defenses to the site.

From Detection to Communication

Surveillance of its surroundings, however, is only one part of a plant's struggle for survival in this harsh world. Plants have the fascinating power to relay the information they collect. Imagine if you were attacked—you might find it quite helpful to call for help and alert others to possible danger. Plants are no different, often releasing volatile organic compounds, lighter-than-air organic chemicals that easily drift and that their neighbors can pick up on. This ability is especially critical for organisms that cannot move. Communication allows both attacked and healthy plants to ramp up herbivory defenses, increase immunity to disease, warn their neighbors and relatives of threats, and prepare for recovery. Plants use communication to learn a great deal about their attackers and subsequently prime a hand-tailored suite of defenses. Since defense investments such as toxic chemicals and structural enhancements get quite expensive, efficient defense priming leaves more resources for growth, proteins, and other necessities. Thorough surveillance allows plants to dispatch only defenses that match the correct attacker. Denser trichomes might work well for aphids, for example, while growth-regulating toxins or antinutrients such as tannins might take care of the next pesky caterpillar outbreak. Thicker cell walls with stronger lignin and silica compositions might stop a wood-boring beetle in its tracks, while antibiotic compounds and CODIT activation might stymie potential disease outbreaks. These are the complex but necessary decisions plants must make each and every day.

So we've explored reasons why a plant might create a message. How about transferring that message from one part of a plant to another? Plant survival systems are fully wired up for communication. Scientists now commonly accept that communication, once a radical and ridiculed concept, is widely used by plants within, between, and outside their cells. This task is made simple because plant cells are completely interconnected. It makes sense that the same connections that constantly transfer fluid, food, defenses, and nutrients can also conduct chemical and electrical messages. A plant's phloem not only is a lifeline for supplies but also doubles as its major transfer and wiring system. Phloem is an optimal conduit because it ranges through the entire plant, and its cells have efficient transfer openings such as sieve plates and plasmodesmata. Messaging molecules, transporter molecules, and electricity flow along these tightly connected cells.

Two different messaging trajectories through cells exist, providing a complex and astonishing internal regulatory system for the type, strength, and direction of communications to be sent. One is the apoplastic pathway, a term that indicates its location along the outside of a cell between its membrane and cell wall. The second is the symplastic pathway, consisting of communication through the cytoplasm fluid inside the cell membrane. Myriad signaling molecules exist, among them the major plant hormones, proteins, chemical compounds, and ions. Specialized transporter enzymes and gates help these molecules navigate through cells and locate their appropriate counterparts: receptors. Plant defense–related responses are often encoded with two signaling molecules called salicylic acid and jasmonates. Salicylic acid is noteworthy because it is the source of aspirin, an anti-inflammatory and pain-killing drug. The chemical draws its name from the willow genus, *Salix*, from which it was first derived. While the vast majority of plants employ salicylic acid in signaling, willow bark happens to be a high-enough natural source that it has been gathered as a painkiller for thousands of years by Native Americans, ancient Greeks, and ancient Egyptians. Both of these chemicals help coordinate defense responses in many plants.

You might wonder who is on the receiving end of all these plant messages. It turns out that almost everybody is, even though we humans are often oblivious to the conversation. Plants respond to their own signals, but so do other plants, fungi, and even animals such as pollinators, herbivores, and predators. Plants might recruit helpful companion organisms such as pollinators and mycorrhizal fungi. Flowering plants spend time heavily communicating with pollinators during reproduction. The scent of a flower can be explicitly geared to a particular animal. Plants select patterns and colors that appeal to a distinct audience. Insects, for example, visually notice ultraviolet patterns, which are often located on the petals of flowers. To us, these petals might look solid white or yellow, but to an insect, they appear as a map, pointing the way to nutritious nectar. Chemical scents can also be used to call in seed distributors. Think of all the vibrant scents of fruits and vegetables you are familiar with. Each of those enticing scents helps animals locate the precious cargo. Plants can also use chemical communication for defense in addition to reproduction. A plant might send warnings to prepare in advance for horrible pest swarms or disease epidemics. Secretly, competing plants have learned to wiretap these signals, listening discreetly for their own benefit. Animals, too, are on the lookout for plant signals to know where to find food such as nectar, fresh leaves, or a hapless caterpillar.

We'll discuss three incredible types of plant communication in greater detail: underground signaling, airborne signaling, and electrical signaling.

Underground Signaling through Mycorrhizal Fungi

Let's dig a bit deeper into the plant communication that occurs underground. Below the soil exists an expansive interspecies communication network where plants and fungi are linked over an expansive area. Each "conversation" can span just a handful of plants or even entire forests. Across this network, distress signals are beamed, water and nutrients are exchanged, and clever plants are selfishly eavesdropping. Underground communication helps plants survive through counteracting resource shortfalls, preparing for disease and herbivore invasions, and passing "head starts" to their offspring.

To send chemical messages through this subterranean pipeline, plants require the help of wide-reaching mycorrhizal fungi, which are symbiotic fungi that colonize plant roots. The belowground ecosystem that occurs between plants and mycorrhizae is called the common mycorrhizal network (CMN). As we'll learn in this book's nutrient stress chapter, mycorrhizal fungi intimately connect with plant roots using tiny, root-like branches called hyphae. The interconnectivity of this network is vast. Mushrooms, the reproductive structures of some fungi, are just the tip of the iceberg. The majority of the fungus, called the mycelium, grows predominantly underground. Fungi create a large web of pathways through the soil, often linking plants along the way. Fungal networks can be exclusive, wherein plants sharing a kin relationship or species also share the same fungus.

Another type of relationship is communal, wherein associations occur between multiple plant and fungal species. To set up a mycorrhizal network, plant roots actively secrete chemical messages that attract and facilitate infection by the fungi. Among these communiqués are plant hormones called strigolactones, which roots pump into the soil. Strigolactones cue mycorrhizal spores to germinate, stimulate fungal branching, energize fungal growth rates, and help nascent fungi locate and distinguish "friendly" plant roots for colonization. Research has shown that the vast majority of plants form mutualistic associations with fungi. This is because each stands to benefit. The photosynthesizing plant provides, among other things, carbon in the form of sugar, while the fungus can get at the hard-to-access water within tiny soil pores using its thin, branching hyphae. Belowground ecology is still a developing field, with promising discoveries no doubt awaiting in the future. To date, scientists have detected rich common mycorrhizal network interactions in spruces, acacias, alders, she-oaks, eucalyptus, pines, hemlocks, poplars, oaks, willows, and legumes.

Plants work hard to establish a common mycorrhizal network because of its substantial survival benefits. Plants are capable of sharing carbon, nitrogen, phosphorus, micronutrients, stress alerts, and allelochemicals (detrimental chemicals that debilitate or slow competing plants and other organisms) through the common mycorrhizal network. Healthy plants serve as source plants, which are often those of greater age or with better access to certain resources. Older plants, likely possessing taller stems and deeper roots, usually have the greatest access to sunlight, water, and nutrients. Some of the carbon produced by these successful individuals fuels the mycorrhizae. The fungi, in turn, barter with some of the nutrients they're specialists at acquiring, such as phosphorus, and give plants expanded root zones. Plants facing stress and in need of common mycorrhizal network supplementation are sinks. Younger, seedling plants are often sinks. The first few growing seasons of their lives are tough ones. Seedlings face herbivory and light competition with other plants, and they are less resilient to environmental stresses and shortfalls. Their less developed defenses may only generate meager levels of protection, while their roots and shoots may not be as productive as those of their older cohorts. Especially in a dense forest system, light for photosynthesis might be in short supply. These networks allow older trees to shuttle carbon to their offspring, giving them a boost on the darkened forest floor. Additionally, a few extra minerals could supplement the construction of new tissues, recovery of wounds, or production of defense chemicals that might help the struggling seedling survive. This is why we see impressive communal sharing networks in dense forest systems, where this kind of adaptation would be particularly advantageous.

Once a plant advertises its distress over the common mycorrhizal network, resources are on the way with the speed of modern global shipping services. In just one to two days, aid molecules from source plants begin arriving at the fungal mycelium—the "body" of a fungus. The fungi then distribute these molecules to sink plants, wrapping up an entire transfer in a mere total of three to four days. Scientists can track this kind of exchange through radioactive carbon isotopes that are "fed" to one plant. The radioactive signal can be tracked as it moves along the tree and through the soil and fungi within the system, eventually making its way into the juvenile plants. Furthermore, the common mycorrhizal network is a constant, mutually beneficial pipeline that can last for life. Care packages and shrewd trading arrangements work their way back and forth between plants and fungi on a daily basis. Admittance within the umbrella of the common mycorrhizal network leads to a bright future with enhanced chances for a seedling's survival, establishment, and reproduction. Often, but not always, the common mycorrhizal network is a way for adult plants to contribute to the health and well-being of any nearby offspring. There is an amazing and complex communication highway underneath your feet each time you hike around a forest.

The common mycorrhizal network is also an alert system about incoming bad weather, disease, and herbivore invasion. Plant hormones, such as ethylene, may be pumped into the ground by roots, warning of incoming floods or even large obstacles. Defensive signaling molecules, such as salicylic acid and jasmonates, often divulge pest outbreaks as they travel along the common mycorrhizal network. A 2010 study connected pathogen-infected and healthy tomato plants with only a common mycorrhizal network, limiting other communication mechanisms such as gases called VOCs. The sick plants, under the attack of early leaf blight (*Alternaria solani*), responded with an array of defenses, including microbe-digesting enzymes. Without direct contact to the disease, the healthy plants began ramping up antimicrobial defenses. Studies involving infestations by aphids and caterpillars had similar findings, with unaffected plants primed into producing toxins and repellents by common mycorrhizal network communication. Across the board, defenses rose if healthy plants remained in common mycorrhizal network connection with embattled pest- and disease-laden plants. Plants have been shown to adjust their defense levels as quickly as six hours after receiving these communications.

Eavesdropping along the common mycorrhizal network is also possible. Monitoring for generalist pest outbreaks, incoming diseases, or natural disturbances gives plants an edge in advance preparation. It is well established that plants monitor airborne communications for these cues, and it is increasingly being found that plants monitor underground communications as well. Many pests (such as aphids and grasshoppers) and diseases (such as botrytis and root rots) can be quite generalist in their host preferences. It behooves a plant to find out sooner rather than later if it might be subject to the same infestation or infection that is harming nearby neighbors. Plants often respond to diseases by killing or digesting them with specialized defense enzymes. Therefore, levels of these telltale molecules are surveilled very closely by other plants.

Information Clouds: Volatile Signaling

Let's take a deep breath before we enter another marvelous hidden world: plant aerial communication. Plants communicate using gaseous molecules called volatile organic compounds. VOCs are organic gases created by plants. Since they are volatile (lighter than air at room temperature), VOCs readily drift and disperse. VOCs can be absorbed through plants' stomata, their chemistry read by receptors, and their elusive messages decoded. A plant's chemical cloud might consist of dozens of different VOCs, each with a specific purpose and target. The blend of gases a plant emits might also change depending on what positive or terrible things happened in its life that day. Some VOCs are messaging molecules or hormones, which illicit growth responses in surrounding plants. Other VOCs, such as the familiar essential oils, can directly disrupt or kill herbivores and microorganisms. Communication via air is particularly advantageous because it offers plants the ability to send complex messages quickly over distance and during stresses that restrict cell-to-cell communication.

The VOC clouds surrounding plants are an exquisite and complex mixture of different messaging chemicals. Volatile plant-speak sounds like it comes straight out of a mad science lab. Each plant whips up its own floaty concoction depending on species and circumstances, but common plant VOCs include esters, aldehydes, acetates, ketones, alcohols, essential oils, organosulfurs, isoprenoids, terpenes, sesquiterpenes, sterols, and methylated hormones. The important thing to realize is that plants constantly set adrift dynamic, sophisticated messages that can achieve numerous goals simultaneously.

Plant hormones are among the most potent messages that can be vaporized and sent through the air. The chemicals that make up plant hormones influence the growth, development, and differentiation of cells and tissues. Since these hormones are active in only minute amounts, plants need only catch a mere whiff of them to institute a response. One hormone, ethylene, is already conveniently produced in gaseous form and thus easy to relay. Ethylene is a general growth regulator and is also responsible for the ripening of fruit, the opening of flowers, and the shedding of leaves. Hormones such as jasmonate and salicylic acid are signaling molecules, which are normally transported inside cellular fluid. Should the need arise to extend their signaling range, plants can chemically modify these hormones into VOCs using a chemical process called methylation, which allows them to become gaseous and drift. This gives hormones the lift they need to reach distant plants or recruit the assistance of pest predators.

VOCs are one of the quickest ways for plants to send messages to their own tissues. The process of emission, reception, and response can take only seconds. Large plants such as trees, for example, might choose volatiles as the best way to communicate with distant branches and leaves or to chat with their neighbors about the weather or local bugs. By communicating with each other, plants can mount a coordinated and much more successful counterattack against massive pest and disease outbreaks. Communication systems are survival systems.

Airborne signals are also capable of reaching a variety of eavesdroppers: other plants, herbivores, predatory insects—even us! We often become aware of plant VOCs only because herbivore repellency is a common

Out of hundreds of orchids, the exotic three-colored vanda (*Vanda tricolor*) is the most alluring because of its amazing and intense fragrance. Orchids are among Earth's more than 352,000 species of flowering plants, which pump wonderfully scented VOCS from their flowers and fruits. A little bit of clever chemistry secretly ropes pollinators and dispersers into helping out with reproduction.

function they possess. Other volatiles, such as those emitted by flowers or ripe fruits and vegetables, are deliberate attractants for pollinators and dispersal agents. Plants secretly use these wonderfully scented chemicals to rope us and other organisms into helping them out with reproduction. Many plant species have strong, distinctive smells because of the volatiles that surround plants or that are emitted when they are crushed or otherwise wounded. Close your eyes and think about some of the plants in your life whose unique scent you can easily conjure. Oranges? Pineapples? Roses? These are examples of attractive VOCs. What about the scent of cut grass, pine, brussels sprouts, or garlic? These are examples of repellent chemistry, where plants make their best efforts to drive attackers away.

These airborne communications are useful because they work regardless of a plant's moisture content. Electricity, hormones, and many signaling molecules and ions are only effective when they can be distributed through fluid transfer, transportation, or diffusion from cell to cell. Taking communication to the air provides an ingenious alternative, and both forms of communication often work in tandem. Severe moisture loss, such as from drought stress, root rot, or mechanical wounding, can impair plants' fluid-based communication channels. Despite such events, plants remain fully capable of emitting gaseous distress signals. Xerophytic, or drought-tolerant, plants such as eucalyptus, tea tree, sagebrush, and rosemary have foliage that is notably pungent because of their predominant VOCs.

Table 5 A Selection of Plants That Produce Essential Oils

Plant family	*Common name*	*Major VOC produced*
Mint	Basil	Linalool
Mint	Mint, peppermint	Menthol
Mint	Thyme	Thymol
Mint	Oregano	Carvacrol
Mint	Rosemary	Pinene, camphor
Mint	Lavender	Lavandulol
Myrtle	Cloves	Eugenol
Laurel	Cinnamon	Cinnamaldehyde
Mint	Catnip	Nepetalactones
Myrtle	Tea tree	Eucalyptol
Myrtle	Eucalyptus	Eucalyptol
Pine	Pines	Pinene
Mistletoe	Sandalwood	Alpha-santalol and beta-santalol
Laurel	Sassafras	Safrole
Laurel	Yellow anisetree	Safrole
Citrus	Citrus (lemon, orange, lime, grapefruit, kumquat)	Limonene, nootkatone
Pine	Eastern red cedar	Cedrene, cedrol
Cypress	Arborvitae	Alpha-thujone, beta-thujone, fenchone
Grass	Lemongrass	Citronella oil
Amaryllis	Onions	Organosulfur
Amaryllis	Garlic	Organosulfur
Mustard	Mustard, cabbage, broccoli, brussels sprouts, kale, turnips, collards	Organosulfur
Mustard	Wasabi	Organosulfur
Mahogany	Neem	Azadirachtin, limonene
Ginger	Ginger	Alpha-zingiberene
Black pepper	Black pepper	Transcaryophyllene, pinene

Speaking of smelly, you're familiar with VOCs if you've ever smelled essential oils. To get the highly concentrated bottles of essential oils we find in stores, truckloads of plant foliage are heated up with steam. The foliage's VOCs are carried by the steam and are then cooled, condensed, separated from the water, and distilled into liquid, bottled VOCs. Leave the bottles open too long, and you'll find out just how volatile their contents are. The essential oils readily turn into gas and drift into the air, which is why essential oils are well known for their distinctive scents.

Plants predominantly manufacture essential oils as natural pesticides and disinfectants. Essential oils strongly affect insects, mites, diseases, and other plant pests. A plant's VOCs can act simultaneously as antioxidants, repellents, digestive inhibitors, growth and reproduction regulators, and outright poisons. Many essential oils have been traditionally used and have a promising future as temporary pest repellents around gardens, mosquito deterrent sprays, and organic pesticides. Numerous essential oils (tea tree oil, for example) have renowned antimicrobial properties. Since both plants and humans are subject to infections by bacteria and fungi, we can hijack and apply plants' defense chemistry successfully.

Not only can our noses detect VOCs but so can our taste buds. Plant VOCs are partially responsible for strong, unusual, or alluring tastes and flavors. Table 5, which lists common essential oils, includes many of our beloved foods and spices. An essential oil can contain dozens of volatile ingredients, but here we list only the predominating VOC.

The Hidden World of the Lawn

The smell of a freshly cut lawn evokes pleasant memories of picnics, springtime, sports, games, cookouts, and other outdoor pastimes. Subconsciously, this scent also delights us because it contains volatiles similar to those emitted by ripening fruits and vegetables. Our olfactory senses have cued into the fragrances of plants to help us survive and distinguish ripe, nutritious plant food from inedible or toxic plant parts. The reason plants concoct these powerful aromas after being mown down, however, is far from peaceful. They want revenge.

What you're actually smelling in the wake of a rampaging lawnmower are the aromatic chemical screams of plants that have just been butchered by a massive, whirring blade. The reaction of plants to mowing injury is virtually the same as if they were chomped by a ferocious herbivore. They're busy warning their neighbors and ramping up defenses. Every time you watch a lawnmower fling shards of cut grass through the air or catch the scent of a mown lawn, you're connecting with a hidden world of plant communication.

As a lawnmower blade slices and dices its way through the outdoors, it leaves a wake of plant communications behind. This cloud covers fair distances and lasts for hours. Incoming VOCs from other plants, fluxes of hormones and internal signals, and elicitors such as shattered cell wall pieces rapidly trigger the lawn's survival system. Even the dying chopped particles of leaves and stems participate. Before they shrivel, their last breaths are vengeful volatiles. While plants are unlikely to succeed against an unnaturally powerful motorized blade, communication allows them to realize that something is gravely wrong and mobilize their toughest defenses. Collectively, your yard's do-or-die opposition to a mower's assault is a united, extraordinary effort capable of driving away all but the strongest biotic attacks.

Plants throw almost every defense they can muster against the steel butcher's chopping frenzy. These defenses combine into the aroma of cut grass, a complex chemical cocktail often referred to as green leaf volatiles. Each volatile released in this process causes its own signaling cascade in the plants that detect it. Green leaf volatiles jump-start wound-healing genes, speeding the damming and repair of multiple cut surfaces and damaged cells. Airborne green leaf volatiles exert antimicrobial pressure so that diseases aren't able to utilize mowing cuts as entry points. As plants bandage and disinfect themselves, they shift focus to preventing further attack. Volatile hormones such as jasmonates increase the mobility and concentration of toxins such as insecticides. Repellent VOCs are pumped into the open air, ensuring that the attacker never returns. One class of these VOCs, the organosulfurs, are present in wild onions and occasionally give cut lawns an oniony smell. Finally, some volatile messages alert predatory insects such as wasps, lacewings, and ladybugs to the possible presence of prey and hosts. The next time you catch the scent of that freshly mown grass, realize it is because your lawn is making an amazing, last-ditch effort to survive a very, very bad day.

Fast as Lightning: Electrical Signaling

Bzzt! The chatter in the plant world is charging with current. Electrical communication is a rapid and convenient way for plants to send signals across their own cells. The speed of electricity outpaces any other mechanism of plant communication. There quite simply isn't a faster way for a living thing to transmit information and implement a response. To transmit electrical impulses, plants create gradients of charged ions such as calcium, magnesium, sodium, chlorine, and potassium. The alteration of these gradients is a very simple mechanism by which plants control the strength and distance of their communications. For example, a weakly charged signal may only pass across a single cellular membrane. Longer-distance versions might speed across the cells in an entire leaf or even from the roots to the leaf system of the plant. This same principle—the difference of electrical potential energy—drives electricity to flow between the positive and negative terminals of a battery and across the neurons in our bodies. In this section, we'll learn how plants use electrical signaling and what kinds of information they send.

It's not that weird to think of plants communicating by jolting themselves occasionally. After all, our cells are running current all the time as well. It is how our muscles move—beating our hearts, puffing our lungs,

adjusting our senses, moving our bodies, and powering our digestion. The world's plants, animals, and microorganisms are buzzing with electricity. Electrical signaling is a feature that cells have possessed for a long time, having evolved billions of years ago when life first drifted through our oceans. Electricity has been responsible for enormous evolutionary events, such as the power of movement. Movement isn't restricted to animals, either—we'll highlight a few amazing plants that feature rapid movements as part of their survival systems.

Some of the most scintillating displays of electrical plant signaling occur in plants that move. Rapid, touch-sensitive movement in plants is called thigmonasty. Amazing plants exist that are famous for their power of movement for herbivory defense or pollination. There are much more sinister moving plants as well. It may surprise you to learn that electrical signals power the ability of carnivorous plants to chomp on various animals. Plant carnivory is an extraordinary survival strategy that allows plants to transform nutrient-rich components into plant fertilizer. These extraordinary plants still photosynthesize to create sugars. However, the nutrient-poor or waterlogged environments that they inhabit make it difficult to acquire macro- and micronutrients such as potassium, nitrogen, and phosphorus. Instead of grabbing those nutrients out of the soil as most plants do, these carnivorous plants capture any crawling, falling, or flying packets of fertilizer nearby, such as insects, leaves, feces, or even small vertebrates. Not all carnivores have moving traps, but the ones that do, such as the Venus flytrap, seem spectacular and otherworldly. For these species, it's only by the power of electrically signaled trapping movements that they can survive the starvation induced by the wet, mucky, nutrient-poor soils of the boggy wetlands they inhabit.

The lickety-split movements performed by the traps of carnivorous plants such as the Venus flytrap, bladderworts, and sundews are fast enough to surprise and capture insects. Bladderwort traps have the claim to fame of having the fastest plant movements on the planet. The bladder traps on these plants are like tiny aquatic pouches. The traps are often filled with air, which helps the bladderworts float at the water's surface. Each trap contains a one-way door that opens to slurp in water and prey. The doors to their underwater traps are triggered to open and shut in as

The lickety-split movements performed by the traps of carnivorous plants such as the Venus flytrap (*Dionaea muscipula*), bladderworts (genus *Utricularia*), and sundews (genus *Drosera*) are fast enough to surprise and capture insects.

little as a thirty-fifth of a second. They are so fast that they can barely be captured with slow-motion cameras. Bladderworts make the larger and more ferocious chomping leaves of the Venus flytrap look like snails. Flytraps, which have two leaves that close to hold any wriggling creature inside, rely on trigger hairs to set this sequence in motion. Once triggered, flytraps use electrical signaling to close their jaw-like leaves. This is where the prey is imprisoned and eventually broken down. Flytrap movement clocks in at a mere three-tenths of a second. That's more than ten times slower than a bladderwort, but it gets the job done with striking panache.

To set and release their traps, carnivorous plants create and unleash large fluid pressure differentials in their tissues. This simply means that the plants build gradients of high fluid pressures in one area and low pressures in another. When released, the pressures flow to equalize each other, creating a substantial mechanical or signaling force in the process. An average Venus flytrap trap takes fifteen to sixty minutes to pump up and prime but only a fraction of a second to release because of the cascade of electrical signaling. The electrical signal causes complex patterns of ions to disperse throughout the leaf cells. Water follows rapidly along these ion gradients, plumping up cells with more ions and deflating those with fewer. As this contrasting expansion and contraction takes place, the entire leaf begins to move. The result of the cascade of electricity, ion transport, and water diffusion is one of nature's marvels: the movement of the hinge-like midrib of the flytrap leaf.

But how does a carnivorous plant know when to fire its trap? Carnivorous plants have evolved trigger hairs—touch-sensitive trichomes—that grow on the surfaces and entrances of their traps. Every time prey bumps these triggers, electrical potential energy builds, allowing the plants to sense that live prey has come within lethal range. In a sense, the traps really do "charge up." For the flytrap, two taps of the trigger hairs within six seconds jolt the leaves into closure. Why two taps? The plant needs to be sure that an organism is the trigger—not a raindrop, a fallen twig, or the brush of a neighboring leaf. Once triggered, the leaf almost completely closes, leaving a gap just large enough to allow any tiny creatures to escape. To flytraps, there is no sense in wasting a leaf on a small meal. A further half-dozen taps of the hairs tells the plant it has successfully caught live prey that is too large to escape the jail cell. At this point, the flytrap leaf seals completely shut and enters a juicy digestive phase for about a week, leaving nothing but an empty bug exoskeleton behind.

The traps of bladderworts are aquatic. Each plant possesses thousands of bladder traps: air-filled pockets with hinged doors surrounded by trigger hairs. Accidentally knocking these hairs fires the electrical trigger for the door to open inwardly. The trap becomes a powerful water vacuum, depressurizing and sucking in any nearby water and prey for digestion. This extraordinary movement may be the last event that unsuspecting tiny animals such as water fleas, baby snails, or mosquito larvae ever see.

In this chapter we've seen some of the most spectacular things plants are capable of. We hope that these revelations stick with you while you're mowing your lawn, stepping around a curious mushroom, watching a special on bug-munching plants, or catching the botanical scents of plants in the air. How much of the true activity of the hidden world were you already aware of? Plant communication is nothing short of mind-blowing. In times of peace, communication strategies help plants acquire natural resources, assist their offspring, and secure pollen and seed dispersers. In times of stress, however, the same communication mechanisms allow plants to mobilize formidable arsenals of defenses. Plants can counterattack, strategize, move, relay intelligence, identify hazardous situations in advance, and recognize attackers—all without making a sound. Perhaps these silent sentinels are a bit keener than we give them credit for.

CHAPTER 8

The Enemy of My Enemy

Plant Protectors

The hidden world of plant protectors is a strange and astonishing one indeed. Plants are assaulted by voracious insects all the time but are anchored to the ground, unable to run away. As we saw in the vascular system chapter, plant defenses can be overrun by herbivores such as aphids. Thankfully, plants have befriended the predators and natural enemies of their herbivores. You heard right—plants can actually call in the predators of their pest insects to help get the situation under control. Provided the right combination of bribery and distress signals, these beneficial insects form a vigilant and effective security and surveillance system. This chapter opens the strange but effective world of indirect defenses, which are structures and chemicals that plants create to recruit predatory and colonizing insects. Indirect defenses rely on amazing mercenary armies rather than the plant itself to kill and remove herbivores.

Indirect defense is a mutualistic relationship between a host plant and its defenders in which both species benefit. In exchange for housing, food, or shelter for their offspring, defenders provide invaluable services to facilitate the survival of their hosts. These services include killing and removing herbivores, planting and protecting seeds, cleaning and disinfecting the host, cutting down and maiming competing plants, pollinating the plants, and depositing nutrients into the host's specialized absorptive zones.

To achieve these symbiotic benefits, plants have become surprisingly adept at bribery. Carnivorous ants, wasps, beetles, and mites sign up as indirect defenders in droves to access the bribes they receive from their hosts. Some plants, such as the diverse grouping known as the ant plants,

This unusual epiphytic plant, called the Malayan urn vine (*Dischidia major*), forms large ant nests throughout tropical forest canopies. The leaves of a *Dischidia* are hollow pouches that are often filled with the plant's own roots. Colonies of ant defenders live inside these domatia, protecting the plant while producing nutrient-rich garbage and carbon dioxide.

are so specialized at indirect defense that most of their morphology is dedicated to supporting their protectors. These noble plant guardians are rewarded with a round-the-clock life on easy street. The ant plant provides interior housing and food to ant colonies, encouraging the insects to move into the plant's interior. Ants are highly territorial and fierce defenders of their homes. By letting ants reside within a plant's internal chambers, both the ant plant and the ant colony gain automatic protection. These plant hosts grow customized, protected shelters called domatia and produce plentiful supplies of food on demand from extrafloral nectaries, food bodies, and elaiosomes. Ant plants aren't the only type of plant to form domatia, but they are some of the most impressive. Each of these intricate structures deserves a closer look.

Epic Treehouses

Domatia can be quite diversely shaped and sized, occurring across a range of plant organs. Some domatia form from hollow leaves, such as in *Dischidia*, *Tillandsia*, and the fanged pitcher plant, *Nepenthes bicalcarata*. Debris-trapping foliage and roots act as domatia in the staghorn ferns, bird's nest fern, oakleaf ferns, basket ferns, and *Pachycentria glauca*. Hollow stems or caudexes (swollen, bulbous stems) house insect colonies in *Cecropia*, *Myrmecodia*, and *Hydnophytum*. Even thorns (*Acacias*) and rhizomes (*Solanopteris* and *Lecanopteris*) have evolved as hollowed-out, defender-bearing structures.

Some of the most successful and striking domatia are produced by plants called ant plants. Ant plants are strange, otherworldly-looking organisms that have spent millions of evolutionary years favoring ants as their preferred defenders. Some of the oldest ant plant lineages, whose origins date back more than fifteen million years, include the genera *Myrmecodia*, *Hydnophytum*, *Cecropia*, *Lecanopteris*, and *Dischidia*. You might consider ant plants with their bulbous tissues grotesquely misshapen at first glance, but a closer look reveals spectacular ant-based defenses within these botanical oddballs. The most prominent feature of these tubby little plants is their intricate, swollen domatia, which serve as housing for entire ant civilizations. The ant plant domatium is a turnkey, rent-free, custom-built home. As the water-laden vascular tissues course through each domatium,

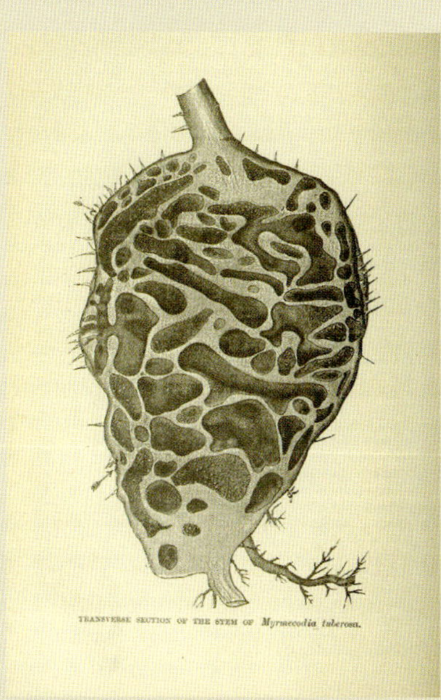

If you or an herbivore were to tear open an ant plant, you'd find it entirely hollow. Each plant grows precisely ant-sized entrance holes that lead to labyrinths of tunnels and chambers. The internal surfaces of these chambers are lined with absorptive glands or specialized roots that uptake and process the colony's nitrogen-rich waste and carbon dioxide into fertilizer for the plant.

Cecropia trees (genus *Cecropia*), seen here, are well-known ant recruiters. A young ant queen will alight on the *Cecropia*'s hollow stems, chew into the soft, temperature-controlled interior pith, and seal herself inside an internodal chamber. She will lay a brood, fight competing queens for sole ownership of the entire *Cecropia* tree and its resources, and stockpile food bodies produced on *Cecropia* leaf stems.

Staghorn ferns (genus *Platycerium*), seen here, are epiphytic ferns that form two types of leaves. The conspicuous antler-shaped fertile fronds bear the plant's reproductive spores. The more subtle shield fronds form the "core" of the plant. The living and dead shield fronds form an overlapping, multilayered, and highly creviced structure that houses ant defenders, traps nutritious debris, and attaches the fern epiphytically to its substrate.

they regulate the chamber's temperature. In essence, the ants receive free air conditioning in the intense tropical heat. Plants even install basic security systems, as many domatia are studded with spines and coursing with caustic latex. If you or an herbivore were to tear open an ant plant, as seen in photo 84, you'd see something that looked like Swiss cheese or a potato with a bunch of holes drilled into it. These plants are mostly hollow. Each plant grows precisely ant-sized entrance holes that lead to labyrinths of tunnels and chambers. The internal surfaces of these chambers are lined with absorptive glands or specialized roots that uptake and process the colony's nitrogen-rich waste into fertilizer for the plant. The ants also puff large amounts of carbon dioxide into the domatium as they breathe, stoking their host plant's photosynthesis and growth rate. As herbivores munch, shake, or otherwise attack ant plants, they unwittingly generate vibrations that call the ant colonies to battle. Regarding damage to their host plants as a direct attack on their home, ants are prepared to fight herbivores to the death.

Domatia are an amazing benefit for ants. Many ant plants are epiphytes, nonparasitic plants that grow on the surface of another plant and glean nutrients and moisture from the surrounding environment. As a result, moving into treehouses allows ant colonies to exploit the resources of the forest canopy. These arboreal ants dodge competition, disease, and predation by ground-dwelling ants, pathogens, and predators. The treehouses are also in close proximity to novel and diverse food resources such as fruit, seeds, canopy insect prey, honeydew, and foliage. Not only do the ants gain a free, sturdy, and protected house, but they can collectively focus on tasks other than digging and construction—tasks such as plant defense! Plants even try to make ants' work easier. Specialized anchor-shaped trichomes on ant plants such as *Cecropia* trees are precision-fitted grips for the hooks on *Azteca* ant legs. Teamwork, carrying prey, fighting, and sawing are all made easier by these trichomes. Amplified sticking power also keeps the ants safer, protecting them from being dislodged by predators, prey, wind, and rain. Once a colony has united with a host, the two organisms become an unstoppable force.

Ants make perfect defenders. Ants, like bees, operate as one superorganism, where everyone is working as a collective unit. As such, ant colonies steamroll problems as a unified whole with overwhelming force. A colony is a living superorganism, instantaneously and simultaneously coordinating daily activities. Ants unite to forage for food, cultivate gardens of fungi, cut leaves, and tend insects such as aphids. Nest construction is a communal affair, as is clearing competing vegetation, cleaning, disinfecting areas of the nest, and disposing of waste. Ants also improve their host's photosynthetic ability by removing leaf debris as well as sawing down vines and any other competing vegetation that grows too near. These other plants are potential highways for the intrusion of predators and competitors, so the ants benefit as well. Brood rearing, collective learning, surveillance, and defense against threats are also part of the ant team's daily mission. Each colony may be comprised of thousands or even millions of ants, each ant fiercely armed with its own biting, stinging, and poison-spewing weaponry.

Ants are intensely attuned to chemical signaling, which is a feature plants have found pretty nifty. While ants evolved chemical sensory perception to communicate quickly within a colony, ants and their symbiotic host plants have evolved the ability to speak the same chemical "language." Like superheroes responding to the bat signal, ants readily tap into the chemical distress signals and identification markers emitted by their plant hosts. Additionally, when ant plants are damaged, they emit volatile chemicals that place ants on high alert. Workers encountering these signals immediately scan for danger. If the scouting mission finds a menacing caterpillar or mammal, the worker paints a chemical path of pheromones back to the closest domatium entrance. Within seconds, a minute militia bursts forth, intent on bringing instant regret to the herbivore for its mistake. Woe betide the muncher of the ant plant!

Botanical Catering

Plants are literal food factories, bursting at the seams with prodigious quantities of sugars, minerals, and proteins. By selectively dispensing these substances or broadcasting the presence of prey, plants utilize the assistance of carnivorous and parasitic ants, wasps, beetles, and mites. All indirect defenders are rewarded day and night with an enormous all-you-can-eat buffet stocked meticulously by their plant hosts. Let's see what's on the menu of this hidden catering service!

The bumps on this blue passionflower (*Passiflora caerulea*) leaf stem are limitless nectar spigots called extrafloral nectaries. Extrafloral nectaries bribe ants and other beneficial organisms to pay great attention to this plant. Rarely a minute goes by without this plant being patrolled by ants, wasps, and other predatory critters.

Most of us are familiar with nectar, but few realize that its evolution in plants far predates flowers. More than three hundred million years ago, ferns, not flowering plants, became the original inventors of nectar. Before pollen, pollination, seeds, flowers, and fruit, nectar's sole purpose was to sustain the process of indirect defense. This defensive nectar is called extrafloral nectar because it occurs on plant parts other than flowers. The limitless botanical spigots for this sweet, vitamin-rich fluid are called extrafloral nectaries. Extrafloral nectaries can be found in a wide array of plants, many of which probably live around you. Scan the nooks and crannies around plant leaves and stems—extrafloral nectaries look like tiny bumps, and much of the time they emit small droplets of nectar. Scientists believe that one possible origin of nectaries may have been inadvertent, occasional "leaks" from early phloem cells, which are highly pressurized with sugary sap. Only by chance encounters with these leaks did indirect defense relationships begin. With increasing protection and animal attention, extrafloral nectaries eventually evolved into the standalone, specialized structures we see today. There are hundreds of species of plants in more than seventy families that produce extrafloral nectar. You may already be familiar with many of these, such as the common garden bean, cucumber, cotton, passionflower, castor bean, basket fern, peach, and elderberry. Extrafloral nectar production is strategically scattered across a plant, and its purpose is to place indirect defenders on constant patrol. Nectaries usually protect flowers, fruit, leaves, and stems, so entire plants may be crawling with ants or alighted on by carnivores or parasites at any moment. Such frequent and attentive surveillance is well worth the cost of donating a few easily produced sugar packets!

To ensure their healthy little fighters take in a balanced diet, plants also bribe them with structures called food bodies. Food bodies vary in their shapes, placements, and even names depending on the species of plant that produces them. The general unifying trait is that food bodies are rich in proteins, lipids, and carbohydrates. After all, fueled up and happy defenders are formidable ones. Trumpet tree leaf stems produce Müllerian bodies, fuzzy whitish patches of a sweet, energy-rich substance called glycogen that entices ants. Beccarian bodies occur on species such as the parasol-leaf tree. These are essentially bands of deep-fried ice cream:

This *Pseudomyrmex* ant is looking for candy. What a trove it has found on this bullthorn acacia (*Acacia cornigera*) leaf! Each acacia leaflet bears a treat at its tip called a Beltian body, custom designed to bribe ants into servitude. These food bodies have nutritional profiles similar to protein bars—they are rich in oils, protein, and carbs.

greasy, fatty bands that encircle each node of the plant's stems. Beltian bodies, which have ratios of lipids, protein, and carbs that put energy bars to shame, are produced on the leaflet tips of certain species of *Acacia*. Pearl bodies, named for their orb-like, shiny, and pale appearance, are produced on the epidermises of leaves, petioles, and stems. Pearl bodies occur in more than nineteen plant families in the tropics and subtropics, including the black pepper family (family Piperaceae). The occurrence, distribution, and richness of food bodies are carefully controlled by plants.

We find these food bodies on the fascinating ant plants as well. After ant plants detect colonization by ants or an attack by herbivores, they escalate their production of food bodies. The nutritional content of the food bodies themselves also increases, creating tastier snacks that strengthen and entice defenders into conducting more frequent and thorough patrols. Food bodies are so delectable that ecologists have noticed ants dropping other food, prey eggs, and even their own larvae to the forest floor in order to stop and pick up a tasty food body. In the world of indirect defense, plants become enchanted, candied lands where all kinds of happy helpers scamper merrily around, plucking delightful little morsels. Along the way, the miniature mercenaries rip, gnash, stab, and sting herbivores to death. What a mind-blowing juxtaposition we find in the hidden world!

Elaiosomes, another type of botanical bribe, are similar to food bodies except that they are always attached to seeds. Elaiosomes are specialized seed appendages that are edible, fleshy, oily, and rich in proteins and lipids. The organism that most commonly craves this tasty product are—you guessed it—ants. Plants have used elaiosomes to addict ants to planting, dispersing, and protecting their seeds—a phenomenon called myrmecochory. These seeds are much too hard-shelled for the ants to eat, but the tasty elaiosomes make excellent baby food for developing ant larvae. Elaiosomes are perfumed and scented with chemicals that identify them to ants as valuable food sources worthy of defense. Ants merrily carry the seeds back to their nests, snipping off the elaiosomes and tossing the seeds aside like garbage. However, being discarded is exactly what plant seeds want. The ants' underground waste pile is a perfect location for germination: it's moist, nutrient-rich, and safeguarded by thousands of territorial ant protectors.

This photograph showcases the elaiosomes of twelve different plant species. Basically a food body attached to a seed, the elaiosome convinces ants to collect, haul, and bury the seeds in their nutrient-rich underground garbage piles. Free food for free dispersal!

Calling in Air Strikes

Part of the web of ecological diversity thrives on a valuable food source we may not even think of: the herbivores that infest plants. Eggs, larvae, nymphs, and adults are all on the menu for carnivores and parasites, which can include predatory insects, mammals, birds, and any other hungry creature. All plants need to do is find a way to advertise their presence through VOCs. Their production and emission through the air surrounding a plant call in an air strike by the natural enemies of herbivores. VOCs can act as repellents to herbivores, but some are highly attractive to higher trophic

Get 'em! A mix of tiny parasitic aphid wasps (circled in red), green aphids, and brown aphid mummies are seen here on a crop of corn. The plant called for help, and the wasps came to the rescue!

levels of the food web. Hiding all around us are these imperceptible clouds of information that stressed-out plants emit to draw the attention of the top predators, competitors, and parasites of their pests. You've smelled these botanical screams and perhaps even enjoyed them. This chemical signaling is found in the smell of freshly cut grass or a bouquet of flowers. In these instances, the plants are emitting a distress signal from being damaged or fed upon by entities such as the herbivorous lawnmower or a gardener's shears.

VOC-based indirect defense begins with elicitation. The breakage of trichomes, the piercing or gnashing of cells, the slobbering of feeding, or the gluing and deposition of new eggs all induce plants to sound the alarm. Plant hormones transmit the order for plant cells to begin off-gassing a broad spectrum of messages. Volatiles such as indoles, aldehydes, ketones, esters, alcohols, terpenes, and nitrogenous compounds are set adrift around the stressed plant. Some studies have concluded that these VOC screams can reduce the population of certain herbivores by more than 90 percent!

It can be almost impossible for predators and parasites to detect camouflaged prey such as these caterpillars. However, plants outsmart cryptic patterning with the emission of VOCs, replete with information about the herbivore's species, location, and population size. Indirect defenses paint a giant smelly target on herbivores' backs, attracting predators no matter how hard the insects try to hide.

Botanical chemistry communicates well not only with ants but also with wasps, beetles, and flies. Usually carnivorous or parasitic, these rapid-response defenders attack from the air, being especially sensitive to airborne distress signals. Aphids, caterpillars, bugs, sawflies, boring insects, and a wide variety of other herbivores have everything to fear from these enemies. Insect larvae often attempt to avoid predators through crypsis—the color pattern or mimicry of their surroundings. Plant volatiles outsmart this strategy completely. While these varmints try to hide away, the plant is busy tattling on them. The type of information provided includes an herbivore's species, population size, and exact location. VOCs leave herbivores nowhere to hide from their hungry predators. The carefully crafted aromas of plants mark their herbivores with massive, smelly targets. We previously learned that structures such as glandular trichomes or fluids such as resin and latex can also mark attackers with scent trails.

Predatory insects are great for munching herbivores out of existence. After an appetizer of extrafloral nectar, a ladybug might snack on some aphids, too. Lacewings might select a stressed plant as an egg-laying site

Carnivorous insects are great for munching herbivores out of existence. Lacewings might select a stressed plant as an egg-laying site for their voracious larvae, seen here, which devour thrips, mites, caterpillar eggs, and scale. Many lacewing larvae assassinate their prey by fabricating camouflage disguises from debris, including soil, plant parts, lichens, and prey carcasses.

for their voracious larvae, which devour thrips, mites, caterpillar eggs, and scale. Hoverflies, esteemed as gentle pollinators for their nectar-sipping behavior, actually have ravenous larvae that devour thrips, aphids, and other small sap-sucking insects. Wasps, beetles, or ants might abduct a hapless herbivore at any moment, gnashing it to pieces. VOCs also get the attention of predatory mites and assassin bugs, which stab and drain the bodily fluids of herbivorous mites and caterpillars. By no means are plants alone in their struggle against herbivores.

Even more gruesome than predator insects are the parasitic wasps, midges, and tachinid flies that flock to plants' chemical messages. These parasites develop their larvae in horrifying ways, laying eggs either on the outside of the host or within it. Idiobiont parasites sting and paralyze their host, lay eggs on the stiffened body, and then pack it in a safe space such as a mud or paper chamber. This process allows their offspring to have an immobile but living host to feed on. Mud dauber wasps (families Sphecidae and Crabronidae), spider killer wasps (family Pompilidae), and potter wasps (subfamily Eumeninae) are all examples of idiobionts. Koinobiont parasites pierce the host with their ovipositors and insert an egg, which develops inside as the host continues to move, eat, and grow. Aphid wasps (subfamily Pemphredoninae) and braconid wasps (family Braconidae) are examples of koinobiont parasites. If you've ever been out in a vegetable garden and noticed small white cocoons mottling a tomato hornworm caterpillar, you've seen braconid larvae draining a poor host caterpillar. In each type of strategy, parasite larvae emerge armed with large jaws to begin eating and draining the living host from within. However, they are careful to avoid damaging their host's major organs in order to allow themselves maximum time for development. There is no light at the end of the tunnel for herbivores—they become weakened, sick, and eventually die when squirming, chewing hordes of larvae burst out of their bodies. Plump and juicy caterpillars are one of the most popular parasite hosts, but pestiferous aphids, beetles, flies, and true bugs are also victimized—there are even hyperparasitic wasps that attack other parasitic wasps! Parasite larvae lack openings in their digestive systems to excrete their wastes. By retaining their excretions, the larvae refrain from polluting the host or warning its immune system of their existence. Knowing the host will eventually die, the larvae simply leave their dump

This tomato hornworm has experienced the wrath of plant defenses. As the healthy caterpillar munches on the tomato, the plant recognizes the caterpillar's location and species and prepares to unleash a sinister resistance. The plant releases a silent alarm: VOC advertisements offgas, reaching tiny parasitic braconid wasps. These wasps require hornworm caterpillars to host the next generation of their life cycle, so they spend much of their adult lives patrolling for these plant scents. Adult wasps flock to the tomato plant, where they insert their eggs beneath the caterpillar's skin. Voracious hordes of legless and flightless wasp larvae hatch, writhe, feast, drain, and eventually kill the hornworm from inside. To pupate, the larvae chew out of the dying caterpillar, spin white cocoons, and finally emerge as legged and winged adults. The life cycle continues as these minute wasps hunt for other distressed plants with fresh caterpillar hosts.

pile behind when they pupate. After eating their way out of the drained and dying shell of the host, the wasps and flies fly away to unleash their own terrifying children upon another hapless herbivore. Some parasitic wasps hold endogenous, symbiotic polydnaviruses that infect a host and disrupt its immune system so that the parasite's larvae can develop safely and undetected. Additionally, it is believed that these viruses can actually change the behavior of the host. Some diseases include insect mind control among their symptoms. In some instances, the virus causes the dying organism to guard and assist the parasitic offspring. For example, the white butterfly wasp (*Cotesia glomerata*) causes parasitized caterpillars to spin protective cocoons and aggressively fight off parasites of the wasp's larvae. Who knew what horror plants and their helpers could unleash at a moment's notice?

Entomologists have long been aware of these fascinating ecological relationships. In agronomic and horticultural situations, we deliberately apply many predatory insects and arachnids on our crops as targeted pest control efforts. Ladybugs, lacewings, aphid wasps, and predatory mites are all examples of what we call biological control agents, which help us keep our crops free from damage and work alongside other methods of pest control. Beneficial organisms remain a powerful tool in our arsenal. Their existence is the result of millions of years of evolution and yet another window into the world of plants fighting their predominant herbivores.

Plants seem like vulnerable organisms passively anchored to the ground and letting whatever happens to them happen. But nothing could be further from the truth. Plants around the world have armed themselves with guards, chemical warfare, mercenaries, and their own colleagues in order to ward off the massive onslaught of attacking pests. It's a tough world out there, and plants work hard to survive in it.

CHAPTER 9

Hiding in Plain Sight

Camouflage and Mimicry

How Does Plant Mimicry Work?

There's one plant survival tactic that's simply out of sight: camouflage! Called mimicry or crypsis, it keeps plants alive by helping them evade the perception of their attackers. Animals cannot eat plants they cannot see. Through millions of years of natural selection, some plants have developed a physical appearance or smell that closely replicates some fascinating and inedible parts of their immediate environment. Plants around the world escape detection by mimicking rocks, dead leaves, useful crops, herbivory damage, and even the smell of rotting flesh. Despite these plants being full of water and nutrients, animals simply pass them by on the way to munching their more easily spotted competitors.

Mimicry is a passive strategy that has taken shape over generations of evolution and natural selection. Unlike the great animal masters of disguise such as the chameleon and the octopus, plants cannot make instantaneous changes to their color or shape. So the patterns and colors they develop through mimicry are a life-or-death test at any moment. The plant mimics we see today are the product of generations of beneficial, natural, and random genetic changes. The plants that adapted a shape, color, or pattern that enhanced their camouflage or mimicry of inedible objects survived and reproduced at greater rates than the plants that wound up on an herbivore's dinner table. Even the most well adapted mimics, if transplanted to distant substrates of a slightly different color, would stick out like sore thumbs. Catching a thirsty or hungry animal's eye places a plant in immediate danger. In deserts, the evolutionary crucible for some of the world's most spectacular mimics, herbivore pressures are extreme because

Living rocks are living plants after all! *Lithops*, the living rocks, are a genus of stone-mimicking plants from extremely dry deserts in South Africa. Here is a colony of *Lithops marmorata* in bloom. Living rocks use a defense called mimicry to avoid being seen. Undetected plants are safe plants.

plants are often the only sources of vital water. Let's see if we can detect how plant mimics became so successful.

In 1811 a British naturalist and explorer in South Africa named William Burchell noticed an odd-looking pebble. He stopped to pick it up but, to his surprise, found it rooted to the ground! Burchell had stumbled upon a group of plants known as the living rocks: the genus *Lithops*. Looking around, Burchell realized that these strange, diminutive plants were growing in every nook and cranny in the desert landscape around him. Coincidentally, he'd also made the first discovery of plants using camouflage.

A perfect match! Can you spot the *Lithops aucampiae* in this photo? Living rocks are superb mimics of rock and soil.

Lithops and related genera such as *Conophytum*, *Titanopsis*, *Pleiospilos*, and others in the ice plant family (Aizoaceae) reside in South Africa, succeeding in some of the driest and most heavily herbivorized habitats in the world. Some of these living rocks have specialized to the terrain of individual mountains, a process that took eons. Their leaf pigmentation and patterning perfectly match the color and shape of the chunky, rocky soils in which they eke out a living. This particular type of camouflage is called background matching. *Lithops* come in a geological rainbow of colors—rusty reds, orange, brown, green, slate gray, or even white to mimic certain types of quartz. Other Aizoaceae members, such as *Titanopsis*, have craggy white leaves that mimic limestone. Additionally, the upper *Lithops* leaf surface is colorfully mottled. While beautiful, it's another visual trick called disruptive coloration. Animal, military, and outdoor camouflage patterns utilize this same principle. When a plant's markings give the appearance of false edges and boundaries, animals have a tougher time noticing the plant's true shape. Rock mimicry has independently evolved in several other genera, including the milkweed, euphorbia, cactus, and crassula family. The exquisite beauty of these forms has led numerous succulent enthusiasts, including the authors of this book, to include a few mimicry plants in their collections.

Titanopsis, from South Africa, is related to living rocks and is seen here mimicking limestone with the white craggy patterning on its leaves.

The baseball plant, *Euphorbia obesa*, is a euphorbia rock mimic.

Various genera of living rock cacti from the southwestern United States and Central America have evolved to mimic dry rocky desert soils. Here, you can scarcely see the foliage of *Ariocarpus fissuratus* poking above the surrounding rock—and neither can herbivores. *Aztekium* and *Frailea* are other genera of rock-mimicking cacti.

Pseudolithos cubiformis is an African rock mimic in the milkweed family. Though not seen in the photo, this plant bears an inflorescence whose stinky scent is yet another layer of mimicry.

This *Sclerocactus papyracanthus* in Arizona has papery, twisted spines that mimic dead grass leaves. For an herbivore, it's not worth taking a bite.

Euphorbia platyclada, from Madagascar, is known as the dead wood plant. Despite it being perfectly happy and alive, its color and shape don't look appetizing.

Another type of visual deception plants employ as a defense is called masquerade, which occurs when the mimic poses as a different and less palatable object. Using shape and coloration, plants have evolved to masquerade as dead vegetation. In the deserts of the American Southwest lives a cactus called the paper spine or grama grass cactus. Paper spine cacti have quirky spines that mimic dead grama grass, a common habitat companion. The spines are the exact same color of brown, freakishly elongated, twisted, and papery thin. Definitely a meal to avoid! Some plants mimic dead sticks, such as the dead wood plant of Madagascar and the milkweed genus *Rhytidocaulon*. Even when they're perfectly happy and alive, you'd barely know it. The dead wood plants produce pale, brown, photosynthetic stems rather than lush, green leaves. Should a desperate animal decide to take a chance on the branches, a bitten dead wood plant will gush poisonous latex. Being pale brown has its cost—it is less efficient at performing photosynthesis because of the missing chlorophyll. Making latex also has an energy and resource cost. However, when you live in a desert and can easily stick out to any hungry herbivore, it is worth the sacrifice. This is why we see these adaptations in limited use around the globe.

Nikolai Ivanovich Vavilov (1887–1943) was a brilliant Russian botanist and geneticist best known for having discovered the centers of origin of common crop plants. He devoted his life to traveling the world, collecting and studying strains of corn, wheat, and rice. Vavilov established the world's first seedbank and among many accomplishments discovered crop mimicry, now also called Vavilovian mimicry. Tragically, this man, whose contributions helped feed the world, was imprisoned and starved to death by Joseph Stalin for his support of basic Mendelian genetic principles.

The Mimics Hiding in Our Crops

The evolution of mimicry has even been growing right under our noses in common crop weeds. In the early 1920s a Soviet botanist and geneticist named Nikolai Vavilov was struck by the uncanny resemblance of certain weeds to cultivated crops. Species such as rice, wheat, lentils, and flax appeared to have weedy doppelgängers scattered through their fields. These interlopers include the rice mimic, early barnyard grass (*Echinochloa oryzoides*). Wheat and barley are mimicked by oats and rye, which interestingly became crops in and of themselves. Common vetch (*Vicia sativa*) mimics the lentil plant, while camelina (*Camelina sativa* var. *linicola*), also now a crop in its own right, copies the look of flax. By masquerading as crops, these weeds have hitched a ride in farmers' pampered, fertile, and irrigated fields for tens of thousands of years. Vavilov's brilliant hunch was the discovery of the hidden world of crop mimicry. Crop mimicry is now also called Vavilovian mimicry in his honor. Vavilov's insight led all the way back to the dawn of human agriculture during the Neolithic period, when a few wily weeds changed their species' futures forever. Unfortunately, this amazing visionary was imprisoned and starved to death in 1943 by Joseph Stalin for his outspoken, scientifically sound stance on these and other basic principles of genetics.

The basic driving force behind crop mimicry is *us*—artificial selection by humans. When farmers scour their fields, rogueing undesirable weeds, they are making visual selections between which plant species stay or go. In the modern age, machinery such as combines, threshers, winnowing machines, and seed sorters use lasers and mechanical means to process our crops. Size, shape, color, weight—these physical features determine how a farmer and his or her tools remove plants and propagules that don't resemble the crop. On occasion, a weed with a similar leaf shape, height, or seed size might slip through undetected and live to reproduce. As these duplicitous survivors multiply, as we know they have for generations, the mimicry intensifies to a point where some crop mimics are almost perfect replicates. In fact, some even go a step further and overpower our crops in their own fields. For example, barnyard grass, assuming the guise of rice, outcompetes rice by every metric, including growth rate, seed production, nutrient guzzling, herbicide resistance, and paddy water slurping.

Rye (*Secale cereale*), now a commercial grain crop, is a happy accident of Vavilovian mimicry. Wild rye (*S. montanum*), a close relative of wheat

Wheat (genus *Triticum*), almost ready to harvest. Its domestication marked the dawn of agriculture during the Neolithic period. Over tens of thousands of years, pampered wheat fields were the crucible for the evolution of crop mimicry.

Can you tell them apart? Rye (*Secale cereale*) is a crop mimic of wheat. Not only does rye look like wheat, but rye also evolved an annual habit to avoid being destroyed by tilling.

and barley, began as an invader of agricultural fields. Wheat and barley, as annuals, produce seed and die each year. Wild rye, with a longer-lived perennial life cycle, was always vulnerable to the annual tilling of fields because it rarely produced seed before its second year. At some fateful point before recorded history, a quite fortuitous wild rye mutant decided to produce seed in its first year, a trait passed along to its offspring. That rye placed a few of these precious seeds and their genetics into ancient collection baskets alongside the season's wheat and barley. Season after season, this new annual rye mutant was inadvertently sown into wheat and barley fields. Further artificial selection, such as better seed retention and flavor, allowed rye to morph into a fully domesticated annual crop plant.

Small genetic changes replicated over generations explain the origin stories of other Vavilovian mimics. To mimic lentils, the common vetch transitioned from a rounded seed to a flattened shape. Camelina, a look-alike of flax, has adapted seeds of the exact weight, shape, and color as flaxseed so that the two are impossible to separate by sorting machinery. Camelina uses this tactic to bypass sorting techniques such as winnowing, which uses air blown through grain seed to remove the lighter chaff and retain the heavier seed. Two plant species with identical seed weights are virtually inseparable with this process. The flax dodder (*Cuscuta epilinum*), a flax parasite, also tosses a few propagules into the collective mix every year, having adapted larger seeds over time to pass through flax sorting equipment. In rice fields, it's common to find a few barnyard grasses (*Echinochloa oryzoides*), a species that has altered its appearance, flowering, and seeding times to match up precisely with rice. Perhaps a few thousand years from now, the weeds we see on the edges of fields today might become the next big crop plants cultivated by our descendants.

Other Extraordinary Fakeries

Mistletoes are another fascinating example of plants mimicking the appearance of common plants around them to avoid being eaten. Mistletoes are parasitic plants from several families, including Loranthaceae, Santalaceae, and Misodendraceae. They live in the canopies of trees in forests throughout North America, Europe, South America, Africa, and Australia. As hemiparasites, mistletoes produce some of their own food but also have specialized roots called haustoria that sap water and nutrients from their hosts. Mistletoe berries are viscously sticky and spread from branch to branch by birds. The foliage of a mistletoe spends its life side by side with the leaves of the host tree. Over many generations, mistletoes have gained an advantage by mimicking the species they parasitize.

The foliage of Australian mistletoes in particular is famous for bearing an unbelievable resemblance to their host trees. The coevolution of Australian mistletoes with their entire ecosystem has been going on for millions of years—the parasites, their hosts, herbivores, and fruit dispersers have all played a role in the development of their mimicry. Needle-leaved mistletoes, such as the species *Amyema cambagei*, are dead ringers for their hosts, *Casuarina cunninghamiana* and *Allocasuarina torulosa*. Another species, *Amyema sanguinea*, has nailed the mimicry of the flat, falcate leaf shapes of *Eucalyptus* and *Acacia*. *Dendrophthoe homoplastica*, in order to look exactly like its host, *Eucalyptus shirleyi*, has rounded, stem-clasping leaves. Even Australian mangroves (*Rhizophora*) have mimics of their thick, fleshy leaves: *Amyema mackayensis*, *A. thalassia*, and *Lysiana maritima*.

You may be wondering why in the world mistletoes would stretch to such evolutionary lengths to duplicate their hosts' foliage. Plant scientists have wondered the same thing, and it all started with the search for which animals might eat these mistletoes and pressure them to such extremes. Scientists first eliminated herbivores that are terrestrial (kangaroos and wallabies), are extremely specialized (koalas), and navigate by sensing chemicals rather than sight (insects). The whodunit mystery ended when scientists realized that the mimicry was designed to fool the ringtail and common brush-tailed possums. The animals were operating nocturnally, using their superb night vision to navigate to food sources. When given the choice in controlled feeding experiments, possums greatly preferred the softer and more nutritious mistletoe leaves. In contrast, many of the mistletoes' host species, such as eucalyptus, were less preferred because of their harder, less digestible, and toxic leaves. By blending in visually with their hosts, a mistletoe on a eucalyptus might as well be a clump of needles in a massive haystack. Plant mimicry helps mistletoes survive by tricking these poor possums into eating the wrong plant. While rummaging in the dark for their favorite food—mistletoe—the possums more often than not pull their *least* favorite food—the mistletoes' host plants. That's the downside of navigating to food by sight and touch—an animal can be tricked by mimicry, leaving the mistletoes able to survive night after night of possum foraging.

Ornithologists have posited a second reason for mistletoes' peculiar evolution. They think the mimicry might help birds efficiently spread mistletoe seeds to new host branches. These dispersers are the Australian mistletoe bird (*Dicaeum hirundinaceum*) and honeyeaters (*Grantiella*, *Acanthagenys*, and *Plectorhyncha*). Because mistletoes have a high host specificity, any seeds birds deposit on nonhost trees or the forest floor are essentially wasted, with no chance of survival. Ornithologists believe that host mimicry strongly influences bird dispersers to spend a great deal of time hopping through the branches of potential host trees looking for mistletoe fruit to eat. Mistletoe mimicry has conditioned seed dispersers to seek out not the mistletoe plants themselves but rather their host trees. Whether mistletoes are already present on the trees or not, these birds equate finding the mistletoe's preferred host tree with finding food. By ensuring their host species are always quite popular with the birds, mistletoes bump up the chances that their offspring can colonize new trees. During their investigation of every host in the area, mistletoe birds and honeyeaters dispense the propagules in their mistletoe fruit–laden diet. By hiding within the foliage of their hosts, mistletoes escape the wily possums and secure the survival of their offspring.

In addition to camouflaging into other plants to avoid predation by marsupials, plants have some pretty smart mimicry tricks up their "leaves" to confuse insects. During their long coevolution with insects, plants figured out that bugs, like us, hate to sit down for a meal someone else has already claimed or eaten half of. The strange patterning of each leaf of some passionflowers, ficuses, and caladiums mimics an intense competition over or absence of food. Scientists have found that plants experience less herbivory if they simulate the visual symptoms of insect visitation, egg laying, and previous herbivory damage. When insects see these cues, they pass these plants by in favor of others with less obvious damage. As it turns out, insect females are actually very careful about selecting where to lay their eggs. We have to keep in mind that insects, like plants, are also competing with one another for the best food sources and offspring-hosting sites.

Butterfly and moth caterpillars have a voracious appetite for foliage, so each egg needs an appropriate amount of space for the developing larva—usually one egg per leaf. Impregnated females often make a "flyby" of a plant prior to laying their eggs. It's a considerable investment to lay each egg. For the best chance of success, females need to get a sense if other eggs have been laid already. If they have, it means another competitor's caterpillar will hatch first, giving it a critical head start that may leave no foliage behind.

Enter the passionflowers, a group of plants that absolutely nailed egg mimicry during their evolution. The genus *Passiflora* contains about four hundred species throughout North and South America. Their primary butterfly herbivores, heliconian and fritillary butterflies, have evolved sequestration as a work-around for one of passionflowers' primary defense chemicals: cyanogenic glycosides, which we discuss in a later chapter. In a coevolutionary arms race, the passionflowers' next line of defense is the fake eggs they produce on their leaves. These are visual dots that look like eggs that a butterfly would lay. Sometimes, each leaf bears half a dozen or more of these fake eggs—a horribly overpopulated situation for the discerning butterfly parent. In every way, these "egg spots" are perfect duplicates, exactly the same yellow-orange color and size. The cells of the leaf even rise up in a rounded, spherical bump to simulate the eggs' three-dimensionality. Passionflowers make the considerable investment in egg spots because every butterfly female that passes by without laying eggs means the plant retains more leaves, energy, and evolutionary fitness.

While *Passiflora* species mimic impending damage, some *Ficus* and *Caladium* species mimic damage that has already occurred. Juvenile *Ficus politoria* plants, native to Madagascar, grow strange, highly dissected leaves that appear to have gotten in a rough fight with some voracious caterpillars. Caladiums have evolved variegation patterns on their leaves that mimic the telltale squiggly lines left behind by leaf miner larvae as they tunnel and eat through foliage. Like butterflies and moths, leaf miner females make similar flybys before laying eggs. If they see the squiggle damage from competitors, they'll ensure their larvae sit down for dinner at another leaf instead. If a plant wisely mimics damage on every leaf, leaf miners are likely to leave the entire plant intact and continue searching elsewhere. Not only do these strategies help the individual plant, but they also lead herbivores to victimize competing plants.

The final visual mimicry strategy we'd like to dig up the dirt on is called decoration. But plants aren't getting festive; rather, the preferred decoration plants use as a mimicry defense is actually sand! Let's take a look at some plants around the world that willingly bury themselves in dirt to survive. They use a type of decoration defense called sand armor, or *psammophory* (from the Greek words *psammos*, meaning "sand," and *phora*, meaning "bearing"). Plants that employ sand armor have evolved leaf and stem surfaces festooned with hordes of sticky, gloopy trichomes. As sand splashes or blows onto the plant's surfaces, it gets stuck. The reason for this is that a coating of sand is perhaps the best camouflage a plant can

Seen here is beach sand verbena (*Abronia umbellata*) in Montaña de Oro State Park in California. This species uses gloopy trichomes to create sand armor!

use—it precisely resembles its growing environment without the need for leaf patterning. Additionally, if sand armorers are uncovered by an herbivore, each bite of the plant also becomes a bite of sand. This may not seem like much, but as we learned earlier in this book, silica and sand are great for grinding down the teeth and mandibles of any herbivore. Sand is indigestible and nonnutritive, and studies have shown that insect and mammal herbivores wind up less active and smaller if they have a steady diet of sand armor. Sand armor is used by over two hundred species of plants across eighty-eight genera and thirty-four families. For an exhaustive list of the world's little sand armorers, see LoPresti and Karban (2016). Examples include but are by no means limited to petunias, tobacco, monkeyflowers, dwarf lupine, sand verbenas, and the honey-scented pincushion plant. In labor-intensive experiments in the wild, UC Davis plant scientists found in 2016 that herbivory attacks were doubled against sand verbenas and honey-scented pincushions that had their sand armor deliberately removed. Fighting dirty might be a great choice after all!

Perhaps the weirdest type of mimicry involves plants that capitalize on mammals' sense of smell. We've known for a long time that many plants,

Imagine finding a dead carcass in your salad! Using their flowers' ability to mimic the smell of rotting flesh, carrionflowers repel mammalian herbivores. This particular species is *Stapelia gigantea*. The uncanniness of this mimicry packs a pungent punch—it has to be smelled to be believed. To mammals, including us, the smell of putrefaction and sewage is gross because it implies that death, disease, and predators are in close proximity.

Table 6 A Selection of Plant Mimics

Common name	*Scientific name*	*Subject of mimicry*
Calligonum comosum	*Calligonum comosum*	Carrion/dung smell
Arums	*Arum palestinum, A. italicum, A. dioscoridis*	Carrion/dung smell
Corpse flower	*Rafflesia* spp.	Carrion/dung smell
Corpse flowers	*Amorphophallus* spp.	Carrion/dung smell
Carrionflowers	Genera *Stapelia*, *Huernia*, *Carraluma*, *Hoodia*, and *Orbea*	Carrion/dung smell
Sand verbenas	*Abronia* spp.	Sand armor
Monkeyflowers	*Mimulus guttatus, M. nasusus, M. lewisii, M. cardinalis*	Sand armor
Honey-scented pincushion	*Navarretia mellita*	Sand armor
Dwarf lupine	*Lupinus pusillus*	Sand armor
European mistletoe	*Viscum* spp.	Host mimicry
American mistletoe, Pacific mistletoe	*Phoradendron leucarpum, villosum*	Host mimicry
Oats	*Avena sativa*	Crop mimic
Rye	*Secale cereale*	Crop mimic
Flax dodder	*Cuscuta epilinum*	Crop mimic
Camelina	*Camelina* spp.	Crop mimic
Common vetch	*Vicia sativa*	Crop mimic
Barnyard grass	*Echinochloa oryzoides* and *E. crus-galli* subsp. *oryzicola*	Crop mimic
Living rocks	Aizoaceae family, including the genera *Lithops*, *Conophytum*, *Titanopsis*, *Argyroderma*, *Ruschia*, *Dinteranthus*, *Monilaria*, and *Pleiospilos*	Rock mimic
Living rock cactus	*Ariocarpus* spp.	Rock mimic
Dead wood plant	*Euphorbia platyclada*	Dead plant mimic
Living rock asclepiads	*Larryleachia* and *Pseudolithos*	Rock mimic
Baseball plant	*Euphorbia obesa*	Rock mimic
Paperspine cactus	*Sclerocactus papyracanthus*	Dead plant mimic
Dead wood milkweeds	*Rhytidocaulon* spp. and *Echidnopsis* spp.	Dead plant mimic
Living rock stonecrops	Genera *Crassula* (*C. deceptor*, *C. fragarioides*, *C. mesembryanthemopsis*), *Cotyledon*, *Tylecodon*, and *Adromischus*	Rock mimic
Australian mistletoes	Genera in multiple families, including *Amyema*, *Dendrophthoe*, *Lysiana*, *Muellerina*, and *Diplatia*	Host mimicry

such as corpse flowers and carrionflowers from Africa, utilize the scent of decaying flesh and dung to attract flies and beetles to pollinate their flowers. Disgusting floral scents are actually beneficial to the plants, luring peculiar but dependable pollinators that are enticed by the smell of rot. However, these crafty plants are actually getting two benefits from one smell. Bad-smelling flowers happen to also play a bizarre but significant defensive function. Mammal herbivores, of which there are dozens throughout Africa, are especially put off by the terrible scents of these potential food plants. Unfortunately for plant munchers, getting their mouths up close enough to chomp these plants also puts their nostrils straight into the grotesque, nasty flowers. Imagine smelling a dead carcass in your salad—would you continue to eat? Let's think about what risks these rotten smells express in an evolutionary sense. The explanation is the heart of the reason we ourselves are put off by the smell of carcasses, refuse, and sewage. Overwhelmingly, bad smells imply the proximity of things that can kill us: hazards, disease, contaminated food and water, parasites, and predators. The next time you're grossed out by a smell, think about how amazing it is that those feelings have evolved to protect you from sickness and danger. Furthermore, it is fascinating that plants utilize our visceral reactions as part of their hidden world of plant defense. Hope your day stinks, mammals!

In this chapter, we've seen plants assume some pretty spectacular disguises to stay under the radar of herbivores. You never know where those crafty plants might be hiding. The supposed rocks, leaves, sticks, bug damage, and smells around you might never be what they appear.

CHAPTER 10

True Colors

Defensive Plant Pigments

The vivid world of botanical color around us doesn't just make plants visually appealing. These hues also advertise additional plant defenses. The coloration of plants also provides protection from herbivores and UV radiation. Behind plants' greens, violets, browns, reds, oranges, yellows, and whites are powerful chemicals called pigments, each of which holds its own meaning and function. Some of the pigments housed in fruits, flowers, and leaves are used for peaceful purposes. They generate energy, imbue plants with protection from excessive sunlight, or call in pollinators and fruit dispersers to come interact with the plant. However, other pigments send a clear warning signal to stay away. In this chapter, we take a deeper look at both ends of the spectrum, looking at pigments that serve a function to plant well-being as well as those pigments that deliver warnings to other organisms to stay away.

The vivid world of botanical color around us secretly advertises plant defenses. Pigmented chemicals deliver warnings to herbivores and imbue plants with protection from excessive sunlight and ultraviolet radiation.

Pigment Power

Plants often use their pigments to advertise how powerful their defenses are, frequently with contrasting patterns or colors. This is called aposematic pigmentation and is a phenomenon found in both the plant and animal kingdoms. These plants aren't lying, either. Usually, strong contrasts, colors, and markings on a plant's leaves signal its high investment in defenses. These may include chemistry such as secondary metabolites, antioxidants, latex and poisons, or defense tissues such as spines, trichomes, and idioblasts. Plants can't move, so instead they warn, stab, and poison. Pigments, which are strong antioxidants, signal that a plant holds immunity or a strong ability to resist and counteract herbivore toxins and attack chemicals. You may have noticed and read these visual messages before, perhaps without fully realizing that they are part of the hidden world of plant survival. Spiny plants such as cacti, euphorbias, roses, and agaves have color-coded their weaponized leaves and stems. Even without spines, poisonous plants can appear threatening using just color. Think, for example, of the stark reddish coloration of castor beans or the strikingly mottled leaves of dumb cane. These plants are loaded with toxins and use their conspicuous patterns to keep hungry herbivores at bay. These brightly colored messages are stern "do not disturb" signs for herbivores.

The familiar phenomenon of fall color is also where you'll notice plants revealing their true defensive colors. Autumn is a time when nutrient-rich pigments (except for tannins) are dissolving and retreating from leaves before freezing temperatures arrive. Chlorophyll, the photosynthetic green pigment strong enough to mask all others during the growing season, is first to break down. Flavonoids, carotenoids, and tannins, the three other major pigment classes behind fall color, were hiding there all along. These pigments usually remain because they were slower to break down, but some are synthesized during fall as temporary protectants. With chlorophyll gone, the lingering and mingling of these pigments craft the amazing, complex beauty of fall color. Reds, oranges, yellows, purples, whites, and browns owe their hues to pigments we'll focus on in the remainder of this chapter. If these secretive pigments were there all along, what were they doing? What is their purpose?

One of science's best answers is that the botanical color code is a defensive display. We can't forget that plants live alongside their attackers,

Every rose has its thorns . . . and its aposematic pigmentation! The stark color difference between the "safe" green leaves and stems and the "harmful" red thorns is a color-coded warning not to munch on this Cherokee rose (*Rosa laevigata*).

Flavonoids and carotenoids power the aposematic pigmentation of this fire croton (*Codiaeum variegatum*), which belies the plant's toxic, caustic latex. Other chemicals within the croton are carcinogenic and induce intense vomiting.

In the Agave family, plenty of species, such as this *Agave parryi*, color-code their weaponized parts. The pigmentation differences between the spiny leaf margins and the leaf blades are highly contrasting, communicating to herbivores that this plant means business!

Yellow and orange pigmentation during fall is created by flavonoid and carotenoid pigments. They are always inside leaves during the growing season but are revealed as the more powerful chlorophyll breaks down for winter. Many herbivores and their larvae are likely to be seen by predators as leaves change color. Eagle River Valley, Anchorage, Alaska.

whose life cycles are also deeply affected by seasonal changes. Often, fall and winter are times of resource scarcity and vulnerability. Usually, this means that insects and other herbivores compete for resources, lay eggs, enter immobile or dormant stages of metamorphosis or hibernation, tunnel into plant tissues to create shelter, or create stockpiles of plant food to last through the winter. During the mad rush to ensure their survival for winter, animals rely on color to strategize where to craft the best shelters, safely enter dormancy, leave successful offspring, or create the largest and safest food supply. Fall is a critically important season in temperate animals' lives. It is a pertinent time for a plant to advertise its inhabitability, unpalatability, or toxicity toward these animals.

Through successive generations, plants figured out that fall is an optimal time in which to strike at their herbivores using pigment alterations. Just the simple act of changing foliage pigments is a severe disruption to the crypsis (camouflage patterning) and mimicry of insects and their larvae. For example, a green caterpillar or egg might be caught off guard and become extremely visible to predators when the originally green leaf underneath it turns yellow or crimson. The color of foliage also affects insect overwintering behavior. Insects seeking overwintering sites in autumn are less likely to oviposit, attach, or nest within plants whose defenses are strongly advertised. In general, yellow and orange provide warnings of mild toxicity, while red is often a cue of high toxicity. In several experiments on pests such as aphids and apple moths, researchers have found that diets of yellow and red leaves profoundly decreased survival, slowed development, and interfered with breeding. However, pigments perform a great deal more for plants than just toxifying herbivores.

Botanical Sunscreen

Plants often find that too much of a good thing, such as sunlight, can actually create harmful problems. Plants do capture sunlight with the powerful green pigment called chlorophyll to create their own food, grow, and reproduce. However, excessive levels of sunlight contain debilitating levels of heat and UV radiation. Plants go to great lengths to regulate and optimize photosynthesis while simultaneously protecting themselves from solar harm. Much of this work is performed by colorful pigments in addition to chlorophyll, such as tannins, flavonoids, and carotenoids. This part of plants' survival systems is called photoprotection. Let's shed some light on how it works.

Pigments are protective: they absorb light and heat, acting as botanical sunscreen. Chlorophyll is but one of these pigments, simultaneously helping a plant generate energy while also shielding its cells from phototoxicity (sunlight-induced damage). However, light, UV and heat damage to cells, nutritional deficiencies, or herbivore and pathogen attacks all erode chlorophyll. When chlorophyll production is impacted in any way, the familiar green color of leaves wanes. The sun protection chlorophyll offers is gone, and leaf cells begin to face deadly challenges. When plants are under stress, the production of secondary pigments such as flavonoids, carotenoids, and tannins ramps up to save their lives. This causes leaves to flare into threatening hues of yellow, orange, and red. Every time you witness these colors, you get an ephemeral glimpse at pigmentary survival systems. Elevated pigments combat attackers, dissipate heat, and counteract the toxic effects of stress. Stresses, if they're not dealt with properly, produce massive quantities of free radicals such as reactive oxygen species (ROS), which are bad news. They are violently reactive, destroying proteins, lipids, and DNA and wreaking havoc on critical cell processes. If ROS are left unchecked, cells will die one by one until the plant itself can no longer survive. Photoprotection, however, is just one of the many marvelous things pigments do.

Flavonoids

Flavonoids are a huge and widespread group of plant pigments that are involved in many of the survival capabilities of plants. They are produced in all parts of plants, functioning within both the leaves and the root zone. Flavonoids help filter damaging UV radiation, recruit and shelter mycorrhizal fungi in plant roots, inhibit and repel animal and microbial attackers, and create beautiful yellow, red, and blue pigments in flowers, fruits, and leaves. You see their cheery brightness in the palettes of citrus such as lemons and grapefruit, pineapples, bananas, daffodils, dandelions, parsley, onions, blueberries, tea, grapes, and chocolate. In fall, the golden hue of flavonoids is apparent in maples, birches, aspens, willows, elms, cherries, tulip poplars, hickories, witchhazel, and many other plants. Flavonoids are expert phytoprotectors, actively scavenging, quenching, and blockading the chaotic movements and reactions of free radicals caused by UV radiation. When used against herbivores, flavonoids can decrease the nutritive value of tissues, decrease digestibility, and even accumulate as toxins. When they meet plant-parasitic nematodes, a common plant pest, flavonoids put a chemical blockage in their digestive systems, have a strong repellency effect on hatched nematodes, and interfere enough with developmental regulation to stop their eggs from hatching. Flavonoids act as an "antioxidant shield" against the damage associated with the attacks of plant diseases and microorganisms. Many ROS are generated during infections, and flavonoids rapidly quench them, allowing the plant's cells to survive as long as possible. Flavonoids are also part of defense signaling systems, flooding through a plant in times of stress in response to a myriad of different types of attacks. They help close vascular tissues during wounding and disease outbreaks, a duty they perform by helping plugs called tyloses to form during the process of CODIT (the compartmentalization of decay in trees). Flavonoids can even deactivate the digestive enzymes produced during the attacks of many microorganisms. On fungi, flavonoids inhibit the germination of spores and growth of tentacle-like hyphae. There are almost a dozen different classes of flavonoids, but we'll look at some of the most powerful defensive ones.

Isoflavonoids, which include chemicals known as phytoestrogens, are one class of powerful flavonoids. These flavonoids are botanical mimics of estrogen, the female sex hormone that all vertebrate animals and some insects produce. Phytoestrogens are a powerful control on the reproduction of animals that eat them. Fewer herbivores running around correlates to a decreased chance of being eaten. Think about the power that this gives plants. Female animals consuming vast amounts of phytoestrogens encounter reproductive problems including reduced fertility, higher rates of miscarriage, and fewer offspring produced. Phytoestrogens enable plants to jump into the evolutionary driver's seat of their herbivores. Animal populations that eat more isoflavonoids reduce their own reproduction and fitness. Those that avoid isoflavonoid-producing plants are rewarded, becoming more populous and successful. In males, phytoestrogen-rich diets have been found to have very few effects, so it appears these chemicals

This soybean plant (*Glycine max*) is being clobbered and drained by thousands of microscopic parasites called soybean cyst nematodes (*Heterodera glycines*). These nematodes burrow into plant roots, inject needle-like mouthparts into the plant, and siphon nutrients. This attack is so draining to this plant that chlorophyll production has shut down almost completely, and the flavonoids remaining have turned the soybean a distressed, conspicuous yellow as they fight the nematodes. Contrast this with the plant's greener and healthier neighbor in the background.

are targeted primarily at female herbivores. It is quite fascinating that plants picked up a way to interfere with animals' bodies in such a manner. Examples of plants producing isoflavonoids include yams, hops, mulberries, cannabis, and many legumes, including soybeans, alfalfa, and red clover.

A second class of flavonoids is called anthocyanins, which have a reddish or purplish hue. Anthocyanins create these hues in leaves, flowers, and fruits. For example, the coloration of mimicry plants such as *Lithops* originates in large part from anthocyanins. In deciduous plants, anthocyanins are produced en masse during late summer and fall as sugars in the leaves are breaking down. One of their primary purposes is to shield the leaf from sun damage. As chlorophyll withdraws during fall, leaf tissue is at risk of light damage on any remaining hot and bright autumn days. Brighter and more intense light during the fall stimulates a correlated photoprotective response from plants. The intensity of the redness in fall color demonstrates the greater levels of anthocyanins produced as a part of plant survival systems. You'll also see the tinge of anthocyanins temporarily protecting tender young leaves as they emerge or on fruits such as apples, blueberries, grapes, cherries, and cranberries. If you would like to appreciate the red-purple glow of anthocyanins, take a look at an oak, maple, tupelo, cherry, persimmon, sourwood, dogwood, or sweetgum on a crisp fall day.

Carotenoids

Red, purple, and yellow aren't the only colors plants are capable of displaying. Orange is another common color we see in the landscape of the planet. The primary purpose of the vivid orange pigments known as carotenoids is to help plants survive the stress from excess sunlight. Carotenoids are amazing heat dissipators, UV absorbers, and antioxidants. Throughout the growing season, carotenoids, like some of the other pigments we have seen, dwell in foliage but are hidden behind chlorophyll. Carotenoids also exist throughout the rest of the plant, which is why roots such as carrots can be so colorful. Like flavonoids, carotenoids provide defensive functions just as easily as they aid fruit dispersal and pollinator attraction. Both pigments color many fruits and vegetables, indicating to dispersers (like us!) when these botanical products are ripe and safe to eat. Similarly, flowers may hold yellowish or orange hues that indicate that nectar is available at select locations of the plant (the flower). Carotenoids are divided into two groups. The first class, called the carotenes, are responsible for the orange and red colors in carrots (from which carotenoids get their name), squashes, cantaloupes, persimmons, citrus,

The primary purpose of the vivid orange pigments known as carotenoids, seen here coloring a Chinese pistache (*Pistacia chinensis*), is to help plants survive the stress from excess sunlight. Carotenoids are amazing heat dissipators, UV absorbers, and antioxidants.

papayas, mangos, apricots, tomatoes, and peppers. Famous pigments such as beta carotene (the orange in carrots) and lycopene (the red in tomatoes) are carotenes. The second class, xanthophylls, range from yellow to orange and red, causing the colors we see in plants such as sunflowers, corn, daffodils, and saffron.

Carotenoids provide two broad-spectrum functions. First, they assist chlorophyll in absorbing and processing the sun's rays. Second, using their massive antioxidant potential, they fight free radicals. Carotenoids are universally useful against any oxidative stress plants face. Many sources of oxidative stress feature in this book, such as drought, cold, heat, salinity, flooding, UV radiation, herbicide injury, pest damage, and disease. Without carotenoids, plant cells risk genetic mutations, death, and overheating caused by stress.

Tannins

Have you ever wondered why dead leaves are brown? In fact, if you look around a forest floor, it's the most ubiquitous color you encounter. That brown color comes from chemicals called tannins, which belong to the same polyphenol class of chemicals as lignin. After all other pigments have drained from the leaves during autumn, tannins remain. Tannins are a pigmented set of chemicals that also help to defend plants against herbivores and microbes. They are so vital to plant defense that, after lignin, hemicellulose, and cellulose, tannins are considered to be the fourth most numerous type of botanical molecule on Earth. Anywhere from a quarter to a half of every plant you see around you is loaded with tannins. They lurk in leaves, fruit, bark, and wood.

There are many familiar signs of the tannins hidden in our world. The dark hues of tea, coffee, chocolate, and red wine are created in part by tannins. Tannins have a bitter and astringent taste that reduces herbivory and protects unripened fruits and seeds. Many foods owe part of their characteristic flavors to tannins: pomegranates, blueberries, grapes, açaí berries, cinnamon, cloves, walnuts, almonds, acorns, red beans, and persimmons. Blackwater swamps and rivers are stained brown by the tannins leaching out of the surrounding vegetation. Most plants hide their high

Tannins are some of the most abundant defensive chemicals on Earth. While the oak leaves seen here are among the most well known, the vast majority of all plants enrich their leaves, bark, wood, and fruits with tannins to fight herbivores and microbes. Tannins are responsible for the brown color that remains in dead leaves after other pigments have decayed.

A classic southerners' trick for unsuspecting visitors is to prank them into eating unripen persimmons. Inedibly rich in tannins when green, persimmons (*Diospyros virginica*) defend their immature seeds with an unpleasant astringency and bitterness. When the fruit turns orange, persimmon seed coats are hardened and ready for dispersal. Once the fruits' seeds have fully developed, the tannin levels inside begin to fade, while their sugar content and fragrance dramatically increase.

tannin contents, however. Some tannin powerhouses include the bean family (acacias, forages such as clover and alfalfa), sumacs (genus *Rhus*), grasses (sorghum, corn, barley), oaks, pines, willows, maples, eucalyptus, quebracho (meaning "axe-breaker," genus *Schinopsis*), red mangrove, bald cypress, and eastern hemlock.

If tannins are so plentiful in the world, why do plants make so many of them? The most valuable contribution of tannins is that they bind with and precipitate valuable nutritive substances such as proteins, starches, minerals, and cellulose. Let's unpack what that means a little bit. Tannins use their binding powers in a plant defensive strategy called digestibility reduction. When tannin-containing plant matter breaks down in an animal's digestive system, proteins and other nutrients begin to dissolve into an absorbable form. Tannins interfere with this process by chemically transforming nutrients, especially nitrogen-rich proteins, into indigestible and unabsorbable forms. To do this, tannins bind with the nutrients directly or "precipitate" them, causing the nutrients to fall irretrievably out of the digestive solution. Tannins are essentially chemical nutrient thieves! Herbivores can be starved of nutrients even when their bellies are full of plant tissue. For insects, tannins work slightly differently. Once inside insect digestive systems, tannins react to form damaging free radicals and reactive oxygen species. Oxidative stress from these dangerous reactions interfere with vital life processes inside the insect, sometimes to the point of death. For both mammals and insects, tannins can exert a powerful effect.

Tannins continue their benefits to plants even past the death or loss of the tissues containing them. Once stolen, much of the nutritive value of tannin-rich food passes undisturbed through animal digestive tracts and reaches the ground. Here, among the decomposers and leaf litter, plants gain a fighting chance to reabsorb the nutrients they lost through herbivory. Tannins are plants' master control switch on the nutrient cycle of their habitats. Plants can be induced to modify the tannin content of their leaves in response to environmental circumstances, and the results deeply affect other organisms. A plant too often munched, experiencing nutritional deficiencies, or threatened with disease will respond by producing suppressively high levels of tannins. Higher tannin content benefits plants in several ways. First, herbivory is discouraged because foliage grows bitter, unpalatable, and indigestible. Second, the nutritional content of animal waste increases, giving plants chances to steal nutrients back as the nutrients in this waste leaches into their root zones. Third, tannins reduce the activity of microorganisms, which is an effective method of defense against diseases.

Tannins' inhibitory effects extend even to decomposers such as microorganisms, arthropods, nematodes, potworms, and earthworms. Through modifying the tannin content of their tissues, plants gain control over the decomposition rates in their habitats. Essentially, the use of tannins gives plant roots a head start in the nutritional race against soil microorganisms. This is important because plants actually compete with decomposers for sources of nutrients such as nitrogen. A plant might be stimulated to elevate its tannins in nutrient-poor environments, during times of nutritional stress, or while rebuilding tissues after an herbivore attack. Soil fauna that ingest tannins might be directly toxified. Additionally, tannins bind and coat molecules such as proteins, nitrogen, and cellulose, rendering them into forms that are simply unusable for microorganisms but perfect for root absorption.

Interestingly, plant tannins are one of the reasons why we make spit. Have you ever wondered why you make spit? What is it made of, anyway? Saliva is a complex chemical fluid that aids digestion and food breakdown. Along this same line, saliva helps animals stay safe from plant defenses such as tannins. Animals that regularly consume tannins, including humans, secrete tannin-binding proteins in their saliva called mucin and histatins. Mucin works basically as a decoy, binding tannins so they cannot catch more valuable proteins. Histatins simply disable tannins by forcing them to precipitate. Insect guts have also evolved ways to tolerate ingested tannins, deactivating their effects with high pH, a protective lining called the peritrophic matrix, and antidotal chemicals such as antioxidants and surfactants. Thus the inevitable arms race between the plant and animal kingdoms continues.

Leather is one common product created by tannins. The process of "tanning" animal hides into leather draws its name from plant defense molecules. The ability of tannins to create leather is a direct but unintended result of plants' defensive intentions. When tannins meet protein-rich animal hides, they bind with and permanently alter the proteins inside the skin. Tannins render the nutrients in hides insoluble to most of the digestive tactics used by decomposing insects and microbes. The result is leather—tough, resistant to microbial decomposition, and colored a deep brownish hue. Tanning is now a highly industrialized activity, but for most of human history, tannins were sourced from the wood and bark of plants.

One of the most extreme and grotesque phenomena created by tannins' preservative powers are "bog bodies." Bog bodies are naturally preserved mummies created by the tannin-rich, flooded, acidic, and anaerobic soils of peat bogs. The skin, nails, and hair of bog bodies are remarkably well preserved because these protein-rich parts bind strongly with tannins. A natural leather then forms, keeping microbial decomposition at bay for millennia. These time capsules are a fascinating way plant chemical defenses provide a glimpse into our own history.

The rainbow of colors held by the flora of our planet not only renders plants beautiful but also provides a full spectrum of pigment-based survival adaptations. Pigments are an incredible group of molecules that

An eastern hemlock (*Tsuga canadensis*) grove in the aftermath of bark harvesting, mid-twentieth century, Herkimer County, New York. Tragically, the timber was too uneconomical to harvest and was left to rot, while the bark made its way to tanneries, usually drawn over the winter snow by horses and sleds.

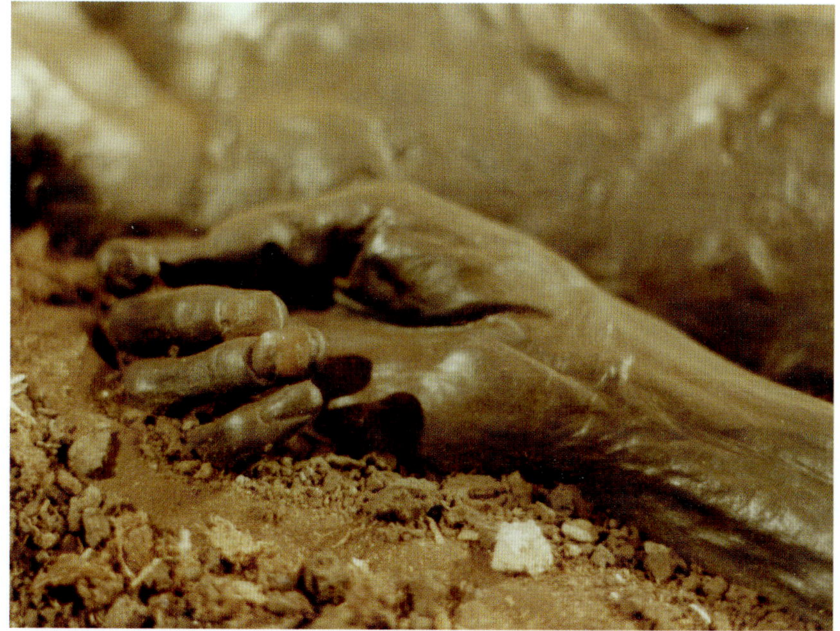

The bog body known as Grauballe Man, shown here, died in a bog in Jutland, Denmark, more than eighteen hundred years ago. His features were so well preserved by tannins that his fingerprints could still be taken. This is because the tannins' antimicrobial effects kept the tissues from decaying.

intimately interact with the entire natural world around plants. The next time you note the tones of the leaves, wood, and bark around you, try to see them as signs of the chemical armor protecting plants from UV radiation, diseases, herbivores, and theft of nutrients. Phytopigments are so common that they might just be in your next beverage, the clothes you're wearing, your landscape at home, the flowers in your vase, or the tastes and colors in your next meal.

CHAPTER 11

Feel the Burn

The Hidden World of Botanical Flavors

The vegetables and spices residing in our kitchens are a minefield of plant defenses designed to turn up the heat on herbivores. Plants manufacture repellents that taste spicy or bitter to protect their vital roots, leaves, and seeds. We confront their defenses when we experience the heat from chili peppers, the spiciness of black pepper, the zing of ginger, the pungency of wasabi, and the sting from garlic. Human preparation and consumption of food might as well be a plant's worst nightmare. For plants and their seeds, our dinnertime represents a horrifying disassembly line where they are chopped, broken, mashed, chewed, and digested. Thus, many plants have adapted these unpleasant tastes to try to ward off our insatiable appetites.

In spite of this repellency, hot vegetables and spices have been culinary mainstays for thousands of years, surviving alongside us for so long that we've domesticated them into integral crops. This chapter explores the stories behind why we love these plants despite the fact that they are trying so hard to avoid being eaten. We'll look at how common tastes in our food evolved as repellent substances and how humans ignored these warning signs and instead included spicy plants in our daily lives. It might seem like an odd choice, but we did this because many of the same botanical chemicals that provide flavor also possess strongly antimicrobial properties that keep plants healthy. Throughout human history, seasoning our food with plant defenses has simultaneously been keeping *us* healthy by decontaminating it of microbial pathogens. We hope this chapter whets your appetite for the hidden world of repellent chemistry.

Our food and the flavors within it are a chemical soup. Many plant-based foods or dishes contain valuable nutrients, protein, and carbohydrates sandwiched between defensive chemicals. Spicy and pungent plant repellents such as capsaicin, piperine, gingerols, allicin, and mustard oil bombs manipulate animal bodies and nervous systems in some pretty compelling ways. As we break up plant material through food preparation, chewing, and digestion, these chemicals are released into our bodies.

These repellents also differ from botanical poisons such as ricin and neurotoxins. To work most effectively, repellents need to be nonlethal so they can teach avoidance through incredibly unpleasant experiences. Herbivores need only ingest spicy or pungent plants a handful of times before developing lifelong aversions. In fact, by leaving herbivores alive, repellents may be diverting pressure to competing plants instead. This secondary benefit, the harshly learned lesson, is more ecologically valuable than killing herbivores outright.

Plants have been concocting substances in the chemistry lab for millions of years, but so have animals. During our own evolution, our internal organs, such as the liver and kidneys, developed detoxification processes for many chemicals inside plants, including most of the repellents in this chapter. Humans have an even further advantage against these chemicals compared to insects, microbes, and other animal herbivores. We've learned that heating, cooking, boiling, and dilution can destroy or remove harmful chemicals before they enter our bodies. Because of these biological and learned adaptations, inherently repellent foods, including chilies, mustard, wasabi, ginger, and garlic, support the cuisine of entire cultures and regions around the world. Let's take a closer look at some of these beloved flavors that are actually repellent in nature.

The Chili

One such repellent plant, the chili pepper, possesses amazing defensive powers that have shaped human history. Chili species (genus *Capsicum*) are native to Central and South America, where they have been domesticated and deliberately cultivated into culinary mainstays for millennia. What should be naturally repellent, the heat and pain generated by a chili pepper, has been willingly adapted into our dietary experience. The intensity of these fruits should instinctually drive us away, but humans have grown to enjoy the pain. When we eat a chili, its defenses trick our nervous system into sending two powerful signals to the brain: extreme pain and heat. Both of these feelings imply life-threatening danger that we should avoid. The reason we feel this burning sensation is because of the alkaloid capsaicin and related compounds called capsaicinoids, the protective chemicals of choice for chili species.

In this particular example, the plant is putting its chemical energy into protecting the seeds, its future offspring. Instead of using components such as shells or hulls to allow the seed to survive the chewing and digestion by mammals or insects, peppers evolved to select for a different dispersal agent. *Capsicum* peppers have thin-walled, fragile seeds. However, when broken, these seeds release a flood of capsaicin. Thus, capsaicin's purpose is to stop animals from crushing the delicate yet critically important genetic package within seeds. Mammals and many insects mash their food using molar teeth or mandibles, a feeding strategy prone to destroying seeds. When the capsaicin-producing veins holding chili seeds are damaged, a flood of capsaicin binds to any animal pain receptors it contacts. Usually, the fireworks erupt in the mouth and throat, but capsaicin can also cause intense burning to cells in the eyes, nose, lungs, and skin. This sensation is almost instantaneous and quite excruciating at the levels present in wild species of chili peppers. Capsaicin seeks out and binds with the animal nerve receptors that sense physical heat and pain, tricking our brains into thinking we are being hurt and burned.

Capsaicin shows no mercy and accepts no apologies. Once herbivores have made the mistake of attacking a chili pepper, the plant launches this noxious napalm at them to protect its offspring. Capsaicin is a tricky little molecule specially designed to spread and avoid being washed off. The key to this protection is capsaicin's long hydrophobic tail, which makes capsaicin too oily to dissolve in water, sweat, or saliva. While we produce these fluids to flush out toxins, capsaicin instead carries along with them to spread the burn over a larger area. Pepper plants are hijacking our own bodily fluids as a vehicle to ensure that their chemical minions coat our mouths, throats, skin, and face for maximum impact. If you have ever cut spicy peppers and then felt the burning on your hands hours after washing or rubbed your eyes after touching hot peppers, then you have experienced this effect. As luck would have it, capsaicin's magic tail has an extremely high affinity for protein receptors on our tongue as well as the mucus membranes in our throat, eyes, and nose. Not only this, but the tail allows capsaicin to diffuse through our cells' oily membranes, meaning the burning effect can spread tenaciously from cell to cell. Capsaicin's effects can last for hours both inside and outside the body but generally clear up after a day or so. The burning may also inflict minor damage to nerves, skin, and corneal cells, but these are capable of healing up within a few days. The overall experience, however, may be tough to forget.

Our brains are so fooled that they engage the production of pain-killing chemicals called endorphins. These feel-good response chemicals may underlie one major reason why some of us enjoy spicy food so much. Endorphins are hormones released by our hypothalamus and pituitary glands in response to pain, stress, vigorous exercise, laughter, and, apparently, enjoying our favorite spicy dishes. Altogether, endorphins are excellent not only at relieving pain but also at creating a general feeling of well-being. Clearly, plants hadn't counted on us finding enjoyment in this process. Perhaps we have some questions to ask of ourselves.

There is also substantial evidence that capsaicin inhibits the growth of multiple species of bacteria and fungi. Chilies evolved in wet, humid, tropical places with substantial disease pressure. Fungicidal and bactericidal properties protect the developing seeds as the chili fruits ripen, but humans have co-opted these effects for medicinal purposes and sanitizing food. Maya people in the Yucatán Peninsula have relied on herbal medicine from *Capsicum* fruits for thousands of years, often treating diseases that have microbial origins. This impact has also been studied in microbiology laboratories. After extracting capsaicin from wild chili fruits, scientists found that chili extracts inhibited the growth of yeasts as well as species of *Fusarium*, *Bacillus*, *Clostridium*, and *Streptococcus*. In experiments from the medical field, capsaicin began killing multiple strains of *Streptococcus pyogenes* bacteria (most of which were antibiotic-resistant) within fifteen minutes, and no strains showed resistance. Further studies in food preservation by industrial food researchers found that capsaicin also inhibited common foodborne pathogens such as *Escherichia coli*, *Salmonella*, and *Listeria monocytogenes*. The antimicrobial function of plant defense chemicals such as capsaicin is a powerful rationale for why spices are so valuable in our cuisine. One study even found that people in hotter, wetter climates, in which food spoilage is a notable concern, utilize more spices per dish overall compared to those in cooler, less disease-prone climates.

The chemicals inside our spices that deliver flavor are often powerfully antimicrobial, protecting not only the plants they are derived from but also ourselves. Chilies are just a fraction of the list of spices with innate defensive chemistry that simultaneously flavors and disinfects our food. Almost all major spices provide this function with different defense chemicals, including black pepper, onions, garlic, cloves, citrus, mint, oregano, vanilla, cinnamon, allspice, thyme, rosemary, ginger, anise, saffron, tarragon, mustard, bay leaf, lemongrass, sage, basil, cilantro, coriander, nutmeg, cumin, and cardamom. We'll highlight a few of the most pungent ones throughout this chapter.

Spiciness also repels insects regardless of how they feed on the plant. Some of the most prevalent pests of peppers include piercing-sucking pests such as aphids, leaf-footed bugs, and seed bugs (such as *Acroleucus coxalis*). Slurping capsaicin up their straw-like mouthparts is so unpleasant that these pests reduce their feeding. Since diseases such as *Fusarium* are so prevalent in wild chilies (infection rates can be as high as 90 percent), and many diseases originate after the wounding from an insect bite, the heat from capsaicin defends against multiple layers of pressure simultaneously. Furthermore, capsaicin coats the ovipositors (egg-laying equipment) of any lady insects that might be tempted to cultivate offspring on the pepper plant. Female weevils, moths, and beetles are deterred from laying eggs anywhere near a chili pepper lest their ovipositors be painfully scorched.

We know capsaicin creates the sensation of heat, but this is an insanely powerful weapon against bugs. Scientists have found that capsaicin causes an intense disruption to the thermosensing and regulation of the body temperatures of a chewing pest called the Colorado potato beetle. Temperature regulation might seem like a strange thing for plants to tinker with, but it exerts a powerful effect upon animal behavior. Insects lack the cooling abilities we mammals have evolved, namely, sweating and panting. Insects ingesting capsaicin experience the same nerve sensation of overheating that we do but possess no natural mechanism to diminish its potency. A flood of capsaicin is eventually fatal to many bugs in two distinct ways. Some insects, under the pepper's "hot flash," may cease feeding for prolonged periods to move to cooler locations, away from the sun-loving plant. Thus, the plant is manipulating bug behavior to the point of starvation. Others may be so confused by the chili's blast that they fail to perceive their own body temperature altogether, with both cold and heat sensation being disrupted. This places them at a dangerous risk of fatal exposure to the elements. With their nervous systems overridden by capsaicin, insects remain in heat and direct sunlight far too long. A battle with a pepper plant could easily result in death via desiccation and actual overheating. Capsaicin turns a hot day into a fatally hot day!

There is one completely immune group of organisms, however, to which capsaicin molecules do not bind. That would be the birds. Curiously, birds lack the pain receptor that capsaicin binds to in insects and mammals, so they'll eat the hottest capsaicin-loaded peppers without feeling a thing. To find the reason behind this, just watch a bird eat. Birds' feeding strategy is much different from the strategies of mammals and insects. Birds have beaks that lack teeth, so they swallow their food whole. To break the food down, they rely on a specialized digestive organ called a gizzard, which is a muscular pouch filled with grit or small stones that grind and pulverize food. In many species of plants, including the chili, bird feeding not only leaves seeds intact but also preps them to germinate by removing the fruit flesh and scarifying the seed coat for the absorption of water. Compared to mammals and insects, birds are the perfect flying, long-range dispersers. Thus, capsaicin's selective repellency is a defense that allows the chili to pick a preferred seed delivery system.

The story of how the chili pepper landed on our tables is fascinating. Perhaps the closest living wild relative of chili peppers is an adorable, diminutive landrace called the chiltepin pepper (*Capsicum annuum* var. *glabriusculum*). *Tepin* is derived from a Nahuatl word meaning "flea." Another common name for the chiltepin is the "bird pepper," which clues us in to its close relationship with bird dispersers. There are two quite noticeable things about wild chilies such as the chiltepin: they're only a teeny fraction of the size of modern chilies such as the jalapeño, and they're also extremely hot. A single chiltepin is barely the size of a pea, yet this little warhead is packed with twenty-three times the capsaicin of a jalapeño.

The ancient domestication of hot peppers likely began in the highlands of Bolivia and then spread through Central and South America. In general, human selection drove chilies to reduce their innate capsaicin-based defenses as well as produce larger fruit. Within the genus *Capsicum*, five

The diminutive warhead seen here is the closest living relative to the wild ancestors of chili peppers. It's called the chiltepin pepper (*Capsicum annuum* var. *glabriusculum*). Despite its size, its heat is full-blown nuclear! The birds that swallow the pepper's fruit whole, not the humans that crush the seeds, are the chili's preferred dispersal agent.

species of chili have been domesticated by humans: *C. annuum* (responsible for bell, wax, banana, jalapeño, and cayenne peppers), *C. frutescens* (tabasco, Thai, and piri piri peppers), *C. chinense* (habanero, ghost, and Scotch bonnet peppers), *C. baccatum* (aji peppers), and *C. pubescens* (rocoto peppers). Bird dispersal appears to be the norm for the *Capsicum* genus, with the majority of species possessing small, spicy fruits. When ripe, chilies turn red because this color is highly visible to birds. To insects, whose spectrum of color vision perceives higher wavelengths than vertebrates do (such as ultraviolet), red is invisible. In addition to the chiltepin pepper, other wild species include *Capsicum chacoense*, *C. cardenasii*, *C. eximium*, *C. lanceolatum*, *C. rhomboideum*, *C. lycianthoides*, and *C. geminifolium*. On occasion, some wild pepper plants do not make capsaicin. This trait has a genetic basis, and these spiceless plants have survived and reproduced for millions of years. Humans, finding these defenseless individuals, utilized their genetics in the domestication of peppers as we know them today. This process had already been under way for thousands of years when Columbus landed in the Americas, because he and his men discovered both large, sweet peppers in dishes as well as smaller, hotter cultivars of chili pepper.

It might seem odd for chilies, which invented capsaicin as a defense, to voluntarily give it up. However, producing capsaicin molecules comes at a high cost for the pepper plant. The "capsaicin tax" demands some of the plant's carbon and nitrogen supply, which ordinarily would be diverted toward growth or other defenses such as lignin production. Capsaicin production also necessitates high water consumption, so plants with hot peppers are more susceptible to drought and wilting than peppers without defenses. This is because capsaicin production correlates to denser stomatal pores in leaves, meaning that hot plants transpire more water. Resultingly, capsaicin production trades repellency against diseases and herbivores (the pressures from both being higher during wet years) at the cost of susceptibility to drought. Because of the substantial cost of defense for the chili, a few wild peppers lacking capsaicin succeeded in dry spells because their drought tolerance allowed them to survive without chemical defenses. The precious genetics within these defenseless chilies have factored heavily into our domestication of peppers. Generations of agricultural selection and hybridization between sweet and spicy chilies have generated the varieties we know, love, and fear today. We have them to thank whenever we eat a bite of food that, although spicy, packs a fraction of the punch of wild chiltepins.

According to the scant written records we have of the cultures of ancient Native American tribes such as the Aztecs, capsaicin had a rich and varied usage in their cuisine, medicines, and even chemical warfare. Bernardino

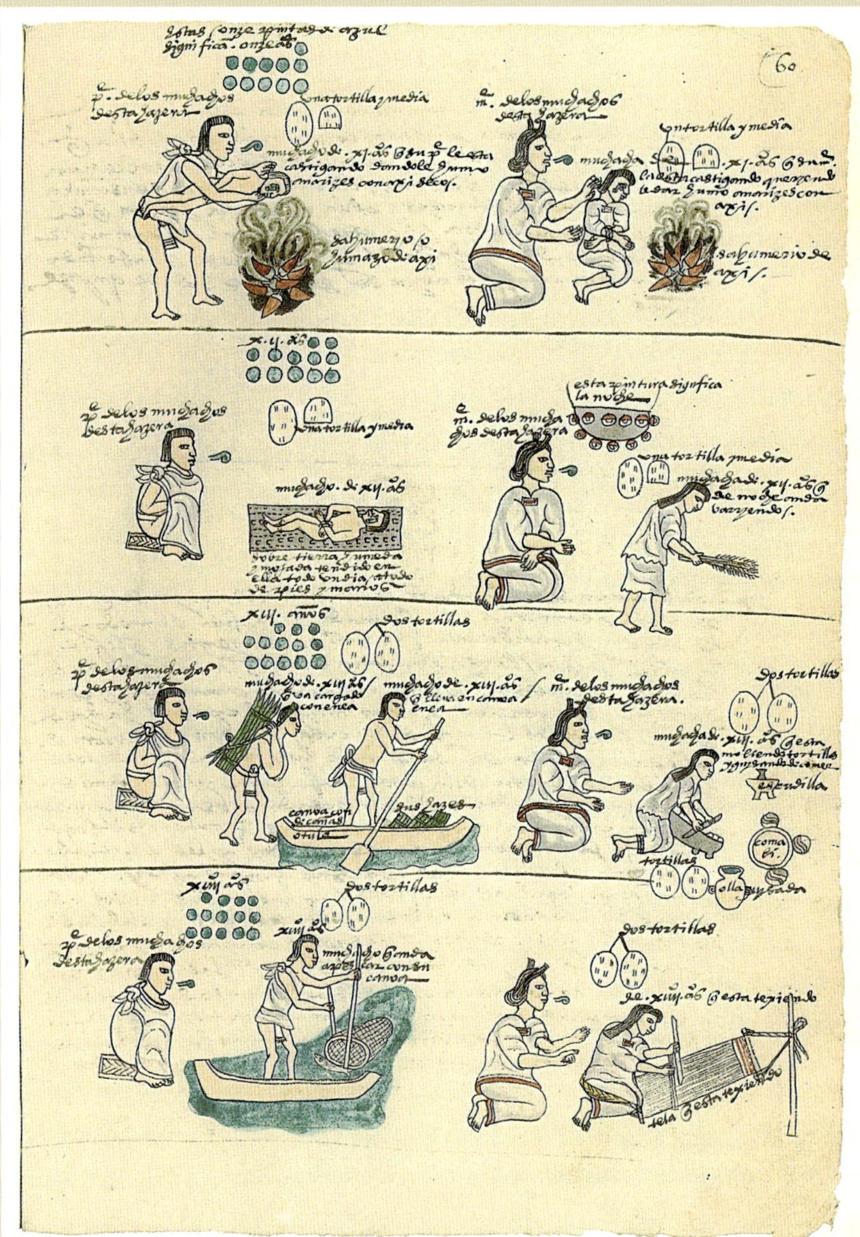

The Aztecs, recorded here in the Codex Mendoza in the sixteenth century, are punishing their children using the excruciatingly painful (but nonlethal) smoke from burned chili peppers. A father and son (*left*) as well as a mother and daughter (*right*) are engaged in the same punishment. The irritating capsaicin inside burning chilis was also weaponized by Native tribes, who used masses of burning hot peppers in assassinations as well as siege warfare.

de Sahagún, a sixteenth-century Franciscan missionary from Spain who learned the Aztecs' language and spent more than fifty years studying their culture in Mexico, found chili peppers deeply embedded in Aztec dishes. Sahagún encountered a rich assortment of chili-seasoned foods, including corn, beans, squash, duck, sauces, chocolate beverages, frogs, ants, locusts, roasted snakes, and even tadpoles! In fact, our words "chili" and "cayenne" are both based on Mesoamerican words for the pepper: "chili" is derived from the Aztec word *chilli*, while "cayenne" comes from the Tupi word *quiínia*.

Chilies found unusual uses outside the kitchen, too. A chili-based assassination was recorded by Spanish conquistadores in the Codex Mendoza. Lords of the province of Cuetlaxtla were annoyed by visits from imperial messengers from the Aztec capital. The lords' response was to seal off the messengers' sleeping quarters and pipe in the smoke from a massive bonfire of chili peppers, causing what must have been an unimaginably painful asphyxiation. Chili fumes were also employed to smoke out defending villagers during siege warfare, and piles of burning chilies were employed as household fumigants for insect and mammal pests. Conquistadores also recorded that Aztec parents used flaming-hot chili smoke as a form of discipline for children. Aztec parents are depicted in the Codex Mendoza as holding their delinquent youths over smoking chilies. Whippings involving spiny cactus pads, another plant defense discussed in this book, were a popular alternative. Let's hope none of these parenting habits make a return anytime soon!

Chilies have come a long way since then both in an evolutionary sense and in a global one. Today, chilies are the biggest spice crop in the world—almost forty million tons of fresh and dried chilies are grown worldwide. Hundreds of cultivars of chili pepper exist in a range of sizes and colors, including white, yellow, orange, red, purple, and even black. Heat varies as well depending on how much capsaicin a cultivar produces in its peppers. The chili has reached most cultures and areas of the world, experiencing almost daily use in a wide variety of iconic dishes, sauces, and dried and pickled products. *Capsicum* peppers are also widely bred and used as ornamental plants for the delightful variation in shape, color, and pest resistance of their fruits. So even though humankind has defeated the chili's mammal-repelling powers through selective breeding, in the end, the chili pepper is the victor, as it reproduces successfully on farms around the world.

The heat of a chili pepper is measured in chemistry labs using a technique called high-performance liquid chromatography (HPLC), which assesses capsaicinoid content. A well-known but antiquated and subjective measure of chili heat uses the Scoville scale, which reports pepper heat in terms of Scoville heat units (SHUs). The scale is named after the American pharmacist who created it in 1912, Wilbur Scoville. To derive SHUs for each type of pepper, Scoville created a test using panels of human testers. Chilies were dried and their extracts diluted with successive concentrations of sugar syrup. When the majority of testers could no longer detect pungency in the sample, Scoville based the official SHU rating for each chili variety on the strength of the dilution. A higher Scoville rating indicates a hotter chili with a higher dilution rate required to render it tasteless. Pure capsaicin measures sixteen million SHUs, while the bell pepper is zero SHUs. Other major points of reference can be found in table 7.

Table 7 Chili Varieties Rated by Scoville Heat Units (SHU)

Chili variety	*SHU rating*
Bell peppers and pimentos	0
Cubanelle, pepperoncini	100–1,000
Cherry, New Mexican	500–1,000
Ancho, poblano, pasilla, mulato, and Española	1,000–1,500
Anaheim, rocotillo, Sandia, and cascabel	1,500–2,500
Jalapeño, mirasol, and hot wax	2,500–5,000
Serrano	5,000–15,000
Chile de arbol, aji, and shipkas	15,000–30,000
Piquin, cayenne, tabasco, and rocoto	30,000–50,000
Chiltepin, Thai, and bird's eye	50,000–250,000
Habanero and Scotch bonnet	100,000–350,000
Red savina	350,000–577,000
Naga morich and ghost	1,041,000
Trinidad scorpion	2,000,000
Pepper spray	2,000,000–5,300,000
Pure capsaicin	15,000,000–16,000,000

The chili continues to be useful far beyond the kitchen both as a medicine and for its powerful repellency. Capsaicin adheres to our pain receptors so effectively that, ironically, it can "mute" the pain signals sent to our brains by other ailments. By superseding these signals, capsaicin blinds our nervous system to other pain. Capsaicin is used topically in creams and ointments. It provides temporary relief for muscle and joint pain caused by strains, sprains, arthritis, bruising, backaches, fibromyalgia, migraines, and surgery.

Pepper spray and tear gas are chemical tools used in policing, riot control, and self-defense. They work effectively against primarily humans but occasionally threatening animals such as bears and dogs. The capsaicin content in pepper spray and some types of tear gas is extracted from crushed and dried chili peppers. HPLC is usually used to determine the true strength of pepper sprays, but different products can range anywhere from two to five million SHUs. The advantages of these plant defense–derived sprays is that they are generally nonlethal as well as instantaneously irritating and effective. By inflaming the eyes, nose, and mouth, pepper spray incapacitates with temporary blindness, breathing difficulties, and severe pain, all of which can last up to an hour. Police officers and riot control personnel can use pepper spray to control and restrain subjects, while people in danger can disable and potentially escape from attackers. There also isn't much subjects can do to resist the effects of pepper spray—capsaicin's oily nature causes it to spread like wildfire. Once pepper spray has permeated our cells, almost no amount of water can wash it off.

Capsaicin-based sprays are also used as pesticides and feeding repellents for crops. Hot pepper extract is one of the most common home garden products against insects, mites, deer, and mice. It can also be used as a coating for seeds to prevent herbivory. Essentially, we're stealing the chili's chemistry and applying it to plants that lack such a long and successful history of chemical warfare.

We've covered the chili pepper's defenses in depth both because of its fascinating story and because of its place at the top of the world's spice rack. However, other spice plants manufacture repellents that toy with pain and temperature receptors. Let's take a look at a few of these.

Black Pepper

Black pepper (*Piper nigrum*) is one of the top ten most valuable spices in the world and the only species in this chapter to make this list. Native to India, black pepper is a tropical vine that produces small fruits called peppercorns. These peppercorns protect the seeds within by exploding with pungency when eaten, an effect created by the alkaloid piperine. Piperine molecules have a similar chemical shape to capsaicin, interacting with many of the same receptors but packing only 1 percent of the heat of capsaicin. Black pepper has a long history of being traded as a spice. At one point during the Middle Ages, black pepper was so valuable that it was used as currency and sometimes called "black gold." The drive to control the trade of spices such as black pepper was one of the forces behind the European Age of Exploration from the fifteenth to the seventeenth century, which included the colonization of the Americas and Australia. When Europeans discovered the teeny chiltepin chili in the New World, they were struck by the plant's similarities to the black pepper they were seeking. Chiltepin and black pepper have fruits of the same pea size, roundish shape, green (unripe) and red (ripe) color, and a fully thermonuclear taste. These similarities, centuries before the development of plant taxonomy by Carl Linnaeus, convinced Europeans to erroneously name the chili the "pepper," a name that has stuck ever since.

Black pepper (*Piper nigrum*) has been one of the most sought-after spices in human history, propelling worldwide voyages of trade, exploration, and colonization. Our obsession lies with the hot, repellent chemical called piperine, which protects its seeds.

Ginger

Another spicy plant, ginger, uses repellents called gingerols, gingeridione, and shogaols. Ginger's rhizomes (roots) are factories for these chemicals, protecting the underground part of the plant from soilborne diseases as well as herbivory. If you have ever bitten into raw or undercooked ginger root, the plant immediately signals you to leave it alone by shooting these chemicals into your taste receptors and activating heat and pain signals. Ginger is native to Southeast Asia and India, but its use and cultivation have spread worldwide. Roughly four million tons of ginger are produced worldwide each year. Ginger has been domesticated for so long that the plant exists solely as a cultigen, meaning no wild relatives of the species exist. Instead, ginger has tagged along with the ebb and flow of human civilizations and migrations.

Ginger first reached western Europe in dried form, being passed along spice routes through Asia and the Middle East. The ancient Greeks and Romans received it from Arab traders, as did Europeans during the Middle Ages. Dried ginger flakes found their way into Victorian-era bars as a flavoring and water-sanitizing agent for wine, beer, and ale, eventually inspiring the stand-alone products we know as ginger ale and ginger beer. Ginger went along with ancient Polynesians across tremendous transoceanic voyages in canoes, reaching as far as Hawaiʻi in the fourth century.

The ginger (*Zingiber officinale*) rhizome, seen here, is the most highly protected part of the plant. It is essential for survival, containing the plant's buds, water, and food reserves. Ginger rhizomes are factories for chemical repellents against animal attacks. Eating them raw is quite unpleasant. These repellents are also highly antimicrobial, helping the ginger survive an underground assault of microorganisms.

Finally, ginger seized its place as a global spice after European colonization connected its centers of production in the Far East with colonies around the world, including the Americas.

Ginger's aromatic and antimicrobial chemistry, aside from culinary use, has produced perfumes, cosmetics, and medicines. In fact, gingerbread and other gingery treats, which most of us are familiar with during the holiday season, began as a sweetened incentive for European children to take ginger as a medicinal supplement for sickness and nausea. Like that of capsaicin, ginger's chemistry fights nerve pain using even more botanically stimulated pain. Capsaicin, piperine, and gingerols all work by activating the same suite of pain and temperature receptors.

Close relatives of ginger also pack defensive chemistry, powerfully repellent flavors, and antimicrobial properties. They include the spices turmeric (*Curcuma longa*), cardamom (*Elettaria cardamomum*), and galangal (*Alpinia galanga*). Despite these plants flashing their warning chemistry when we ate them, humans found the benefits of using them too great to ignore.

Alliums

Another potent plant repellent is called allicin, which is found in the monocot genus *Allium*. Within this genus you'll discover species that have flavored cuisine around the world: onions, garlic, chives, leeks, shallots, and scallions. Some species native to the Americas were heavily cultivated by Native Americans. These include wild onions (*Allium canadense*) and ramps (*A. tricoccum*). There are even ornamental onions blooming in our gardens (*A. cristophii*, *A. hollandicum*, and *A. giganteum*). All of these species have substantial bulbs.

The bulb of an allium *is* the plant and must persist at all costs for the plant to stay alive. Much like other underground organs such as the rhizomes of ginger, bulbs contain all of a plant's food and moisture reserves for hard times. Bulbs remain belowground throughout the year, insulated from aboveground herbivores and bad growing conditions. The subterranean environment is still a life-or-death battleground full of soilborne diseases, microorganisms, burrowing insects and mammals, drought, flooding, and cold. If damaged, dug out, or infected in any way, these crucial allium bulbs begin an avalanche of the defensive chemicals they have stockpiled, namely, allicin.

The bulb of an allium, like this garlic, must persist at all costs for the plant to stay alive. The subterranean environment is a life-or-death battleground full of soilborne diseases, microorganisms, burrowing insects and mammals, drought, flooding, and cold. If damaged, dug up, or infected in any way, these crucial allium bulbs begin an avalanche of the defensive chemicals they've stockpiled.

You won't find allicin in pure form inside an allium bulb, though. There are two main reasons for this. First, allicin is much too toxic to the plant's own cells, so alliums instead carry separate compartments chock-full of allicin's ingredients. Second, allicin, a chemical called an organosulfur, has a steep cost of production, which includes some of the plant's nitrogen and sulfur. Alliums solved both problems by finding a method to deliver these garlicky defenses only on demand: they allow animals and microbes to mix their own poison! Allium bulbs are like an armed warhead, containing cells brimming with the raw ingredients for allicin. Breaking or digesting any cell of an allium sets off a chain reaction. As the cellular fluids mix, the allicin warhead detonates. When the precursor of allicin, called alliin, mixes with its activation enzyme, alliinase, the reaction concocts a huge amount of allicin within ten to sixty seconds after the plant experiences damage. Allicin is among the most powerful repellents and disinfecting chemicals plants can manufacture.

Allicin is a violently reactive, unstable chemical because of the sulfur within. Allicin fractures almost immediately into a barrage of other defensive organosulfur chemicals, which are simultaneously distributed through a variety of means. Degrading allicin can be emitted as volatile gases or spread as fluids in order to permeate an attacker's cells. Alliums are awash in defensive chemistry that is immediately recognizable. Allicin's by-products are at work behind the potent smells and flavors of chopped onions, garlic, and chives. The smells furthermore carry on our breath and skin and exude through our sweat as our bodies detoxify the organosulfurs. Cooking these alliums renders their taste palatable, but direct consumption of raw onion and garlic tastes and smells painfully repellent because of allicin. Allicin is behind the flood of volatile gases that invades our eyes and sinuses when we damage onions. As we chop their survival structures into bits, they fight back by stinging our eyes and provoking tears and mucus. When we cook with alliums, the last defensive gasps of their dying bulbs fill our kitchens with that familiar oniony, garlicky odor. Mmm . . . sounds delicious!

Ever teared up while chopping an onion? As you damage the plant's precious survival structure, a defense called allicin is released. Allicin is violently reactive, fracturing into subchemicals that impart the characteristic tear-gas repellent effect, smell, and flavor of alliums such as the onion, garlic, chives, scallions, shallots, and leeks. Eating any of these plants raw packs a wallop of plant defenses for your taste buds!

Allicin allows plants to escape the gnashing jaws of animals, but what about microbes? Allicin is one of the most broad-spectrum antimicrobial and nematicidal compounds possessed by any plant, assaulting bacteria, fungi, viruses, nematodes, and protozoans alike. Allicin effortlessly penetrates common microbial defenses and enters pathogen cells with ease. One such defense, a shield of cellular slime produced by many microbes, is thwarted easily by the chemical. Once the allicin is inside, a surgical strike on basic life functions follows. Allicin works as a disinfectant by degrading key proteins and enzymes within microbes, causing these attacked cells to die. Many of these proteins are so basic that they also exist in the plant's own cells, which is why alliums take such great pains to compartmentalize allicin until attacks are imminent. Allicin has successfully conquered a mind-boggling array of microbial life in the petri dishes of science labs across the world. Some of these microbes are even ones that alliums never fight in natural settings but are especially prevalent pathogens in the agricultural, medical, and food-processing industry pipelines. Since allicin targets enzymes and proteins that occur across most of the microbial world, this plant defense possesses extensive efficacy on microorganisms. Allicin protects our food by killing and inhibiting foodborne and waterborne illnesses such as *Escherichia coli*, *Listeria monocytogenes*, *Salmonella*, dysentery, and cholera. Allicin demonstrates a powerful ability to keep us safe in everyday life and hospitals by killing antibiotic-resistant *Staphylococcus aureus*, *Psuedomonas*, *Helicobacter pylori* (stomach ulcers), *Cryptococcus*, *Streptococcus* (including pneumonia-causing strains), gingivitis, *Mycobacterium* (tuberculosis and leprosy), *Microsporum* (ringworm), *Trichophyton* and *Epidermophyton* (athlete's foot), *Candida* (yeast infections), and influenza. Even the blood and urine of patients on garlic supplements showed many of the antimicrobial properties of pure allicin. One feeding study on rodents found that garlic extract eliminated almost 40 to

60 percent of these animals' gut flora in just four hours, with a fivefold decrease after eight hours. Allicin's antimicrobial effects were extreme enough to decimate both the good and bad digestive flora of these rodents, an effect serious enough to lead to debilitating nutritional deficits. The ability to nuke an herbivore's gut ecosystem with disinfectants is a powerful defense for alliums, adding another reason why they have evolved such outstanding germ-killing prowess. In the example of the unfortunate test mice, the herbivores on a sustained high-garlic diet wound up smaller and developmentally delayed compared to control groups fed a normal diet. Plant pathologists have also experimented with allicin, using it to successfully kill awful plant diseases such as *Alternaria*, *Botrytis*, downy mildew, and rice blast fungus. Not surprisingly, allicin was most effective against pathogens that regularly threaten allium bulbs in their natural underground battlegrounds. These are major root rot pathogens such as *Phytophthora*, *Pythium*, *Rhizoctonia*, *Fusarium*, and *Thielaviopsis*. Allicin, more powerful than any other plant disinfectant, allows alliums to defend against a tremendous army of virulent microbes. Sounds like a perfectly reasonable trade for the cost of a handful of sulfur and nitrogen molecules.

Humans discovered the medical benefits of allicin so early that allium use predates recorded history. Crops such as onions and garlic have been used for thousands of years, primarily as foods and food preservatives but also as medicines, repellents, and pesticides due to their antiseptic and repellent qualities. Garlic-based prescriptions exist in ancient medical texts from cultures around the world, including the Sumerians, Egyptians, Greeks, Romans, Indians, and Chinese. Garlic juice and onion juice have been rubbed on the skin to prevent insect bites, purify wounds, and prevent infections. They have also been taken internally for a wide variety of ailments, including internal infections, parasites, and circulatory problems. Additionally, because of their powerful antimicrobial properties, alliums have been topically applied and eaten to treat diseases of microbial origin, including skin infections, dysentery, eye ailments, ear infections, and oral sores. As mentioned above, allicin has exterminated almost anything researchers decide to put it on. Who knew that such atomic weaponry against microorganisms was hiding within the scents and flavors of the foods in our kitchens?

Mustards

The last type of repellent we'll highlight in this chapter is the mustard oil bomb, scientifically known as the glucosinolate-myrosinase complex. Mustard oils, a group of chemicals also called glucosinolates, are the predominant defense in the mustard family, Brassicaceae. You'll find the defenses most prevalent in the plants' roots, seeds, and leaves. Like allicin, mustard oils are organosulfurs, meaning they are violently reactive as well as nutritionally draining on the plant for sulfur molecules. These defenses run so deep within the mustard family that most representatives possess them. All of the repellents mentioned in this chapter activate the same kinds of pain and heat receptors in animals, but mustard oils go a step further and also activate bitterness receptors. The explosion of mustard oil bombs explains the pungent and bitter tastes and smells behind mustard, turnips, radishes, wasabi, horseradish, arugula, cabbage, broccoli, kale, bok choi, cauliflower, brussels sprouts, rapeseed, canola, and watercress. Infamous childhood aversions to these veggies have resulted because we simply heeded the plants' repellent warnings and stopped eating them. Hopefully, we've now vindicated a few readers once mislabeled as "picky eaters." Millions of cell-sized mustard oil bombs are exploding right and left in our kitchens every time we chop, crush, cook, and eat plants in the mustard family. Thankfully, we're able to detoxify these compounds using cooking, boiling, and the protective filtering of our kidneys.

Without activation, glucosinolates are essentially like caged tigers—they have trouble crossing cell barriers. This means not only that they are temporarily suspended from wreaking havoc inside the plant itself but also that they are incapable of causing harm to herbivores and pathogens. The mustard oils bide their time safely housed in the mustards' idioblasts and vacuoles. Similar to allicin, once mustard oils meet activation enzymes, called myrosinases, they attain their full form, which poses a high risk to all living cells, including the plant's own. The myrosinases reside in a special, Brassicaceae-specific type of idioblast called the myrosin cell, whose vacuoles accumulate large quantities of these activation enzymes. Once activated by damage from crushing or chomping, mustard oils and their breakdown products spread easily. In this form, they can permeate cells, volatilize into the air, increase in toxicity, or serve as repellents. Mustard oil bombs are concocted only when the different types of mustard crazy cells are fatally damaged by attackers, exploding and mixing their contents. Each plant produces dozens of different glucosinolates. When jump-started by myrosinases, the noxious mustard oils only last a short time before reacting violently with everything in their surrounding environment. Herbivores and pathogens alike face a barrage of hundreds of different toxic breakdown products formed as the mustard oil bomb detonates.

As this shotgun of mustardy chemicals activates, it is capable of wreaking a lot of havoc. First, mustard oils race for the eyes, mouths, and sinuses of attackers. If you've ever been punched in the nose by wasabi or horseradish, you're already well acquainted with the strength of mustard defenses. Like the other repellents we've discussed, mustards protect themselves by shocking our nervous system with pain and heat, adding their own bitterness signals to this mix. This tactic works just as well on insect herbivores such as aphids, beetles, moths, and butterflies, burning both their mouthparts and egg-laying equipment. To reach herbivores fast and pervasively, mustard oil metabolites become volatile gases or lipophilic (meaning they can freely spread across membranes into living cells). These gaseous compounds are also chemical messages perceived by other plants, which ramp up defense production, and predatory insects such as predatory wasps that race to the tasty bug buffet waiting at the mustard plant. Once inside the cells of attackers, mustard oils sabotage them by binding irreversibly with key proteins and enzymes vital to life.

Perhaps their most potent ability is that mustard oils signal the living, healthy cells of microbes and herbivores to commit apoptosis, also known as programmed cell suicide. Apoptosis is a series of biochemical events that cause changes and eventually death in the living cells of multicellular organisms. The process of programmed cell death is entirely natural and quite common across most living things. For example, it is one of the mechanisms by which a plant's hypersensitive response reacts to pathogen invasions. We experience this, too. The average human adult loses between fifty and seventy billion cells per day as a result of apoptosis during normal growth and development. A cell might be triggered to die because it senses cell stress or receives signals from neighboring cells. Some of the deleterious changes apoptotic cells undergo include shrinkage, the breakdown of cell membranes or walls, indiscriminate destruction of proteins, and the degradation of the cell's genetics held in its nucleus, DNA, and mRNA. These mechanisms of self-destruction are so powerful and irreversibly damaging that the process of apoptosis cannot be stopped after it has been initiated. For this reason, apoptosis is one of a living thing's most highly protected and regulated processes. The ingenious yet humble members of the mustard family use mustard oil bombs to commandeer this deadly, irreversible biological process and force it upon their attackers.

While these defenses are powerful indeed, we have to remember that plants exist in a web of ecology. Over evolutionary time, specialist herbivores such as the cabbage aphid (*Brevicoryne brassicae*), flea beetles, and pierid butterflies have turned the tables on their toxic food plants. These bugs turn *themselves* into walking mustard oil bombs. Like the monarch butterfly with cardiac glycosides, pierid butterflies have evolved the ability to safely sequester the defensive chemistry of these mustardy plants. They store these chemicals within their own body tissues as they feed. In some cases, these crafty insects even produce their own myrosinase to detonate the bomb! Specialist feeders such as these have several tricks for utilizing the mustards' own defenses. The first advancement these little chemists made was evolving the ability to detoxify the glucosinolates outright or simply prevent them from meeting activation enzymes and exploding. Some of the butterflies even go a step further and convert the glucosinolates into cyanide against their own predators. That way, if anyone feeds on these little bugs, they get a powerful surprise. Mustard oils have become so key to the survival of these specialist herbivores that many of them evolved attraction to wounded mustards' defensive gases. When they detect the sulfurous explosions of mustard oil bombs, pregnant female cabbage whites and loopers actually rush toward the fray in an attempt to lay eggs.

Insects aren't the only ones who've found mustard oils useful. Mustard family members populate our produce aisle and contribute flavor and nutrition to cuisine around the world. The hidden world of peppery defenses imparts the smell and flavor of a variety of cruciferous veggies in our diet. From the crunch of a radish, to the sting of wasabi, to the bite of comfort foods such as mustard or kale greens, these glucosinolate-rich veggies are found across tables of the world. Mustard oils are also being explored as biofumigants. Because these chemicals are so effective in gaseous form, we might use them one day to disinfect our food or the soil our crops grow in, ridding fields of plant pathogens in a safe and environmentally friendly way. Imagine using these natural pesticides on harvested crops to eliminate or deter germs, plant pathogens, spoilage, and hidden insects. Studies in food science have lent this use considerable weight. As we've found with other biochemicals such as allicin, mustard oils have proven effective against foodborne pathogens such as *Escherichia*, *Klebsiella*, *Salmonella*, *Bacillus*, *Serratia*, *Staphylococcus*, and *Listeria*. They also fight plant and spoilage diseases such as anthracnose (*Colletotrichum acutatum*), *Botrytis*, *Alternaria*, brown rot, and blue mold disease (*Penicillium expansum*). The mysterious world of plant chemistry is a powerful natural alternative to synthetic antimicrobials and pesticides. We are slowly learning how to co-opt the defenses of plants to increase the safety, quality, and shelf life of our food.

Some of the most common spices and flavors in our food originated through the survival struggle of plants. Each bite of chili, black pepper, ginger, onion, garlic, and mustard is pumped full of concealed defensive chemistry. Using powerful repellent chemicals, plants are able to trick the nervous systems and brains of their attackers with powerful signals of pain and burning. These repellent flavors sting and singe us in an attempt to ruin every mouthful of food we prepare. Ultimately, we have decided that these strangely flavored plants are worth keeping and tolerating, harnessing their disinfecting power to protect us from foodborne diseases. Because of plant defenses, spice *is* life. As chefs endeavor in the creative act of flavoring food, they are simultaneously keeping us safe and alive. So eat up, because after you have read this chapter, your enjoyment of the flavors in your food might take on a whole new meaning.

CHAPTER 12

Energizing and Paralyzing

Caffeine

Our Caffeinated Planet

There's one plant defense chemical that helps power the world: caffeine. Intended as a protective chemical, caffeine has inadvertently and deeply interwoven itself into human history. It's produced in the seeds, leaves, and fruit of over sixty plant species, which utilize it for four powerful purposes. Caffeine paralyzes and kills herbivores, enhances pollination, damages competing plants, and possesses antimicrobial properties. By harnessing these benefits from one chemical, caffeine-producing plants increase their rate of survival. Humans have cleverly turned the tables on this defense by adapting processing techniques that dilute caffeine's toxicity. By doing so, we have lowered the dosage to levels that can potentially provide numerous benefits to the body. Today, caffeine has become a daily part of life for many cultures.

By a long shot, caffeine is the most popular natural stimulant on the planet. It is a ubiquitous part of the daily grind worldwide and a shot of energy throughout the day. According to the FDA, almost 80 percent of U.S. adults consume a form of caffeine daily. The average American drinks roughly eighty-nine gallons of coffee and thirty-four gallons of tea annually. Worldwide, humans drink about two and a quarter billion cups of coffee a day. Our appetite for caffeine doesn't stop there. It's also found in chocolate, soft drinks, energy drinks, foods, supplements, medicines, and a variety of other products. Clearly, this one chemical is an important and valuable component of our global dietary and pharmacological industries.

Looking back through time, it's obvious that humans have a long history with caffeine. Whenever a caffeinated plant popped up across the globe,

Coffee fruits are known as cherries because of their ripe color. The seeds, or beans, are loaded with caffeine and roasted into coffee.

humans found a use for it. For example, Native Americans utilized a caffeinated shrub in the holly family called yaupon (*Ilex vomitoria*), a species whose range includes most of the southeastern United States. They toasted and brewed the leaves into a caffeinated beverage called the black drink. To these cultures, this drink was regarded as one of the most special gifts to offer friends and visitors. Today, many of us plant this humble hedge in our landscapes for its evergreen leaves, dense form, and bright red berries. Yaupon is a dependable landscape plant because caffeine also helps it resist herbivory, pests, and diseases. Unfortunately, modern gardeners may have all but forgotten its long history of consumption.

We tend to think of carefully tended landscapes as a modern occurrence. However, yaupon was just as deliberately cultivated in carefully pruned gardens by many southeastern Native American tribes. The naturalist William Bartram, while traveling through the Southeast from 1773 to 1777, observed firsthand how Cherokee, Creek, and Seminole villages maintained acres of tended yaupon shrubs that resembled tea plantations. Additionally, botanist Francis Harper, who from 1917 until the 1950s researched the work of John and William Bartram, noted disjunct populations of yaupon far inland from their natural range that correlate to sites of heavy Native American

Yaupon holly (*Ilex vomitoria*) is a North American native plant that contains caffeine. Native Americans brewed its leaves into a caffeine-rich beverage called the black drink, or *asi*.

occupation. Archaeological evidence shows that southeastern traders from North Carolina down to Florida and all the way to Texas dried, packed, and shipped the leaves across North America. Some reached ancient mound cities such as Cahokia near modern-day St. Louis, Missouri. Early travelers such as Hernando de Soto, Álvar Núñez Cabeza de Vaca, and William Bartram reported the serving of the yaupon-based black drink in conch shells during purification rituals, social gatherings, village councils, and ceremonies. This caffeinated legacy has been reawakened recently because several U.S. companies have ventured into producing commercial tea products from yaupon. Perhaps one day the noble yaupon will find its way from our landscaping and into the teapot once more.

Camellia sinensis, the tea plant, is another of the approximately sixty plant species worldwide that have evolved caffeine as a defense.

Why Plants Espresso Themselves

We know that many plants produce caffeine, but why do they do it? Caffeine is a psychoactive addictive alkaloid produced by some of the most prominent crop plants in the world, including coffee, tea, yerba maté, chocolate, citrus, kola nut, guarana berry, guayusa, and the yaupon holly. Caffeine is so inextricably linked to coffee that the word is derived from the French word for coffee, *café*. Its slightly bitter taste imparts some of the flavor of coffee and chocolate but also indicates its toxic effects and evolutionary history as a poison. Generally, the more toxic a compound is, the more bitter our taste receptors perceive it to be. This bitter taste is a brief sample of the coevolution between mammals and plant defense. Interestingly, caffeine's effects vary based on dosage. Receive a low dose, and the effects are safe and stimulating to an animal's nervous system. Ingest too high a dose, and the animal's nervous system goes into overdrive, which can cause paralysis and even death.

Cacao pods hanging from a chocolate tree (*Theobroma cacao*). A white, sweet pulp inside facilitates the dispersal of the hard seeds, which are protected by caffeine. Humans process the seeds into chocolate.

Caffeine-bearing plants didn't spend millions of years of evolutionary effort just for our benefit. Each caffeine molecule produced utilizes valuable nitrogen that could be used for growth. Thus, there is a cost associated with production. The reason these species tinkered with chemistry is because it rewarded them with four powerful competitive benefits. Caffeine, it turns out, is a natural pesticide, a pollination enhancer, an herbicide, and an antimicrobial agent. It is such a flexible and useful chemical that coffee and tea plants evolved the ability to make it independently of each other, using two distinct genetic mechanisms. This phenomenon of an evolutionary adaptation showing up independently in different genetic lines is known as convergent evolution. Caffeine is a botanical invention that has paid rich rewards to these species in terms of evolutionary fitness. Let's look at some of those benefits now.

Caffeine's effects are dosage dependent. At mild dosages, caffeine offers numerous neurological benefits to animals. Within fifteen minutes of consumption, it enters the bloodstream and brain of an animal and begins stimulating the central nervous system. A dose of caffeine has a half-life of around six hours. This means animal bodies take six hours to clear 50 percent of a given amount of caffeine. It takes a full twenty-four hours for it to be eliminated completely. Things begin harmlessly enough: in low doses, this chemical powerhouse increases heart rate, respiration, blood pressure, memory, and adrenaline. Sugars are released into the muscles, and mood-improving neurotransmitters such as dopamine and serotonin increase. Additionally, receptors that cause the brain to feel fatigue and sleepiness are blocked. Mostly due to its effects on mood, caffeine creates an addictive feeling of reward. Altogether, we feel a burst of energy and happiness.

The benefits mentioned enable plants to utilize caffeine as a pollination enhancer. Small insects are just as affected by caffeination as we are. Flowering plants need animals to transfer their pollen to another plant for reproduction. Drugging insects with caffeine allows plants to manipulate the behavior of their pollinators. Coffee and citrus plants have turned their flowers into miniature coffee shops for just this purpose. They lace their nectar with nontoxic doses of caffeine. It causes a lot of buzz from bees, moths, and butterflies, enough to keep their pollination and reproduction rates above those of uncaffeinated plants. The powerful chemical reward of caffeine stimulates pollinators to visit flowers more often for longer time periods, fly farther or faster, and remember the location of specific plants.

As the dosage of caffeine increases, however, the chemical's effects morph into a whole different beast. The concentration of caffeine in the seeds, leaves, and stems—the parts damaging herbivores are most likely to feed upon—is higher than in flower nectar, which is utilized by beneficial pollinators. Caffeine then becomes a life-threatening pesticide that can paralyze the central nervous system. Humans aren't the only animals with nervous systems, nor are they the exclusive animals to interact with coffee, tea, and cacao (a.k.a. chocolate) plants. Any animal that can move has a nervous system, which powers a variety of muscles that keep the organism alive. Therefore, caffeine-bearing plants can interact with and potentially poison any attacker. Caffeine is capable of paralyzing insect feeders at concentrations present in plants. This can impact their ability to move and feed and eventually leads to complete paralysis and death.

Mammals also utilize muscles in a variety of ways. Think about your own body's muscles and how important they are for your survival. The operation of your heart and lungs and your ability to move, digest, avoid danger, and utilize your senses all rely on muscle movements. At high concentrations of caffeine, mammals can run into problems. Fatal caffeine poisoning, though, is uncommon for larger browsing mammals because the counterreaction to caffeine's bitter taste and nauseating effects is to purge it from the body. For example, humans would need to rapidly drink fifty to one hundred cups of coffee, more than could fit in the stomach, to achieve a lethal dose. However, caffeine overdoses can be a risk with artificially enhanced caffeinated products. When an overdose occurs, it results in a mix of unpleasant mental and physical symptoms. Symptoms of escalating caffeine poisoning include nervousness, irritability, anxiety, nausea, restlessness, insomnia, gastrointestinal disturbance, and muscle twitching. Fortunately, the protective systems of our bodies usually provide these signs of trouble before mortality occurs. If caffeination exceeds these warnings, severe overdoses can induce heart palpitations, respiratory problems, disorientation, depression, hallucinations, muscle death, paralysis, and death.

Oxidizing, roasting, and mixing coffee beans with water dilute the caffeine content to non-threatening levels for human use.

Thankfully, humans have learned some creative ways to avoid caffeine poisoning for most naturally occurring caffeinated products. With these techniques, we can reap the good benefits of the chemical and reduce caffeine's toxic effects. Roasting, oxidizing, and dilution with water are common techniques that prep the caffeine dosage for safe consumption. Dilution by water helps moderate the punch of the chemical and is one reason we brew plant parts into caffeinated beverages. These techniques have quite an ancient origin; after all, we've been interacting with caffeine since at least the Stone Age. In Africa the kola nut has been chewed for energy since ancient times. This traditional practice still exists, but humans also use the kola nut in soft drinks. In fact, Coca-Cola derives its name from this caffeinated nut. Ancient peoples began steeping tea leaves in Asia, coffee in Africa, yerba maté in South America, and yaupon holly in North America. Oxidation, a process where the raw plant parts are left to react with oxygen in the air, also removes a bit of caffeine. Chocolate beans, coffee beans, and black tea are prepared with this technique. Coffee beans are also roasted, ostensibly not only to affect the taste but also because the heat degrades even more of the caffeine prior to consumption. Each time you safely consume caffeine, you're the recipient of the knowledge and progress our species has made during our extensive interactions with plant defenses.

The kola nut (genus *Cola*) is a caffeinated African seed used to flavor soft drinks like its namesake, Coca-Cola. It has been chewed for energy since ancient times. Caffeine not only protects seeds from herbivory but also impairs the root growth of competing plants.

Sometimes, a defense can kill two birds with one stone. Caffeine not only affects animals but also is toxic to other plants. This ability to release chemical compounds that hurt botanical neighbors is called allelopathy. This chemical warfare can take various forms. In this case, caffeine's effect is to slow the germination and growth of competing seeds typically by impairing their root development. When a coffee seed germinates, the surrounding soil is bombed with caffeine. Even as a seedling, a coffee plant gains a substantial head start. Leaves from adult plants contaminate the surrounding soil with caffeine when they fall and decay. It's extremely difficult for other vegetation to grow beneath or next to a coffee bush, much less a whole plantation of them.

The last benefit caffeine offers plants is disease protection. Amazingly enough, coffee can even affect microbes. The chemical has powerful antimicrobial properties that we have only begun to understand within the past few decades. Scientists have pitted caffeine against numerous bacteria such as *Escherichia coli*, *Staphylococcus aureus*, *Salmonella*, and various other microbes responsible for economically destructive plant diseases. In these petri dish battles, caffeine has demonstrated the ability to suppress or kill the growth of these microorganisms. In an interesting twist, several microbes, including numerous plant pathogens, have evolved methods of degrading and detoxifying caffeine. These include bacterial strains such as *Penicillium* (responsible for penicillin), *Serratia*, *Rhodococcus*, *Stemphylium*, *Pseudomonas* and the fungi *Rhizopus*, *Aspergillus*, and *Phanerochaete*.

There are so many reasons for plants to be admired for their invention of caffeine. This single molecule is responsible for a treasure trove of evolutionary fitness traits. Caffeine has the ability to become either a friend or a foe, which provides producers with an amazing flexibility of benefits. Caffeine literally imposes mind control on pollinators. Humans have become so addicted that we have spread caffeinated plants across the globe. As a defense, caffeine assails foes from the smallest microbes to insects and mammal herbivores. Neurologically, caffeine provides warning shots of energy that lead to a danger zone of muscle paralysis. We hope that a glimpse of the powerful effects of caffeine explains a bit of the buzz behind this awesome chemical.

CHAPTER 13

Clean Enough to Kill

Soapy Saponins

There's one plant defense that can literally clean attackers to death: soap! Plants make natural soap compounds called saponins, originating from the Latin word for soap, *sapo*. Saponins are toxic compounds that distinctively create a soapy, cleansing foam when they are mixed with water. Hundreds of plant species around the world use soap to protect their leaves, roots, stems, bulbs, flowers, fruit, and seeds. Upon attack, saponins form a deadly bubble bath for microbes and insects. They're also unpleasantly bitter and repellent when munched. Organisms that consume soap are "scrubbed" both inside and outside their bodies. Saponins erode the membranes of the cells they contact, rendering soap's cleaning power universally effective against germs, plant-parasitic nematodes, and a wide variety of herbivores such as bugs, mollusks, mammals, reptiles, and birds. Once a cell's membrane breaks, it begins leaking fluid and dies by drying out. Soap is particularly destructive to cells in the digestive system, circulatory system (especially blood cells), and skin. You've experienced this effect firsthand if your hands have ever become dry or cracked after repeated washing. While this external drying effect is unpleasant, we're lucky to be able to grow replacement layers of skin cells to protect us from permanent damage. For smaller, fewer-celled organisms such as microbes, the cells that saponins rupture may be the only ones they've got! Any breach in the cell membranes of single-celled organisms is essentially a death sentence.

Soapy saponins are a plant defense! They create foam, trap and wash away particles, taste unpleasant, and can literally clean an attacker's cells to death.

This diagram demonstrates how saponins work. En masse, they form a spherical blockade around particles of dirt, animal and germ cells, oils, and even air. The trapped objects are then easily carried in water. If gas is the trapped molecule, it forms foams and bubbles.

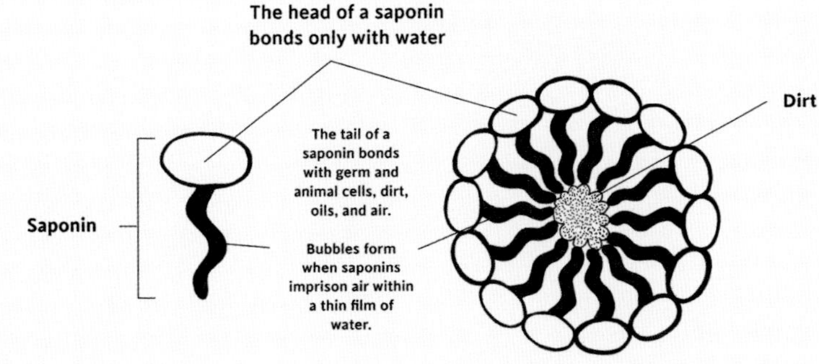

Fortunately, mammals like us have evolved a few methods of resistance to the cellular effects of saponins. Externally, we simply replace broken cells. Internally, counterchemicals such as cholesterol bind saponins, catching them before they absorb into blood and gut cells. Ironically, saponins are a constituent defense in many crop and medicinal plants, such as beans and quinoa. Using simple processing techniques such as cooking and boiling allowed humans to remove the soapy toxins so that we could consume these foods safely. Saponins are by no means a waste product, though. They're added to common beverages such as beer and root beer to create thick heads of foam. Additionally, saponins have a long history of collection and use as soaps and medicines. In the modern chemical industry, saponins, both natural and synthetic, find usefulness as detergents, surfactants, adjuvants, wetting agents, emulsifiers, foaming agents, and dispersants. Saponins can do some amazing things, and now we'll explore more of the science behind how they work and why plants find them so helpful.

Saponins protect plants in multiple ways. As a first line of defense, they're intensely bitter—so much so that foods with high levels of saponins are almost inedible. This type of plant defense is called a repellent. Herbivores simply pass these plants by in favor of more palatable vegetation.

The second function of soap as a plant defense is that it literally cleanses away attackers. Basically, soap has the unique chemical ability to bind with molecules or cells, carry them in water, and wash them away. Soap damages the cells of any attacking organisms on contact, be they part of a bacteria, fungus, nematode, insect, mammal, mollusk, reptile, bird, or even fish. There is enough water in plant cells that the soap is instantly activated if any part of the plant is damaged. What we perceive as cleaning is, in fact, the chemical degradation and removal of a wide variety of microscopic debris and organisms.

The cleaning effects of saponins result because of their molecular shape and properties. Each saponin is made up of two parts, which preferentially bind with different types of molecules. Only by working together in distinct ways can saponins interact with water and remove debris. Their first part is made of sugar and preferentially bonds with polar solvents such as water and cellular fluids. It might be thought of as the "head" of the saponin, and this bonding affinity with moisture is called hydrophilic. The second part is called a sapogenin, often trailing from the head as elongated "tails." Sapogenins do *not* bond to water (called hydrophobic), preferring instead nonpolar molecules such as oil, gas, dirt, hydrocarbons, and other cells. Molecules that combine this remarkable dual personality—the capability to form multiple intricate bonds simultaneously—are called amphiphilic. Depending on what two substances they bond to, amphiphilic molecules display amazing versatility, utility, foaming, and cleansing abilities. Saponins also reduce the strength of the cohesive bonds between water molecules, which results in a lower surface tension. Chemicals with this property are called surfactants. An aqueous solution containing surfactants such as saponins can link with microorganisms and dirt particles more easily. Without a surfactant, these particles would just loosely float in the solution and be less likely to completely wash away. In a typical situation inside an invaded or damaged plant cell, saponins, fluids from ruptured plant cells, and attacking organisms would all mix, bond together, and wash away. Inadvertently, the powerful bonding properties of saponins cause bubbles to form when air is trapped. From a plant's perspective, the creation of soapy foams is almost accidental. Bubbles are simply the phenomenon resulting because saponins arrange themselves in spherical shapes around trapped particles and gases. Within the bubble, saponin tails bond to and imprison gas molecules. To form a bubble's outer layer, the sugary saponin heads bond with a thin film of water molecules.

Human Beings' Discovery of Natural Soaps

Saponins occur naturally in hundreds of plant species across a diverse array of plant families. Some of these are listed in table 8. As we uncover more of the hidden world of plant defenses, we find in many cases how the stories of plant evolution and ethnobotany weave together in fascinating ways.

The best archaeological evidence we have shows that humans have derived soaps from plants for thousands of years. The earliest recorded vegetable soaps occurred as residues in ancient Babylonian pottery dating to 2800 BCE. These versatile molecules have since become part of our daily lives in many intriguing ways. In fact, some of the ways we use soap have nothing to do with cleaning. Let's take a moment to explore the many ways soap has cleaned up our world.

One of the most familiar garden plants that produces soap is the European native soapwort (genus *Saponaria*). Its leaves and roots contain

Soapwort (*Saponaria officinalis*) is an attractive plant with a long history of use of its roots and leaves as botanical soap.

Table 8 Plants Relying on Soap-Based Defenses

Family common name	*Family Latin name*	*Example species*
Pink	Caryophyllaceae	Soapwort
Bean	Fabaceae	Most edible beans, clover, alfalfa
Soapberry	Sapindaceae	Soapberry
Nightshade	Solanaceae	Tomato, potato
Amaranth	Amaranthaceae	Quinoa, beets, spinach
Horse chestnut	Hippocastanaceae	Buckeyes
Ginseng/ivy	Araliaceae	Ginseng, English ivy
Soapbark	Quillajaceae	Soapbark
Sapodilla	Sapotaceae	Sapodilla
Amaryllis	Amaryllidaceae	Onion, garlic
Morning glory	Convolvulaceae	Morning glories
Yam	Dioscoreaceae	Wild yam
Mint	Lamiaceae	Peppermint
Brazil nut	Lecythidaceae	Fish poison tree (*Barringtonia asiatica*), Cream nut (*Lecythis pisonis*)
Lily	Liliaceae	Mondo grass, onion, garlic, leek
Asparagus	Asparagaceae	Asparagus, yucca, agave, hosta
Figwort/snapdragon	Scrophulariaceae	Mullein
Verbena	Verbenaceae	*Duranta erecta*
Honeysuckle	Caprifoliaceae	Coralberry
Arum	Araceae	Jack-in-the-pulpit, dieffenbachia
Greenbrier	Smilacaceae	Smilax
Plantain	Plantaginaceae	Foxglove
Grass	Poaceae	Oats
Tea	Theaceae	*Camellia oleifera*
Buckthorn	Rhamnaceae	*Noltea africana*
Olive	Eleagnaceae	Buffaloberry
Brackenfern	Dennstaedtiaceae	Brackenfern

a natural soap with a long history of use as a cleaner for skin, laundry, and utensils. The Greeks and Romans planted soapwort near their baths and in home gardens where it could be used daily. Monks later introduced the crop to England, and European colonists brought it with them to America. Early settlers such as the Shakers cultivated it for washing, and it remained a garden utility until the development of synthetic detergents during World War I. Modern gardeners may be unaware of this story, seeing soapwort only as an attractive, pest-resistant perennial. Though somewhat obsolete, soapwort finds a niche use today as one of the most delicate cleaners for priceless antique fabrics and tapestries.

Saponins have also appeared in many members of the bean family, where they protect the raw seeds from herbivory and microbes. This includes most edible beans, such as chickpeas, soybeans, lentils, limas, favas, green peas, and black-eyed peas. You can also find saponins in *Phaseolus vulgaris*, the common bean. Varieties from this species include black, pinto, navy, kidney, and green beans. Cooking beans into a safe, pleasant-tasting, and edible product is a learned human ability. To wash away the bitter saponins, we soak and cook beans in multiple changes of water. If the chemicals aren't completely leached, they begin interacting negatively with gastrointestinal cells. While not generally poisonous at these levels, leftover saponins can induce fairly infamous bloating and flatulence as they bond with gas molecules inside the body.

In the hidden world of the amaranth family, you can find saponins in foods such as beets, spinach, and quinoa. Quinoa, a particularly soapy-seeded plant, was domesticated as a food crop by indigenous South Americans. In their natural state, quinoa's saponins are so bitter that they render the seeds inedible. In cultivation, quinoa's natural defenses keep birds, rodents, insects, and microbes from stealing or blighting its seed crops. Humans learned to remove the saponins with a thorough soaking and water change while cooking, rendering the seeds palatable, nontoxic, and nutritious. South Americans in rural villages still employ the rinse water from quinoa processing as a detergent for laundry and cleaning and as an antiseptic wash for skin wounds. Modern processing techniques remove the saponin-rich seed coat of the quinoa grains. Further breeding work is under way to reduce the saponin content of future cultivars. As one might expect, research has shown that removing this innate defense

A field of quinoa (*Chenopodium quinoa*) in Bolivia, South America. The leaves and seeds have a long history as a staple food, provided the bitter saponins are removed in water. The soapy rinse water is still used today for cleaning.

increases the susceptibility of new quinoa cultivars to herbivory and spoilage. Perhaps nature knows best in cases like this.

Saponins also feature as a defense in a North American plant in the olive family called Canada buffaloberry (*Shepherdia canadensis*). Buffaloberry has a wide range throughout North America but is plentiful in the Pacific Northwest and Canada. The fruits of this plant contain high concentrations of saponins and are intensely bitter, yet they have been used to create some fascinating traditional products. If eaten in large quantities, buffaloberries can cause diarrhea and cramps. However, Native people of the Pacific Northwest and Canada discovered that the saponin contents of the fruit could be reduced after exposure to several freezes, drying, or boiling in water. Tribes harvested the berries by using a stick to beat fruiting

Buffaloberry (*Shepherdia canadensis*) is a saponin-bearing plant from the Pacific Northwest. Its fruit has a long history of being processed into numerous fascinating foods and medicinal products.

bushes over pieces of hide or canvas. After processing, buffaloberries were rendered into valuable foods, cleaners, and medicines. Buffaloberries were highly esteemed and traded to Native groups outside the range of buffaloberry populations. One type of a jerky-like food called pemmican involved mixing dried berries with buffalo meat. A second food, Alaskan ice cream, is a dessert traditionally made of whipped fat, meat, and berries such as cranberries, blueberries, and buffaloberries. The berries were also cooked into beverages, stews, puddings, syrups, sauces, cakes, candies, and jellies. A particularly special treat using buffaloberry fruit is still served today at Native-themed restaurants in the Pacific Northwest. It's called indigenous ice cream, or *sxusem* (pronounced "hoo-shum"), and it relies on the foaming effects of the plant's saponins. Indigenous ice cream is a bittersweet, frothy dessert. It's produced by whisking buffaloberry fruit to activate the foaming effects of its saponins. To counteract the bitterness of the defensive chemicals, chefs add generous amounts of sweeteners such as sugar, common camas (*Camassia quamash*) bulbs, or western hemlock (*Tsuga heterophylla*) cambium. Other flavorings include raspberries, cedar (*Thuja*) cambium, Rocky Mountain maple (*Acer glabrum*), and thimbleberry (*Rubus parviflorus*) leaves. American as well as Canadian trappers, pioneers, and settlers learned these techniques and continued the tradition of using buffaloberries as a wild food.

Soapberry or soapnut trees, in the genus *Sapindus*, bear fruit with soapy pulp and seeds. Species in this genus occur worldwide, notably in western and southeastern North America, the Caribbean, China, India, and Hawaiʻi. The fruit of *Sapindus saponaria*, a species from the Americas, contains around over a third saponin content and was widely used by Native Americans to create cleaners and soaps. Interestingly enough, soapberry and other saponin-producing plants may offer us a new way to fight mosquitoes. Scientific research has shown that extracts from *Sapindus* species could inflict heavy mortality on all life stages of at least three different species of malaria-carrying mosquitoes. Remember how saponins exhibit incredible functionality in water? That happens to be the naturally preferred breeding place of mosquitoes. Mosquitoes are aquatic for most of their lives, including the eggs, larvae, and pupae. Only the adults are winged and capable of flight, and only adult females bite and spread diseases. Female mosquitoes create eggs from the proteins in blood and then deposit them to grow in water. It'd be lovely if mosquitoes could create these proteins themselves. Unfortunately, because they cannot, lady mosquitoes are armed with hypodermic needle–shaped mouthparts to steal these proteins from animals. But rest assured: one day, we may utilize plant soap as a new weapon to end mosquitoes' reign of itchy terror.

Several other North American plants exude soap as part of their survival systems. The soaproot plants (genus *Chlorogalum*) and most *Yucca* species produce saponins as protectants. Some familiar species include Adam's needle (*Yucca filamentosa*), soaptree yucca (*Y. elata*), the Joshua tree (*Y. brevifolia*), and Spanish dagger (*Y. schidigera*). Native Americans in California and the Southwest crushed the stems and roots of these species for soap and shampoo and as poisons to help them catch fish. Yes, that's right—fish!

It may sound strange, but soap from plants can actually make it possible to catch fish. Soap defenses are universally effective on animal cells. While mammals have adapted resistances, fish lack these adaptations because they aren't typical herbivores of soap-bearing plants. For fish, saponins are especially bad news. Saponins dissolve readily in water, where they're rapidly uptaken by the gills as fish breathe. Prehistoric fishermen discovered that they could hijack this plant defense to easily harvest large quantities of fish. Saponins are relatively effortless to use compared to hooks, poles, and baits. All

Yuccas, like this soaptree yucca (*Yucca elata*), produce soap as a defense against disease and herbivory. Any cells damaged by herbivory or disease will gush water and soap onto attackers. When humans mash yucca parts into water, they create soap for cleaning or gathering fish.

that was required was adding the soapy substance to the water. Saponins work by impacting breathing, which stuns or kills fish. They then float en masse to the water's surface, where the fishermen can scoop them into nets. Thus, humans co-opted plant defenses as an ideal and unavoidable way to deliver toxicity to massive colonies of fish within just a few minutes. Saponins are so effective that legal bans exist on their use in many countries to prevent overharvesting from crashing fish populations.

Saponins stun fish by interfering with their ability to breathe in two ways. The first is oxygen deprivation. When saponins foam at the water's surface, they create a stable bubble barrier that prevents oxygen from entering the water. The second is cell destruction. Below the water, dissolved saponins enter the gills and reach the fish's bloodstream within minutes. They set to work scrubbing pores in the membranes of cells in the gills, blood, and organs, eventually causing them to leak and die. Plants with this property are called piscicides, or fish poisons, after the Latin word for fish, *piscis*. However, even though the effects of the chemical are potent, poisoned fish are capable of reviving if quickly released into uncontaminated water. For humans, the saponin levels used aren't a problem, and the fish remain perfectly safe for consumption. This is because our cells have evolved the capability to detoxify them, and saponins are also easily removed from food using washes of clean water or cooking.

To fish with these plants, indigenous tribes collected large quantities of vegetation and pulverized them into water sources until they formed a thick foam. The location was also cleverly chosen to keep the water as slow-moving as possible, which prevented clean water currents from diluting the poison. Stagnant pools, tidal pools, ponds, slow-moving streams, dams, and weirs were preferred sites. Gathering the stunned fish was usually done by hand or with baskets, spears, and nets. All continents except Antarctica possess saponin-bearing plants that were discovered and used independently by indigenous peoples. In North America, saponin sources included buckeye seeds (genus *Aesculus*), pokeweed leaves and berries (*Phytolacca americana*), Jack-in-the-pulpit roots (*Arisaema triphyllum*), and the aforementioned *Yucca* and soaproot plants. In China, ground seeds of the tea-oil camellia (*Camellia oleifera*) are used extensively in aquaculture to rid prawn ponds of unwanted fish and insects. The Aborigines in Australia, Gond in India, San and Baka in Africa, and Yanomami in South

America are examples of other peoples who have their own plant-based piscicidal recipes.

The saponins produced by plants haven't become completely obsolete in the modern world. For many food, industrial, chemical, and pharmaceutical processes, we rely on extracts from a plant called the soapbark tree (*Quillaja saponaria*). Soapwort (*Saponaria officinalis*) and *Yucca* species are also commercially cropped for their saponin contents. We produce a bunch of powerful and tasty stuff from soap!

Bubbly goodness! The foaming heads of root beer, cream soda, beer, and ginger beer are created by saponins in quillaia extract, the bark of the soapbark tree.

The soapbark tree (*Quillaja saponaria*), native to Chile, is a highly concentrated natural source of saponins. Extracts from its bark are utilized in a wide variety of products, including foods, beverages, medicines, and chemicals.

Soapbark is an evergreen tree native to the dry forests of Chile. The bark has one of the highest naturally occurring concentrations of saponins, which render the plant almost immune to rot pathogens. Soapbark can be collected and further purified into a powder containing over two-thirds saponin content. The bark in its natural state is a product called *quillaia* (pronounced "qwill-ay-uh"). The word *quillai* is derived from the Mapuche word *quillean*, which means "to wash." Soapbark has been a traditional medicine in the Andes for many generations. Andean people still use this bark for soapmaking as well as treating chest ailments. Soapbark extracts might have already found their way to your table. They're responsible for the foamy heads of beverages such as beer, root beer, ginger beer, and cream soda. Glance at the ingredient labels in common products and see if you notice soapbark adding a bubbly bit of awesome to your life.

Delicious beverages aren't the only thing we have to thank plant defenses for. While plants employ saponins as pesticides, humans have used them in life-saving inventions such as the modern foaming fire extinguisher. This type of extinguisher uses purified soapbark extracts or synthetic saponins called aqueous film-forming foam (AFFF). AFFFs deploy a blanket of water-based foam over fires. Foam fights fire in several ways simultaneously. First, bubbles create an air barrier that smothers flaming material, depriving the fire of oxygen and trapping combustible gases. This layer can even float on water or poolings of spilled fuel. Second, the thick layer of foam is an effective heat insulator. This traps the fire's heat, reducing spread and protecting firefighters. Third, synthetic foams remain stable for long periods of time, which prevents reignition as well as spot fires caused by lofted embers. Fourth, since the foams are water based, they can soak into and cool solid material such as wood. Finally, foam can treat liquid flammables such as gasoline by floating as a barrier between the combustible liquid and the fire.

Saponins from the soapbark tree are also saving lives in the modern pharmaceutical industry. The important property in this case is that saponins have the affinity for scrubbing pores into cell walls. By deliberately introducing pores into specific places in the body, saponins can create pathways for beneficial drugs and chemicals to penetrate into cells. Saponins have also been explored as a way to destroy cancer cell membranes and have been chemically linked with medicines and vaccines to create a cellular delivery tool.

Ironically, soap has given plants a clean way to fight dirty with their attackers. It's an amazing survival adaptation that can create the last bath an offending microbe or insect will ever take. With their bitter taste and cellular scrubbing power, saponins repel and disrupt diseases and herbivores. As an odd side effect of their chemical properties, they also foam intensely. When humans discovered these versatile molecules, we adapted methods of removing them and even began relying on them for economically important products. When you use soap to fight disease, you are connecting with a long history of plants doing exactly the same thing.

CHAPTER 14

Herbicide Rain and Superweed Reign

How Plant Defenses Create Herbicide Resistance

Farming in the Shadow of the Superweeds

Throughout human history, farmers and the weeds in their fields have been locked in a bitter struggle. Humans have escalated this fight from simple hand pulling to industrially mechanized control and, most recently, chemical herbicides. The survival systems of weeds are beginning to win this battle. Superweeds, genetically resistant to many commonly used herbicides, are rising across the world. These crop fiends are steadily becoming one of the most critical agricultural problems of our time, and it's all because of the hidden world of defenses they employ against farmers' chemical arsenal. Superweeds are a daunting predicament, competing with crops for water, nutrients, light, and space. Herbicides and their application cost U.S. farmers alone more than $6.6 billion per year.

Our chemical warfare with weeds has escalated over time. Early farmers employed sea salt as one of the first known herbicides. This affected weeds by drying them out or rendering their roots less able to uptake nutrients from the salinized soil. The salt would accumulate in fields, however, slowly concentrating to potentially threatening levels for agricultural crops as well. After 1750 burgeoning Industrial Revolution sectors such as mining, smelting, and petroleum refining created various by-products that were found to successfully kill weeds. Farmers began spraying fields with slurries of diesel oil, sulfuric acid, copper sulfate, iron sulfate, creosote, calcium arsenate, and sodium arsenite. These chemicals killed every bit of plant life they touched, caustically burning living tissues away. Destructive environmental pollution and inadvertent poisonings of people and animals were inseparable from these early herbicides and the toxic heavy

Chemical herbicides are a critical part of modern agriculture, but their usage has promoted the evolution of resistant superweeds.

Herbicided fields, like this one in rural Georgia, are an evolutionary crucible for herbicide resistance in wild plant populations. If you look carefully at the middle of the photo, within the sea of brown, dead grasses lies a healthy, living, green, and likely herbicide-resistant weed that survived the spray. These types of genetic selection events occur worldwide every time herbicides are used.

metals they contained. Nevertheless, Industrial Revolution era herbicides made their way across the world. European farms were the first sites coated with these poisons, but herbicide use soon became commonplace along roads, railroad tracks, and utility rights-of-way as well as tropical plantations for sugarcane and rubber.

In the twentieth century, the cauldron of World War II chemical research led to the synthesis and discovery of new chemicals that could destroy the crops of enemy farmlands. Chemicals that never before existed in nature were birthed in beakers and created an agricultural war zone. The year 1945 was a hallmark year because of the development of three of the first selective herbicides. This meant that these chemicals could kill a particular type of plant while sparing another. These were

Operation Ranch Hand, from 1962 to 1971 during the Vietnam War, was a defoliation campaign designed to reduce enemy cover and cropland. Aerial applications with more than nineteen million gallons of herbicide were dropped over Vietnam—the dried, dead landscapes were later ignited with napalm strikes, causing immense firestorms and casualties. Agent Orange, an herbicide blend containing 2,4-D, made up eleven million gallons of the total herbicides used in Operation Ranch Hand.

After the Dust Bowl of the 1930s, American farmers learned some hard lessons about the environmental, agricultural, and economic impact of soil erosion. With the advent of mechanized tilling and plowing, rich farm topsoils simply washed and blew away when no plant roots remained as a binding agent—yields and profits dramatically declined along with the dirt and its valuable nutrients. Seen here is part of the Black Sunday 1935 dust storm in Oklahoma and Texas. No-till agriculture, which necessitated the use of herbicides, minimized soil disturbance and was heralded as the best new practice to preserve farm soil.

2,4-D (pronounced "two-four-dee," short for 2,4-Dichlorophenoxyacetic acid), 2,4,5-T (2,4,5-Trichlorophenoxyacetic acid), and IPC (isopropyl *N*-phenylcarbamate). The first two selectively killed broad-leaved or dicotyledonous weeds but left grasses (which are monocots) intact. Grass crops such as corn, sugarcane, sorghum, millet, wheat, oats, rice, and turfgrasses would continue on while weeds below were withering. IPC selectively killed grasses, benefiting broad-leaved crops such as greens, cotton, soybeans, tomatoes, peppers, potatoes, fruit orchards, alfalfa, and many others. Collectively, post–World War II technologies such as chemical pesticides (including herbicides), mechanization, fertilizers, and crop breeding led to a successful agricultural age known as the Green Revolution. Societies reveled as bumper crops rolled in, but a few weed species were still clinging to life on the fringes of fields across the world. This was the beginning of the evolution of herbicide resistance.

It is helpful to understand why people would go to such lengths to poison weeds. While weeds can be mechanically controlled—ripped from the soil by pulling or tilling—chemical control offers quite a number of advantages. Economically, herbicides save time, fuel, and labor, which translates into increased profits and higher yields for the farmer. After the Dust Bowl of the 1930s, American farmers learned some hard lessons about the environmental, agricultural, and economic impact of soil erosion. With the advent of mechanized tilling and plowing, rich farm topsoils simply washed and blew away when no plant roots remained as a binding agent; yields and profits dramatically declined along with the dirt and its valuable nutrients. The reason farmers embraced tilling was, in part, to get rid of weeds by folding them under the soil, where the lack of light would starve adult weeds and prevent their seeds from germinating. We soon learned that soil conservation was essential, and the idea of "no-till" agriculture became a recommended farming practice. No-till farming, developed in the 1940s, keeps the soil intact by minimizing the movement of earth by mechanized equipment. The downside, though, is that farmers still had the weeds to contend with. Resultingly, no-till agriculture necessitated the use of novel herbicides, which began coming online during the Green Revolution. The benefits of the new knowledge and products of the Green Revolution spread quickly. Within just twenty-five years of the registration of the first selective herbicides in 1945, over one hundred other novel chemicals were synthesized, registered, and placed into general use as herbicides. The first university-level programs in weed science developed, as well as the first international and interstate conferences on weeds. In 1956 the Weed Science Society of America met for the first time and published *Weed Science*, the first scientific journal for weed biology and herbicide research. Financial support for scientific weed research in 1962 was six times greater than in 1950. All of this education and funding created a pipeline of bright scientific minds that catalyzed the Green Revolution. The number of U.S. personnel involved in full-time weed research in 1962 was twenty times higher than in 1940. Keep in mind that all the while, chemists were also experimenting with and developing insecticides to use in agriculture. We were soon dosing our crops with a wide range of products with different efficacies and environmental impacts.

During the Green Revolution, we made a tremendous investment in the science and mechanization of eradicating weeds. Ironically, plant defenses had already evolved ways to beat herbicides decades before synthetic chemical herbicides were even invented. Herbicides are an evolutionary selective pressure, and they aren't always fatal to every plant they contact. In any genetically diverse plant population, a few individuals may have genes better suited to survive. Genes for herbicide survival were lying in wait within weed species around the world. Herbicide-susceptible plants die, while resistant ones survive and reproduce in increasingly greater numbers. Mutations in successive generations may stack and increase the level of these tolerances, creating highly resistant superweeds. As the rate of herbicide use has increased, so has the evolution of resistance. By their very nature, weeds are known for their prolific seed counts. Each resistant individual that survives can fling tens or hundreds of thousands of seeds into fields, topsoil, crop seed, livestock digestive systems, boots, muddy tires, and harvesting equipment. Our increasingly industrialized and globalized world was ripe for the picking, wide open for superweed colonization.

Herbicide resistance predates the invention of synthetic chemical herbicides by decades. Plant scientists know this from analyzing the DNA of old herbarium weed specimens. Even though these plants are long dead, dried, pressed, and glued onto sheets of paper, their invaluable DNA tells us their level of herbicide tolerance. The oldest herbicide resistance we know of was uncovered in a 2013 study of blackgrass (*Alopecurus myosuroides*) herbarium specimens. Blackgrass is an infamous modern superweed that has possessed a genetic mutation for herbicide resistance since at least the Victorian age. One very special blackgrass specimen, collected in 1888 in Bordeaux, France, and shelved for more than a century, held a mutation counteracting an herbicide that had not even been invented yet. That fortuitous yet random genetic mutation rendered immunity to a class of herbicides invented in the 1970s called ACCase inhibitors, which disrupt plant cell membrane construction. Blackgrass has since stacked even more types of herbicide resistances, becoming just one in a global pipeline of superweeds. In 1957, along a quiet, nondescript roadside in Canada, a wild carrot (*Daucus carota*) population became the first documented case of herbicide resistance. For the first time in the history of the planet, just twelve years after their 1945 discovery, chemical herbicides that never before existed in nature were being counteracted by the defense systems of ubiquitous weedy plants.

In the 1970s another herbicide revolution began with the development of commercial glyphosate products. Glyphosate continues to be one of the most influential agricultural herbicides of all time. Glyphosate is inexpensive and less environmentally persistent than many other herbicides because it degrades quickly in contact with light and soil microbes. Soil particles strongly absorb glyphosate, so the chemical is also less likely to move into water sources, contaminate groundwater, or linger in the soil to poison successive crops or native plant communities. When a glyphosate spray lands on the leaves of a plant, it begins inhibiting a critical enzyme called EPSPS, which plants use to make proteins and amino acids for growth. As EPSPS "shorts out" in the cells, the entire plant begins to collapse and eventually die. At the time, glyphosate was quite simply unmatched by any other herbicide in its efficacy, ability to kill an incredible spectrum of plants and safety for the environment (relative to other chemicals of the time). Glyphosate's success led to overreliance upon it by farmers, foresters, and rights-of-way maintenance crews. Glyphosate's other major downside was that it killed crop plants in addition to weeds. This meant glyphosate had only a short window preplanting and postharvest when it could be applied to entire fields. However, biotechnology was determined to change this.

Since 1974 over 8.6 billion kilograms of glyphosate have been applied worldwide, the vast majority of it since 1996. This is because in 1996 glyphosate-resistant lines of genetically modified corn, cotton, and soybean varieties hit the market, with alfalfa, sugar beet, and canola to follow later. All were immediate successes. Glyphosate-resistant soy dominated 90 percent of U.S. plantings of the crop in 2010, along with 70 percent for glyphosate-resistant corn and cotton. Since these crops became available, global glyphosate use has increased more than fifteenfold. Glyphosate-resistant crops contain innate, biotechnologically engineered immunity to glyphosate; entire living fields of crops could now be sprayed for weeds during the growing season. Herbicide-tolerant crops led to massive increases in the acreage under a rain of glyphosate, bringing expansive wild plant populations under the selective influence of herbicides. Additionally, growers now had a significant incentive to abandon other herbicides, chemical rotations, and even mechanical weeding such as tilling and plowing in favor of glyphosate. Instead, monocultures of herbicide-resistant crops were planted without rotation, and the weeds within these fields received copious doses of glyphosate. This created the perfect conditions

Across much of the world's arable land, endless, weedless, monocultural rows of crops such as these rows of corn and soybeans are created with the use of herbicides. Since the 1990s, this degree of weedless perfection has largely been achieved using glyphosate sprayed over crops that have genetically engineered herbicide resistance. This management strategy has had a lot to do with the evolution of herbicide-resistant superweeds.

for plant defense systems to engage repeatedly against herbicides. In the last few decades, twenty-four species of glyphosate-resistant weeds have been identified in more than eighteen countries around the world, including the United States, Brazil, Australia, Argentina, and Paraguay.

Meet the Green Revolutionaries: How Herbicides Work

There are many chemical classes of herbicides, all with different killing mechanisms—or modes of action. The mission of an herbicide plays out just like a spy movie. Herbicides need to penetrate successfully into weed cells, locate the target they sabotage within the cell, accumulate and bind to the target, and cause fatal cell damage. Foliar herbicides, such as 2,4-D and glyphosate, are sprayed onto the shoot system of plants. In comparison, soil-active herbicides, such as the triazines, are applied into soil and then uptaken by root systems. Preemergent herbicides, such as prodiamine and trifluralin, prevent or kill germinated seeds.

Let's explore more specifically how some common examples of herbicides actually kill plants. The synthetic version of a plant growth hormone called auxin, 2,4-D was engineered to trick plants into growing at an unbalanced, unsustainably fast rate. Affected plants' hastily constructed tissues quickly run out of resources and die. For example, 2,4-D might cause starvation or wilting by inducing a proliferation of stems and leaves. Such unchecked growth outpaces the supply of water and nutrients provided by the plant's proportionally smaller root system. Other herbicides, such as glyphosate, surgically strike at critical plant enzymes and pathways. Atrazine, a soil-active triazine infamous for contamination of groundwater, is slurped up by plant roots and transported to their leaves, where it inhibits photosynthesis and induces death by starvation. Trifluralin, one of the most widely used preemergent herbicides, disrupts the root growth of newly germinated seeds to kill weed seedlings.

There are now hundreds of herbicides that we send as a chemical barrage to manage our fields, transportation and utility networks, landscapes, forests, and even rare plant populations. Modern chemical herbicides are not without significant effects on health, the environment, and nontarget plants. However, when compared to the metal-based herbicides from bygone centuries, modern selective herbicides have the advantage of being generally less persistent in soil and water and pose fewer hazards to animals. Unfortunately, some of the solutions pitched to counteract herbicide-resistant plants include returning to the costly earlier types of herbicides. After all, it's total war, and we seem quite prepared to accept these risks as collateral damage.

We are irreversibly engaged in an escalating chemical and biotechnological war with plant evolution. More than 250 weed species in sixty-seven countries have developed dozens of different strategies for herbicide resistance. Some species, such as horseweed (*Conyza candensis*) and Palmer amaranth (*Amaranthus palmeri*), have populations capable of resisting multiple herbicides at once! The most prominent plant families with herbicide resistance, from most to least species, include the grasses, asters, mustards, sedges, amaranths, snapdragons, and chenopods. Inevitably, new superweeds will continue to be discovered each year as long as agriculture relies on chemical herbicides.

How have superweeds learned to beat us so fast? There are two general strategies they use, called non-target-site resistance (NTSR) and target-site resistance (TSR). One species of superweed might have numerous

Horseweed (*Conyza canadensis*), seen here, is one of the strongest and most pervasive superweeds in the world. Native to North America and growing three to six feet tall, horseweed thrives in no-till farmlands and along roadsides and railways, producing up to two hundred thousand wind-dispersed seeds per plant. Populations of horseweed have been found with resistances to more than eight chemical classes of herbicides, including 2,4-D and glyphosate. This railway right-of-way in Athens-Clarke County, Georgia, was just herbicided, but the horseweed survives and rains seed atop the dead competing species nearby.

mechanisms of resistance across both of these categories (called multiple resistance) or a single adaptation that confers resistance to different herbicides simultaneously (cross-resistance). Altogether, some of the world's most threatening superweeds, such as Palmer amaranth and horseweed, can resist more than half a dozen different herbicidal modes of action.

Let's talk about NTSR. Resistant weeds with NTSR diffuse an herbicide before the chemical causes lethal damage. As the herbicide takes its cellular

Blackgrass (*Alopecurus myosuroides*) invades a barley field. The thin, dark, straight seedheads in the foreground and background are blackgrass, while the thicker, green, and long-awned seedheads are barley. Blackgrass ripens and rains its seeds upon fields before wheat and barley are even ready for harvest. Furthermore, blackgrass has evolved resistance to a wide range of herbicides. One mechanism it uses is called neutralization—any toxic effects of herbicides are neutralized with peroxide-producing enzymes.

Blackgrass (labeled under the now-antiquated synonym *Alopecurus agrestis* in this historical work). Note the shape of its seed heads and see how many you can spot in the field of barley above. A French herbarium specimen of this species from 1888 was discovered to hold genes for herbicide resistance. Decades prior to the development of the first synthetic chemical herbicides, plant defenses had already beaten us.

journey to the site where it is active, superweeds with NTSR redirect it, erode its concentration, or eliminate its toxicity until the herbicide is incapable of killing the plant. Superweeds such as perennial ryegrass (genus *Lolium*) have evolved reduced penetration: thicker, slicker cuticles that act as a barrier or repellent to herbicide sprays. Horseweed takes a different approach when outsmarting glyphosate. Both horseweed and perennial ryegrass resist glyphosate using altered translocation. They redirect the herbicide away from valuable enzymes, instead locking it harmlessly away in their vacuoles or cell walls. Velvetleaf (*Abutilon theophrasti*) is one of many weeds that employ detoxification. Detoxifying weeds produce enzymes and other chemicals that destroy, metabolize, or alter herbicides, rendering them harmless. A final type of NTSR is called neutralization, and this works by nullifying toxic effects produced after an herbicide has succeeded in chemically reacting with its target. Better late than never—neutralization acts as though the weed learned to provide its own life support. Blackgrass, a major superweed of wheat and barley, possesses neutralization as well as multiple other resistances. One example of neutralization occurs when an herbicide's attack creates lethal levels of reactive oxygen species, which wreck havoc on vital life processes inside plant cells. Reactive oxygen species are highly reactive, unstable molecules that are also known as free radicals. Even though the herbicide has technically hit its target successfully, herbicide-resistant plants survive by neutralizing these deadly effects. To do this, weeds have souped-up levels of peroxide-producing enzymes (called peroxidases) and enormous capability to manufacture other cellular antioxidants. Both are the antidote to reactive oxygen species. Altogether, there are quite a number of weak points where superweeds can stop herbicides along their deadly missions.

Superweeds can also make an herbicide's target more plentiful, elusive, or unreactive. This type of herbicide resistance is called target-site resistance (TSR). Overproduction is a type of TSR during which weeds produce excessive levels of herbicide target molecules. Palmer amaranth is a superweed that brushes off glyphosate with this tactic. Glyphosate targets the EPSPS enzyme, but resistant populations of Palmer amaranth produce 5 to 160 times more EPSPS than they need. When glyphosate enters Palmer amaranth, it fails to completely deactivate the massive quantities of EPSPS molecules floating inside the cells. The amaranth survives because plenty

In target-site resistance, superweeds render an herbicide's target more plentiful, elusive, or unreactive. Palmer amaranth (*Amaranthus palmeri*), seen here, is a superweed that brushes off glyphosate with TSR. Specifically, Palmer amaranth overproduces the enzyme targets by glyphosate. Plenty of EPSPS enzyme remains for the amaranth to use after even the strongest glyphosate attacks. Palmer amaranth also uses another TSR tactic called target mutation to render itself immune to ALS inhibitors.

of EPSPS remains after herbicide attacks. Target mutation is the second type of TSR. Using simple structural modifications to an herbicide's target molecule, resistant weeds can render themselves chemically immune to an herbicide. Target mutations are the most common type of herbicide resistance—more than nine different chemical classes of herbicides have been overcome with this tactic. Even slight changes to an herbicide target can confer partial or complete herbicide immunity. Let's take one example, the herbicide chemical class called ALS inhibitors. ALS inhibitors bind to and deactivate an enzyme called ALS that helps weeds build DNA. ALS inhibitors have been defeated more than seven different ways by resistant weeds such as Palmer amaranth, annual ragweed (*Ambrosia artemisiifolia*), blackgrass, giant chickweed (*Myosoton aquaticum*), Indian hedgemustard (*Sisymbrium orientale*), annual meadow grass (*Poa annua*), hairy crabgrass (*Digitaria sanguinalis*), and barnyardgrass (*Echinochloa crus-galli*, the crop mimic of rice). These plants have independently learned to custom build their ALS enzymes so that ALS inhibitors can't react with them. Superweeds make "switcheroos," substituting the ordinary ALS enzyme with just a handful of amino acids that the herbicide cannot bind with. Herbicides pass these custom molecules by without damaging them. This principle is akin to mechanically changing the shape of a lock such that a key fails to fit anymore.

In this chapter we've met some plants that stay alive in the face of a global extermination campaign using chemical herbicides. We hope that this secret world of war between superweeds and synthetic chemicals may become food for thought. Many parts of our daily life have become dependent upon the success of herbicide-resistant crops such as corn, soy, cotton, canola, sugar beets, and alfalfa. The cotton in textiles, the corn and canola oils in biofuels, and the foods we and our livestock consume result from herbicidal victories in this war. As fast as we work to biotechnologically enhance our crops and chemical arsenal, plant defense evolution is keeping pace. Economically and ecologically, the cost of maintaining the escalating war against superweeds may become too much to bear in the future. Evolution is a powerful force to reckon with.

Part Two

Survival against Natural Forces

Lacking the option to physically run from fire, trees and other plants have prepared resistance and recovery adaptations. These specially adapted species are called pyrophytes. This photo, taken in Alberta, Canada, features a type of wildfire called a crown fire. The blaze has climbed branches like ladders to attack the canopy, or crown, of these trees.

CHAPTER 15

The Burning Question

How Do Plants Survive Wildfires?

As a stifling smoky haze fills the air, a raging inferno creeps along the horizon. The desperate creatures of the forest are running for their lives. Winged insects and birds whizz by like bullets. Frogs and toads hop desperately alongside rodents and reptiles scurrying to hide from the blaze. Deer, raccoons, possums, and even the nocturnal creatures awaken and form a furry stampede toward safety. Yet the plants of this community remain, stolid and quiet as this terrified commotion blows past them. Plants in biomes that regularly experience wildfire have faced this threat before and will again. Lacking the option to physically run from fire, trees and other plants in fire-prone regions have prepared an unassuming set of entirely different defenses. While even resistant plants can succumb to fire, the ones that survive pass on their traits. These specially adapted species are called pyrophytes after the Greek *pyros* (fire) and *phyton* (plant). Let's take a look at this hot topic.

To survive and even benefit from fire, plants seem to have several options. One is simply to adapt resistance to heat and ignition by growing protective tissues or altered chemical compositions. Other plants recover quickly after burns or focus on prolific reproduction, relying on the regeneration of their tissues or seeds for success. Pyrophytes, plants that have evolved to actively tolerate fire, can also possess traits that encourage the incidence and severity of fires, which are beneficial to them but damaging to competing plants. Plants with these characteristics have evolved the resiliency necessary for living in places where intense, frequent, and widespread fires exert significant threats to survival of plants and their offspring. Fire regimes, the natural spatial and temporal patterns of wildfire in ecosystems, vary according to a region's climate, available fuels, and topography. Ecoregions around the world with significant fire-adapted plant species include the longleaf pine–wiregrass forests and sandhills of the southeastern United States, the California chaparral and woodlands, dry Mediterranean ecoregions in southern Europe, the Australian Outback, the Cerrado of Brazil, grasslands such as the prairies of the American Midwest, the fynbos of South Africa, the pampas of Argentina, and extremely active volcanic islands such as the Canaries.

Fire is destructive to plants in a number of ways. The heat of fire is the mechanism that causes plant cells to die and become desiccated, combustible fuel. As a fire burns, the radiating heat forces plants to cool themselves as best they can using transpiration. If they cannot, even plants outside of the reach of flames can simply wither and desiccate to death. If fire actually reaches the plant, living cells in the leaves, stems, and roots are at great risk from burning to death. As plant cells reach critically high temperatures, their proteins denature. What this means is that these proteins, despite being necessary for most basic life functions, coagulate and cease to operate. This is quite like how eggs change from goopy to solid when cooked. The water inside a plant cell can also boil and steam away, placing the cell's organelles and membranes under enough pressure that they can explode or leak. As flames roast live plant tissues near the base of a tree, the water supply to the canopy can be boiled off, killing the entire plant. Even if most of a plant escapes being entirely engulfed in flames, the death of these critical tissues, such as roots, cambium, or vascular bundles, can amplify across other parts of the plant.

Fire Resistance

Bark is often a plant's first line of defense against a wildfire. Among bark's many significant roles in plant survival systems, it may amaze you to learn how specifically adapted it is against fire. In almost every possible way, the thickness of a tree's bark can save its life during fire. On fire-adapted trees, bark grows incredibly thick, sometimes as much as 10 percent of the trunk's diameter. The thickest bark in the world currently belongs to the giant sequoia, whose extremely fire-resistant bark can grow up to three feet thick at the base of the trunk of very old specimens. Bark is comprised mostly of dead, sacrificial cells full of lignin, waxes, tannins, and the plant's waste products. Thicker bark insulates the precious living cambium cells from heat, slowing down the time it takes for a fire to create deadly temperatures. Thicker bark also allows plants to retain more moisture, which is critical to maintain cooling, sustain life processes, and avoid heat-induced desiccation or hydraulic failure. Many barks are built with a flaky, plate-like design that traps microscopic layers of air, each helping to maximize heat insulation. Bark may seem like a ubiquitous part of our world, but it can be a supremely high-tech material when matched against fire.

When flames attack flammable plant carbohydrates, such as starches and cellulose, these fuels combust into substances such as water vapor and carbon dioxide. Since these products are volatile gases, they quickly boil away. This process leaves no physical protection or insulation between the plant and the ongoing blaze and is completely the opposite with bark, whose protective properties dramatically insulate against fire when bark is burned. As bark chars, it transforms into a solid black carbonized mineral called graphite. As plants already know, this material is one of nature's best fire retardants! In fact, fire-resistant plants lace their bark with chemicals such as lignins and tannins that ensure maximum graphite production during a fire. These chemicals play a multitude of survival functions, but here we'll focus on how they contribute to fire resistance. Lignin and tannins *favor* reactions that leave purified walls of insulating graphite between the fire and the plant. At the same time, they *restrict* reactions that produce gases and water vapor, which would normally allow fire and heat the freedom to drive deeper into a plant's live tissues. To generate this almost pure carbon and diminish fire damage, bark's chemical composition is superb at binding harmful oxidants, radicals, and other impurities generated during combustion. A plant has the best chance against fire if it can generate a sufficiently thick "purified graphite shield" in the outer layer of its bark. In fireproof species such as eucalyptus, pines, Douglas fir, and redwoods, up to 60 percent of their bark may be carbonized by fire into graphite. For plants whose ecosystems rarely see fire, such as sugar maples and rainforest trees, bark may be quite thin and prone to combustion. In these cases, very little solid material remains between the fire and the plant's living cells. These species are much more susceptible to being damaged by fire.

But what is it about graphite that makes it so invaluable during a fire? Graphite is a stable, relatively inert, and crystalline form of carbon with

Bark is often a plant's first line of defense against a wildfire. Bark plays many significant roles in plant survival systems, including a specific adaptation against fire. Plant bark contains chemicals that deliberately form graphite when charred. Graphite is a layered mineral that protects live plant tissue during wildfire. It is both one of nature's best heat conductors (along its surface) and among the best insulators, as heat energy cannot easily pass through its layers.

amazingly protective thermodynamic properties. You might also know it as the dominant molecule in charcoal, diamonds, pencil lead, and electronic components. Graphite remains one of the best-known refractory (or heat-resistant) materials and a valuable conductor of heat and electricity. We use it in electronics, batteries, crucibles, metallurgical molds, and linings of blast furnaces.

Like phyllo dough or a stack of paper, graphite has a shape comprised of numerous thin layers. Each layer consists of graphene, which is a lattice-like sheet made of strongly bonded carbon atoms. As a result of these bonds, electrons—generated by wildfire's heat or electricity in our devices—flow very easily across graphite's layers. When fire hits the graphite surrounding a tree, the material's high thermal conductivity transfers heat away from the living tissues beneath the bark. Ideally, this conduction relieves severe hot spots, evenly distributing heat up, down, and around the tree in a way that gives the greatest chance for survival.

While heat energy can move *across* graphite's planes with ease, it has a great deal of trouble transferring *between* graphene sheets, which contain insulating air layers in between them. In this case, we say that graphite has low thermal *diffusivity* because heat cannot flow easily from the hot side to the cool side. Whether or not graphite conducts or resists heat energy is therefore dependent on the direction the energy is trying to travel. This is a life-saving property for plants. It means that the flow of heat generated outside the plant by wildfire is restricted as much as possible from reaching the living core of the plant. By micromanaging the combustion of their bark into the supermaterial graphite, plants enable themselves to both shed and resist heat.

Self-pruning limbs have also evolved in fire-resistant trees as a way to resist fire. In what are called crown fires, fire can ascend into, spread through, and incinerate the

In order to reach lifespans that can approach two thousand years, redwoods prepare themselves for severe fires. Immensely thick bark, the shedding of lower limbs, and the production of epicormic buds and lignotubers are parts of redwood survival systems.

canopies of entire forests by climbing limbs and vines as if they were ladders. Trees such as pines, cycads, redwoods, and eucalyptus routinely shed their lower branches to prevent this disastrous possibility. To shed a branch, the tree uses hormones to signal cells supplying the branch to cease. These cells begin filling with callose, forming an abscission layer that results in the branch dying and falling away. Branch shedding is called cladoptosis after the Greek words *klados* (branch) and *ptosis* (falling). It is similar to autumn leaf fall, when deciduous trees use the hormone abscisic acid to direct the fall of their own leaves. Once these branches are gone, it is much harder for fire to climb upward.

In a third fire-resistance strategy, plants bolster their moisture supply. One way they do this is by growing succulent, moisture-retentive tissues to surround their buds. An excessive extra supply of water is always handy to have when a fire gets going. Succulents' water stockpile cools down and insulates them from the fire's heat, protects tissues from damage, and allows vital life functions to continue. A great variety of succulent

These South African ice plants (genus *Delosperma*) hold moisture as a method of drought and fire survival. Succulent plants' water stockpiles cool, insulate, and hydrate them during fire and excessive heat.

plants in the following families protect their stems and buds in this manner: the agave (Asparagaceae), cactus (Cactaceae), aloe (Asphodelaceae), protea (Proteaceae), ice plant (Aizoaceae), milkweed (Apocynaceae), and stonecrop (Crassulaceae) families. These succulents are the dominant vegetation in fire-prone dry regions throughout the world, such as southern Africa, the arid southwestern United States, and Central and South America. Another method of resisting fire via moisture is performed by broadleaf deciduous trees. Many eastern North American trees such as beeches, oaks, maples, tulip poplars, and sweetgums aren't necessarily the best at resisting fire via their bark, which is woefully thin. Not only do these species live in areas with low and infrequent fire regimes, but their adaptations actively suppress fire. The shape of a plant's leaves is a part of the hidden world of fire survival. Flat leaves generally suppress fire, whereas we'll soon see that needle-shaped leaves tend to be highly flammable. As flat leaves fall to the ground, a thick, insulating, and water-retentive mulch forms. This wet, leafy blanket stifles air flow so well that it can smother fire before it ever reaches the tree's trunk or roots. Each little circle of leaves beneath a tree is thus a fire-free zone for its trunk and feeder roots, protecting both from ignition and heat damage.

Fire Recovery

Sometimes, fire regimes can be so fierce and frequent that plant resistance simply isn't possible. Even with the best bark and water supplies, plant survival systems can still be overcome by massive conflagrations that incinerate hundreds or thousands of acres. In this case, a plant's best hope is to simply preserve a few critical tissues that can recover quickly after the fire has laid waste to the surrounding environment.

One of the best recovery preparations plants make for catastrophic wildfires is the ability to quickly resprout. Epicormic buds, clonal spread, and lignotubers are three types of fire recovery adaptations.

Epicormic buds are dormant, fire-activated buds located underneath a plant's bark. A plant can thus lose all of its leaves and even many branches, but if even one of these tiny epicormic buds survives the blaze, the plant's survival is assured, because it can completely resprout. Fire, disease, or coppicing (the cutting down of an entire plant at once) can trigger the

The longleaf pine–wiregrass savannah ecosystem, seen here, is one of North America's most fire-tolerant habitats. Longleaf pines and associated herbaceous vegetation rely on frequent wildfires to germinate seeds, burn away competitors, and enrich the soil. The pines possess thick fire-resistant bark and buds protected by their long namesake needles. The fine-textured, resin-rich fuels from pines and wiregrass (*Aristida stricta*) help promote regular low-intensity fires across the landscape.

Pando, a single quaking aspen plant, is one of the largest and oldest organisms on our planet. The forest you see here is created from Pando's repeatedly cloned trunks. Extensive cloning is an adaptation for fire survival. Over more than eighty thousand years, Pando has spread over more than one hundred acres in Utah.

sprouting of epicormic buds. The signals that give epicormic buds the green light to grow are combustion products, nutrients, and higher light levels left behind after a fire. Alternatively, the stresses from damage and disease can also stimulate epicormic bud growth. Epicormic buds can be seen in extremely fire-resistant species such as Coastal Plain scrub oaks, cork oak, eucalyptus, banksias, proteas, pines such as the Canary Island pine, and quaking aspen.

Fire can also stimulate the production of large clonal colonies of a plant. When fire removes a plant's stems, it also takes away growing tips, which are little factories for the plant hormone auxin. Among their many functions, high auxin levels induce strong vertical shoot growth, known as apical dominance. The lower auxin availability postfire strongly influences plants into sideways clonal growth. Redwoods, acacias, sumacs, creosote, saw palmetto, gallberry, and wax myrtle are just a few plants that resist fire with prominent clonal growth. Quaking aspens are particularly known for forming huge forests of clones via suckering through a famous example: the single male quaking aspen named Pando. This massive specimen grows in Fishlake National Forest in Utah. We're not joking about massive, either. Genetic testing of the plant tissue has proven that this one plant has cloned itself enough to cover over one hundred acres of forest. Each of Pando's more than forty thousand tree-sized stems—an estimated more than thirteen million pounds of biomass—is a chance for the plant to survive even the most catastrophic wildfire nature can whip up. The tree's giant, interconnected root system has been dated to around eighty thousand years old, also making it one of the world's oldest known living organisms. Pando must've borne witness to more than its fair share of fire during that epic span of time.

The subterranean lignotuber of this Australian mountain devil shrub (*Lambertia formosa*) is a survival insurance policy against fire. While a recent wildfire scorched away the plant's aboveground tissue, fresh buds and sprouts now erupt from the well-protected lignotuber.

Soil is an amazing insulator against the heat from fire, so a lot of plants find ways to hide from fire belowground. Lignotubers are swollen, woody, fire-resistant tissues that protect recovery buds on the roots of plants. Imagine a potato wrapped in a thick layer of bark, and you're not too far from a lignotuber's appearance. Redwoods and Australian plants such as eucalyptus and banksias are particularly good at making them. Herbaceous plants often duck out of a fire's path by relying on deeply buried rhizomes, bulbs, tubers, corms, or taproots. Brackenfern, pine, and fire lilies and grasses such as wiregrass, cogongrass, and pampas grass are just a few of the many examples of this strategy. Some species of rare or endangered plants are so dependent upon fire for their growth and reproduction that they languish during infrequent fire intervals. Trumpet pitcher plants, Venus flytraps, Michaux's sumac, hairy rattleweed, and Georgia aster are just a few examples of some species' dependence upon fire disturbance. Fire-dependent plants such as these require frequent fire regimes to complete specific parts of their life cycles. A fire regime might include clearing away taller woody vegetation, burning down less fire-resistant competitors, clearing bare mineral soil for seed germination, triggering reproductive growth, or providing cues for seed germination.

Fueling the Competition

Once preparations for their own survival are assured, some fire-resistant plants go a step further with adaptations that burn away competing plants. They become pyromaniacal, actively contributing to and fueling more frequent and severe local fire regimes. It might surprise you to realize just how much fuel plants like to throw at their neighbors.

The needles and other tissues of conifers such as pines (especially longleaf pine), cypresses, arborvitae, firs, and junipers are coursing with flammable resin. These trees also drop their needles, which pack together loosely on the ground like a game of pick-up sticks. Plants with long, thin leaves such as conifers and grasses contribute what fire ecologists call fine fuel. Fine fuels conduct fire extremely well. Their loosely packed structure permits ample oxygen flow, and their thin, fine texture dries easily. Fallen needles, especially of pines, are also adept at catching and hanging on the branches of neighboring plants. If you've noticed these hanging needles

This *Eucalyptus camphora* leaf looks like a starscape when backlit. See the numerous clear dots? They're the flammable oil glands of this pyromaniacal genus.

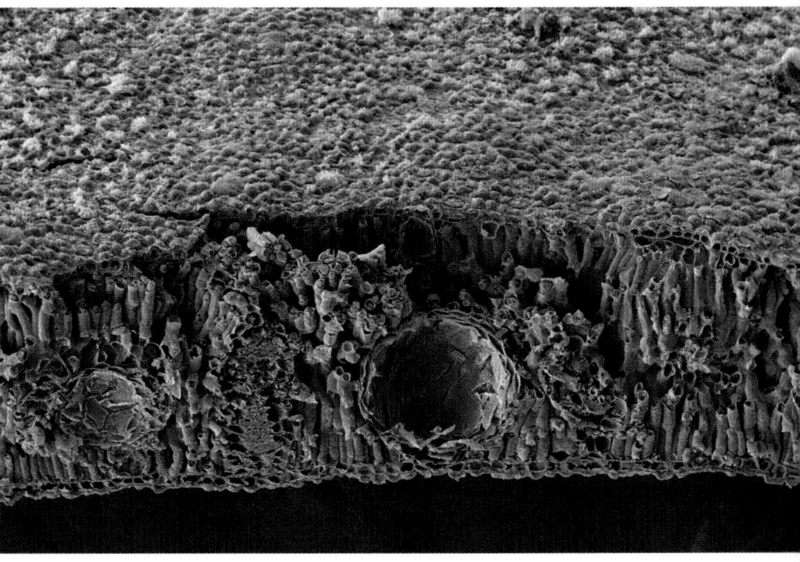

A *Eucalyptus cinerea* leaf side view, showing one of the many huge oil glands located throughout the plant. These glands produce eucalyptol, an essential oil both antiseptic and highly flammable.

or piles of needles on the forest floor, you've seen a landscape primed to burn. Grasslands are another fire-carrying habitat, full of species such as miscanthus, pampas grass, cogongrass, and wiregrass with fine, flammable fuels. Coarse fuel is the term given to larger-sized biomass such as trunks and branches, which take longer to dry, ignite, and combust.

In addition to vegetative fuel, a second common botanical ignition source is oil. Plants with oils inside and outside their foliage, wood, and bark can create hotter and more severe fires with a higher likelihood of killing less fire-adapted competitors. Some of these pyromaniacs include mint family herbs such as rosemary, oregano, thyme, catnip, tea trees, bottlebrushes, tamarisk, sagebrush, creosote, acacias, greasewood, gas plant, Scotch broom, manzanita, and eucalyptus. The flaky strips of bark dangling naturally from eucalyptus species, called stringybarks and candlebarks, promote the formation and lofting of embers. Embers are red-hot, airborne fire starters formed from actively combusting biomass. Each tiny particle burns with the chance to start a spot fire on any unburned fuel it lands upon. The massive convection of a wildfire actually generates its own powerful wind, which lofts flocks of embers far distances. This chance is all the more intensified by the presence of trees that generate thin, flaky, or papery fuels. Ingeniously, many flammable plant oils are highly antimicrobial and decay-resistant. This might seem like an odd trait to mention in a chapter on fire, but when antiseptic leaves drop to the ground, they do one thing very well: they pile up for months or years because they resist decomposition. Masses and masses of this oil-soaked fuel, designed for maximum burn damage on every other plant in range, are simply biding their time until the next big wildfire. As you might have noticed, many of the fire-loving plants take on multiple strategies to survive fire. By beefing up their own defenses and eliminating competition, these plants have the best chance to dominate in these extreme environments.

Rising above the Ashes

It may sound like profligate self-harm for plants to encourage wildfire, but the ones that do so are supremely adapted for high fire regimes. Not only do they pack a hidden world of resistance adaptations, but there are many species whose ability to reproduce and colonize is stimulated by fire. Flowering, seed germination, and seed dispersal can all be triggered by fire, allowing fire-adapted plants to rapidly capitalize on this massive disturbance. Postfire colonization also lowers the chance of fire damaging these delicate features. If a fire has just been through an area, it is a perfect time to put out dainty petals or tiny seedlings.

A fire leaves behind valuable nutrients in its ash, usually withdrawn from the biomass that fueled the blaze. Potassium, magnesium, and phosphorus are notable components of wood ash that fertilize postfire soils. Plant survival systems have adapted to take great advantage of this resource, using chemical cues from fire to trigger flowering. There's no better time to send out delicate buds than when fire has just rolled through. The lack of fuel in the system means these plants are unlikely to experience fire soon after. Flowering is future reproductive growth, leading to seed production, the expansion of territory in fire-resistant species, and survival insurance against future fires. Some of the most rapid fire flowerers are the aptly named Australian fire lilies. The beautiful but diminutive red

Australian grass trees (genus *Xanthorrhoea*), their trunks blackened with soot, fill this fire-prone landscape in the Pinnacles Desert, Nambung National Park, Western Australia.

blooms of fire lilies have been recorded blooming just nine days after severe, landscape-scorching fires, even as smoldering and smoke linger in the backdrop. The cue for fire lily blooming lies in the chemical signature of smoke. When smoke drifts into air pores in the soil or is leached there by precipitation, many plants such as the rain lily initiate their flowers. Australian grass trees (genus *Xanthorrhoea*) and North American plants such as wiregrass (genus *Aristida*), pinegrass (*Calamagrostis rubescens*), and turkeybeard (genus *Xerophyllum*) are similarly adapted. Fire-induced flowerers like these rely upon fire to clear new space for their seedlings.

Fire can also stimulate reproduction by encouraging plants to disperse their seeds. The survival systems of some plants involve building fire-activated seed-bearing structures, a trick called pyriscence. Seed release triggered by any environmental factor is called serotiny. When serotinous cones, capsules, and other types of seed pods mature, they're normally

Lemon bottlebrush (*Callistemon citrinus*), from Australia, has serotinous seed capsules. Plugged with resin, these fire protection chambers keep seeds safe.

These lodgepole pine (*Pinus contorta*) cones have been hanging out in the canopy, full of seeds, for years. Serotinous cones like these won't release their seeds until the extreme heat of a wildfire melts the resin gluing the seed chambers closed.

Everlasting snow (*Syncarpha vestita*) is a fire-germinated species from South Africa.

glued shut with strong, heat-degraded waxes and resins. Only large-scale fire has enough heat to melt these "locks" and unleash a rain of thousands of seeds from these capsules. Using this strategy, plants ensure that their offspring don't meet a fiery, untimely end if they drop prematurely. It also triggers them to open after the system's fuel has been consumed and the risk of fire has passed. This protects the fragile offspring and gives them time to build defenses. Serotinous cones can be found on pines such as the lodgepole, ponderosa, jack, sand, and table mountain pines. Other pyriscent plants include spruces, cypresses, redwoods, eucalyptus species, bottlebrushes, banksias, and proteas. It's quite a smart strategy. Storing seeds in a traditional ground-based seed bank is risky business in high fire regimes. Rather, pyriscent plants can retain a bank of seeds in their canopies for many years, largely protected from all but the most severe canopy fires.

Not all pyrophytes store their seeds in the canopy, however. Seeds can also be adapted against lower-severity fires. They feature thick, insulating hulls with the same graphite-inducing chemistry as bark. However, seeds have one problem: How do they know when fire has passed through and it is safe to germinate? Fascinatingly, seeds of many pyrophytes have adapted to chemically detect cues left behind by fire. In thermal scarification, heat can degrade, crack, or scarify (nick or soften) seed coats, rendering them capable of absorbing water and hastening germination. Fireproof seed coats are usually extremely hard and capable of retaining germination inhibitors within the seed. The fire-facilitated leakage of these inhibitors is another mechanism ensuring that pyrophytes choose their germination opportunity carefully. Smoke and ash are further triggers, each comprised of a complex chemistry that pyrophyte seeds can recognize using specialized receptors. Chemicals from both smoke and ash can carry as aerosols via smoke or leach via water into the soil. When enough of these compounds have accumulated, plant hormones such as gibberellic acid convince seeds that the time is nigh to begin growing. Fire-stimulated seed germination is most commonly found in plants of western and southwestern North America, South Africa, and Australia. North American species include fire poppies, snowbrush, coffeeberry, longsepal globemallow (*Iliamna longisepala*), whispering bells (*Emmenanthe penduliflora*), coyote tobacco (*Nicotiana attenuata*), and redberry (*Rhamnus crocea*). South African plants include the beautiful proteas, leucadendrons, leucospermums, restios, and everlasting snow (*Syncarpha vestita*). Australian plants include fascinating botanical wonders such as triggerplants (genus *Stylidium*) as well as carnivorous plants such as tuberous sundews (genus *Drosera*) and rainbow plants (genus *Byblis*).

Fire remains a tremendous threat to the survival of plants across many biomes of our planet. You could say that plant evolution is a constant trial by fire. The ongoing natural selective pressure and disturbance in the wake of every blaze have forged an impressive array of fire-resistance adaptations in plants. To these stalwart pyrophytes, some of the most horrendous wildfires humans and animals can ever imagine amount only to temporary setbacks. What inspiring fortitude is shown by these tiny noble green sprouts among the black.

CHAPTER 16

Every Drop Counts

Drought Survival

Solar rays beam through the atmosphere, packed full of the hot energy that charges our planet with life. Passing through a virtually cloudless sky, the rays make landfall in a scorching, quiet, and stony place: a desert. Deserts, arid regions defined by their scarce and unpredictable precipitation patterns, are formidable ecosystems imposing incredibly hostile conditions upon plant and animal life. Extreme temperatures, drying winds, and consistently low humidity are associated challenges. Contrary to popular belief, however, these seemingly barren areas are teeming with an astoundingly specialized diversity of plant life. Xerophytes, the drought-tolerant vegetation of dry places, have eked out a parched living for many millions of years. While some of the most spectacular and extreme xerophytes we'll meet live in deserts, xerophytic adaptations are interspersed throughout the plant kingdom.

Plants, like all of Earth's living things, cannot survive indefinitely without an adequate supply of moisture. In fact, not drying out was one of the biggest challenges both plants and animals faced when coming onto land. Drought leads to severe challenges, impairing a wide variety of necessary cellular processes. Plants can overheat, starve, or develop fatal vascular embolisms. Photosynthesis, transpiration, nutrient transport, defense production, resistance to pests and diseases, and general growth and reproduction all fail without water. Let's imagine a plant drying out and think about what happens within it.

The most immediate threat during a drought is the failure of plants' cooling systems. Through a process called transpiration, plants utilize water for cooling. The cost of cooling gets quite expensive, particularly in dry

This desert landscape in Joshua Tree National Park, California, isn't barren. Instead, it's full of amazing plants that can survive long droughts.

environments. Would you believe that only around 1 percent of the water a plant absorbs is utilized for the vast majority of its biological functions? Almost 99 percent of a plant's water budget is allocated toward *cooling*. Water is absorbed from the soil and pulled through the xylem into the foliage, and then it simply evaporates into water vapor as it passes out of stomata in the leaf. This action carries away heat energy, cools the plant's tissues, and ensures the constant movement of water through the plant's system. All plants have diligently used water to cool their tissues each day of their lives, a biological necessity for a terrestrial existence. When the plant's cooling system shorts out, cells begin to overheat. Though less severe than fire damage, drought heating can kill cells by cooking and denaturing their proteins, processes that shut down vital reactions. Open areas of desert soil have been recorded reaching up to 180 degrees Fahrenheit on blisteringly sunny days—just 30 degrees shy of what it takes to boil water! Thankfully, the water-storing plants of arid areas can keep their cooling systems going far longer than most other plants, handling daily temperatures up to 140 degrees Fahrenheit without too many issues. We'll explore some of these water-banking adaptations shortly.

At the same time plants are dealing with cooling issues, a drought soon begins to double as a famine. Among the first reactions to drought stress is that leaves close their stomata. This action cuts water loss via transpiration but comes at a cost. Stomata cannot absorb carbon dioxide for photosynthesis when closed, so energy production begins to fall. Additionally, water is no longer being pulled in at the roots. As previously discussed, water is also a key ingredient for photosynthesis. Without it, plants cannot generate sugars. Leaves are a plant's food factories. Most plants store sugars and starches for short-term shortages, but an unexpectedly long drought can catch them off guard. In extreme drought, leaves wilt and die, absolutely crippling the plant's food supply. Drought impairment of the vascular system can also induce energy shortages. For carbohydrates, minerals, and nutrients to flow and distribute properly, they must be dissolved in water. A supply failure in the vascular system usually means that tissue death is soon to follow. This can have major impacts, even on large, long-lived specimens such as trees or other woody plants. A tree's ability to conduct water through its vascular system is called its hydraulic conductivity. Scientists estimate that, on average, tree mortality during a drought usually occurs when around 60 percent of the plant's total hydraulic conductivity has been impaired.

We've talked a lot about water thus far, but how about air? Desiccation and overheating cause air and vapor bubbles to form in a plant's vascular bundles. This becomes an ancillary threat called cavitation. These air pockets form when the dissolved air in vascular fluid expands due to heightened tension. Air bubbles may not seem awful, but large ones can form an embolism, a situation where air completely interrupts the flow of water throughout the plant because the air severs the chemical cohesive bonds between the water molecules. Unfortunately, it's this attractive force that powers transpiration. The chains of water molecules in a vascular system must remain uninterrupted to work properly. Each molecule, from a plant's roots all the way to its evaporation from the leaves, uses chemical cohesion to tug its successors through the plant. Every drop of water counts in the hidden world.

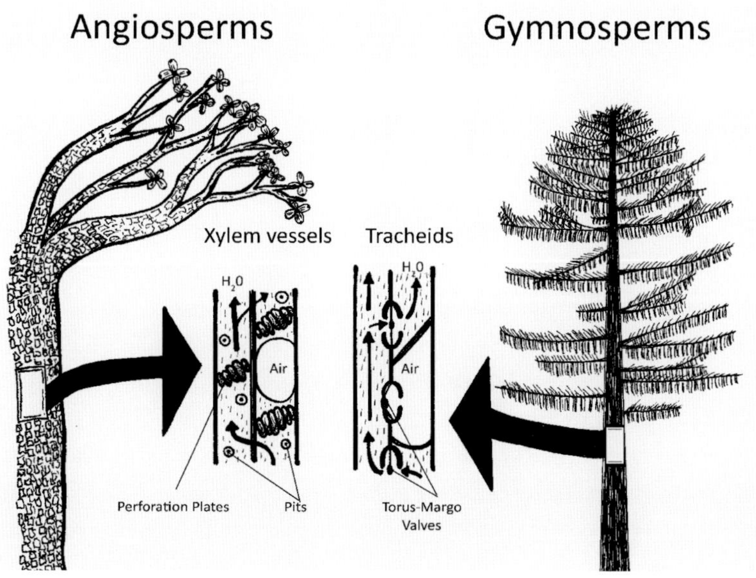

Desiccation and overheating cause air and vapor bubbles to form in a plant's vascular bundles—a threat called cavitation. Air bubbles may not seem awful, but large ones can form an embolism, a situation where air completely interrupts the flow of water throughout the plant. To deal with the threat of cavitation, conifers evolved special valve-like structures on their vascular cells called torus-margo structures. As they dry out, vascular cells depressurize, sucking these valves closed and preventing embolism formation. Angiosperms force air bubbles out with the forceful flow of their sap and build wider and more plentiful transport tissues than they need—a drought insurance plan.

Plants have some ingenious ways to deal with the threat of cavitation. Conifers evolved special valve-like structures on their vascular cells called torus-margo structures. As they dry out, vascular cells depressurize, sucking these valves closed and preventing embolism formation. Angiosperms such as oaks, vines, and maples prevent cavitation by forcing air bubbles out with the forceful flow of their sap and building wider and more plentiful transport tissues than they need—a drought insurance plan. Bubble-blocking valves, sap pressurization, pores between vascular cells, and overnumerous vascular tissues are all effective against embolisms. Both gymnosperms and angiosperms can survive minor air blockages in their vascular systems that might impair individual pipes here or there. Perhaps a few leaves or a branch might suffer as a result. If large or widespread enough embolisms accumulate, however, they threaten to severely interrupt a plant's hydraulic conductivity and cause a complete fatality.

If the direct threats to a plant weren't enough, hordes of herbivores can detect drought stress in plants and initiate an invasion. Plant food remains among the most reliable sources of water for many animals, especially when surface water sources, dew, or precipitation are scarce. Furthermore, as water volumes decrease inside plant cells, the sugars and nutrients dissolved within begin to concentrate into a sweet and seductive "bug candy." When hydrated, plants deliberately try to starve herbivores and diseases by dissolving their valuable minerals, sugars, and proteins in excessive volumes of water. This is not possible when plants are drought stressed. Adequate water is also necessary to keep plant cells under extremely high pressures. When properly hydrated, plant cells become so pressurized that insects feeding on them might as well be drinking from fire hoses. Thus, severe drought begets a whole domino effect of invasions and stressors for plants.

The availability of water, concentrated food, and high plant temperatures during drought allows pest growth and development to skyrocket—and bugs know it! Insects cue in to drought stress the same way sharks detect blood in the water. As cavitations occur in woody plants, the immense negative pressures developing inside plant cells cause them to emit unique ultrasonic acoustic signals. To us, these unavoidable teeny pips and squeaks are an imperceptible part of the hidden world. Insects, however, can easily hear the high-frequency signs of cavitation from more than fifteen feet away. If this weren't enough, plants begin emitting distinctive VOCs during drought such as ethanol and alpha-pinene, compounds that act respectively as an emergency energy source and a growth inhibitor. Evolution has equipped herbivores such as boring insects with receptors that actually detect these signature chemical vapors. Bark beetles and the like flock to drought-stressed plants, knowing full well their defenses may be weakened. Dry plants are often defenseless plants. Without adequate water, plants have a tough time distributing many of their major defensive chemicals, since many are dispensed via cellular fluid. Other defenses, such as the protective oleoresin in bark and wood, cannot be produced without a supply of water. The next time you notice a plant wilting, take pity upon it. Its situation has become dire for many reasons, not the least of which is the voracious descent of everything on the planet that wants to eat it. Hopefully, the plant will receive some much-needed hydration before it reaches the wilting point—the point of no return.

Sounds scary, doesn't it? Thankfully, plant survival systems across the planet have come to the rescue with a multifaceted approach to overcoming a drought. Xerophytes have altered their morphology, physiology, and biochemistry to emphasize water conservation. The solutions they have crafted span two general categories: either plants endure a drought or they avoid it entirely.

Shaped for Survival

When you think of the plant life in a desert, you might conjure up the image of cacti and other unusually shaped and sized plants. Xerophytes purposefully take on a characteristic appearance that enhances their drought survival. A plant's physical features, including their shape and size, are known as its morphology. In arid climates, plant morphology is cleverly adapted for water conservation.

In arid climates, plants are perfectly shaped for water conservation and heat reduction. The spherical shape, thick cuticle, ribbed and photosynthetic stem, and spiny leaves of the golden barrel cactus (*Echinocactus grusonii*) are all part of its drought survival system.

The first thing you might notice is that desert plants have evolved succulent tissues. While all plant cells are filled with water, succulents take water collection to the extreme. The humble houseplants many of us have on our windowsills are among the best water storage devices on Earth. We love these succulents for the plump, fleshy, water-packed storage organs that also keep them alive. Depending on the species, succulent storage may be channeled to leaves, stems, swollen basal structures called caudexes, or protected underground structures such as taproots and bulbs. Cacti, carrionflowers, and euphorbias, for example, have plump stems that hold plentiful stores of water. Ice plants, the stonecrop family, aloes, and agaves, on the other hand, predominantly use their swollen leaves for water storage. An unrelated grouping of plants called caudiciforms are famous for their burgeoning, water-storing trunks. These include the baobab tree, desert rose, Madagascar palm, Buddha belly plant, and ponytail palm. Xerophyte cells are amazing attractors of water, utilizing tricks such as the hydrophilic attractance of salt molecules. An above-average level of salts in a root cell, for example, exerts a strong electrochemical attraction on water molecules, even in saline or dry soils. Xerophyte cells also produce abundant water-conducting proteins such as aquaporins,

Where are all the leaves? This prickly pear cactus is a prime example of how drought evolution drives the shape of xerophytes and succulents. Cacti such as this tick all the right boxes for survival: reduced surface area, flattened photosynthetic stems called pads, leaves reduced to mere spines, fleshy tissues, thick cuticles, CAM photosynthesis, and widespread surface roots.

The ice plant (*Delosperma nubigenum*) is a leaf succulent. Other members of its drought-tolerant plant family include the living rocks.

which help shuttle water through cell membranes. More than twenty-five plant families include succulent members, some of the most recognizable including the stonecrops, cacti, living rocks, euphorbias, aloes, agaves, and milkweeds. Some of the lesser-known families with succulent members might even surprise you—there exist succulent relatives of grapes, passionflowers, lilies, cucumbers, pineapples, figs, and many more.

A desert's precipitation can be fatally extreme and unpredictable. Flash floods can occur after years of drought, or a few light sprinklings can be all a habitat receives over an entire year. To endure these ordinarily fatal fluctuations and scarcities, succulents are prepared to quickly absorb and store as much moisture as their tissues allow. Even a single good rain allows xerophytes to accumulate enough water to survive years of drought. The giant saguaro cacti of the American Southwest are famed for their water absorption ability. Mature ones are known to be capable of packing away over two hundred gallons from a single storm. As succulents desperately flood their tissues with any available water, internal pressures can soar high enough to cause leaves and stems to burst. To reduce this threat, succulents often feature strengthening ribbing and fluting along their stems. This shape contrasts with the mostly cylindrical stems of plants from less drought-prone habitats. Xerophytic ribbing allows plant stems to expand and shrink like an accordion. In contrast to a spherical construction, a ribbed or fluted construction allows a plant to rapidly expand in order to hold an absurd volume of water. During severe drought, ribbing allows the stem to shrink and retract, minimizing sun exposure. Ribbed xerophytes can perform these rapid expansions and contractions with a reduced risk of bursting, whereas normal plants lack the ability to swiftly and safely store vast amounts of water. Plants in which an accordion design is most apparent include succulent cacti, euphorbias, and stapeliads.

Spines, often found in desert plants, not only are protectors but also double as fog collectors. Spines and trichomes often provide superb surface area for moisture condensation. This *Parodia leninghausii* cactus has collected a thick dew on its spines, dripping this precious moisture straight to its roots.

Our favorite drought-tolerant plants are members of the ice plant family in South Africa. Called living rocks or mesembs (many genera, *Lithops* being the best known), these cryptically colored, diminutive plants can survive for years without a single drop of rain. Each plant, humbly looking like a small pebble at the soil's surface, consists of only two leaves a year. This is among the most extreme reductions of leaf surface area for water conservation. Before the heat of summer, the old set of leaves withers into a papery husk. During this intricate process, the old leaves channel every molecule of their moisture and food into the new leaves, which are protected as they develop inside the fleshy core of the plant. This is because *Lithops* leaves are adept light diffusers. *Lithops* leaf cells hold huge amounts of water as well as prismatic, reflective calcium oxalate crystals. As intense desert light, heat, and UV radiation pass through a *Lithops*'s large outer leaves to its core, the plant's cells cool, diffuse, and refract the intensity. *Lithops* and their close relatives aren't the fastest-growing plants around, but this conservative approach has survived the test of time in some of the driest places on our planet, such as the Namib Desert in southern Africa.

As it has with living rocks, the xerophytic way of life has led to extreme reduction of the surface area plants expose to the sun. Certain leaf and stem shapes recur in deserts across the world because they enhance a species' drought fitness. Rounded or upright shapes such as spheres, cones,

Living rocks or mesembs are cryptically colored, diminutive plants that can survive for years without a single drop of rain. While many genera of plants are called living rocks, the best known genus is *Lithops*, seen here. Each plant, humbly looking like a small pebble at the soil's surface, produces only two leaves a year. Their slow growth and extreme leaf surface reduction help them conserve water. A living rock can survive for years without precipitation.

pads, and columns absorb less heat and use less water than the flat or horizontal leaves of more mesic plants. Numerous desert plants, such as cacti and euphorbias, follow this design. Cacti have gone a step further, reducing their leaves almost entirely to nonphotosynthetic spines. Living rocks display perhaps the most extreme reduction, lacking discernible stem tissue and possessing just two recycling leaves per growing season. In the genus *Ephedra*, leaves are virtually absent (a phenomenon called aphylly), and stems perform the majority of photosynthesis. There are also several intermediate morphologies. Microphylly, where leaves are tiny, fleshy, and rounded, occurs in plants such as the elephant food bush (*Portulacaria afra*) and creosote. In sclerophylly, leaves are narrow, thickened, hardened, spiny, and borne tightly on short internodes. Sclerophyllous plants include tea trees (genus *Melaleuca*), bottlebrushes (genus *Callistemon*), and many plants from dry, scrubby regions.

Some drought endurers can reduce their exposed surface area on demand instead of making permanent changes to their size or shape. They use simple tricks to avoid unnecessary sunbathing. Wilting, a common plant adaptation, occurs when water loss drains the turgidity of leaf cells. Wilted plants look sad, but wilting is a potentially life-saving adjustment that helps shrink the photosynthetic area available for the sun to beat down upon. Wilting isn't always a good thing, however. Each plant has a

This Arizona cypress (*Cupressus arizonica*), native to the dry southwestern United States, resists moisture loss with microphylly (leaves reduced to tiny scales) and a white, sunblocking coating of wax.

This Australian lemon bottlebrush (*Callistemon citrinus*) is sclerophyllous. Its leaves are small, thick, hardened, and defensively spiny. They're also oily and flammable, but that's another story!

This *Eucalyptus camphora* likes to keep a low solar profile. Its droopy leaves hang vertically, offering very little surface area for the sun to beat down upon. This is especially true during the hot midafternoon when the sun reaches its highest angles or when the plant becomes water-stressed.

limit, called its permanent wilting point. If the moisture level in the soil dries beyond this point, a wilted plant cannot recover and will die even if given moisture afterward. This is because the chain of water moving through the plant has been permanently broken and cannot grab any free water molecules. Monocots, including families such as the grasses, have a slightly different version of wilting. Drought-stressed monocots invoke specialized cells in their foliage called bulliform or motor cells. When these hinge-like cells become dry, they roll monocot leaves into low-profile tubes. Heliotropic leaves are a third adaptation, seen in leafy, evergreen xerophytes such as creosote and eucalyptus. In these plants, the leaves appear wilted to different degrees and angles throughout the day. Heliotropic leaves might look sad and droopy, but the reason is entirely different from wilting. Rather, heliotropism allows leaves to track the sun's angle each day and shift their orientation accordingly. These micromanaging plants can avoid sunshine at will by keeping their leaves parallel to the sun's rays. To minimize water loss, the leaves droop almost vertically in the midafternoon, when temperatures and the sun's angle are highest.

We've taken a pretty good glance at drought-adapted plants from a distance, but let's get a bit closer and examine their surfaces. These surfaces are highly adapted to reduce moisture loss. The thick, strong, and waterproof cuticles on xerophyte cells help waterproof them and hold in the enormous pressures of water storage. Coatings of wax act like sunscreen, building up into a heavy white or gray layer that reflects light and UV radiation. Dense spines and trichomes not only are herbivory defenses but also dissipate excess heat by acting as reflectors and heat sinks. Desert plants with these

Seen here is a prickly pear cactus (genus *Opuntia*) pad with its cuticle and water storage tissues open for examination. Note the thickness and strength of its papery cuticle as the halves are pulled apart. Succulents like this can heal their wounds and root into new plants after replanting.

This *Eucalyptus cinerea* has stems and foliage densely coated in wax for drought protection, earning it the common name silver dollar gum.

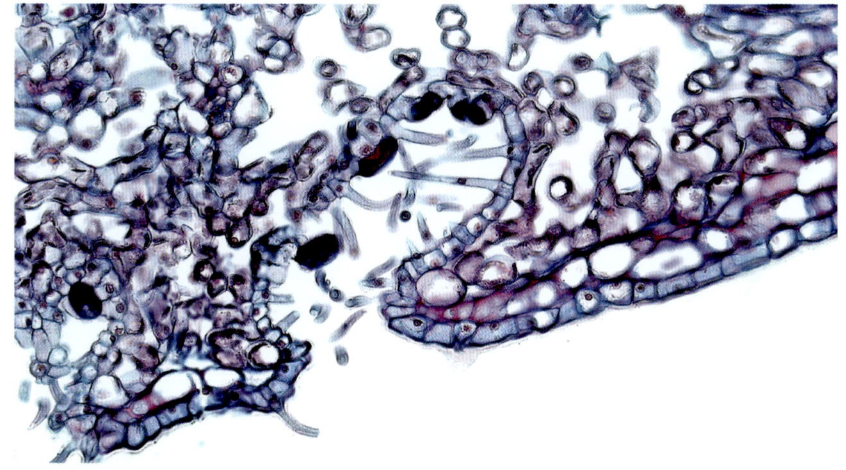

Seen here is the underside of an oleander (*Nerium oleander*) leaf. The protected depression here, studded with dark-colored stomata, is called a stomatal crypt. No matter how dry the air or severe the winds, these little protected bubbles of air are always humid, shaded, cool, and pleasant places. The stomata can open for photosynthesis with little fear of moisture loss.

adorable, almost furry coverings are creating a humid, cool, and shaded microclimate around their tissues, even in severe wind. Below these fuzzy or spiny microcanopies, a plant's surface temperature can remain up to fifty degrees Fahrenheit cooler even in direct sun. Woolly cacti, such as the feather cactus, are a great example of self-shading morphology. Another protective structure can be found underneath a xerophyte's leaves. If a xerophyte chooses to maintain foliage over dry spells, it helps to sink its stomata inside special protective pits called crypts. Stomatal crypts are tiny depressions underneath leaves. Packed with stomata, crypts are shaded from the sun's relentless radiation. Crypt entrances are tightly shielded from heat, dry air, wind, and debris by tiny trichomes. In this protected microenvironment, the stomata lose very little water when open, even in very high winds. Species you'll commonly find crypts on include oleander, conifers, and banksias.

Xerophytes also use the shape of their surfaces to channel and collect any moisture that might become available, no matter how little. This strategy can be so efficient that entire plant species have dedicated their collection efforts toward fog and light drizzles. While fog collection has its challenges, this is sometimes the only source of precipitation available in some of the world's driest deserts. The Atacama Desert (South America), Baja California Desert (Mexico), parts of the Arabian Desert (Middle East), and Namib Desert (Africa) are regions where fog collection becomes a common part of plant survival systems. A hidden world of microscopic

This feather cactus (*Mammillaria plumosa*), unlike most cacti, has become cute and cuddly. Its spines aren't puncturing devices but instead are soft, white, and shaped like umbrellas. They're designed to reflect and dissipate heat, shade the plant's surfaces, and collect fog.

The tiny spines on prickly pear cacti (genus *Opuntia*) are microscopically barbed, spirally arranged, and laced with water conduction grooves. The spines are superb puncturing devices, and they collect and drip moisture to the plant's surface roots. This water conduction works so well that, regardless of a spine's direction, droplets always flow toward the spine's tip. Drops will even fight against gravity to do so!

adaptations enables a plant to condense fog into water droplets and then channel them straight to the plant's thirsty surface roots. The spines and trichomes of fog-harvesting cacti (genera *Opuntia*, *Mammillaria*, *Parodia*, *Copiapoa*, *Oreocereus*, and *Cleistocactus*) are intricately crafted with microscopic barbs, a spiral arrangement, and tiny grooves and channels. The same adaptations that puncture herbivores simultaneously conduct water droplets so well that they can move them against gravity! No matter how these tiny spines are positioned, moisture will condense on them and always flow toward the structure's tip.

Water Conservation

If we look inside xerophytes, we'll find even further layers of drought defense. Drought endurers have discovered a special set of metabolic adaptations that help them tightly regulate water use. They've also developed special proteins that help them relieve drought stress.

Crassulacean acid metabolism, or CAM photosynthesis, enables plants to avert the problems caused by keeping their stomata closed for long periods. As temperatures soar, CAM plants close most of their stomata until nightfall. CAM photosynthesis is like holding a deep breath for hours, with the goal being to clamp down on water loss via the stomata. Thick cuticles, low surface areas, water storage, and other adaptations ameliorate the heat buildup this closure generates. Carbon dioxide gas is necessary for photosynthesis, so the stomata instead open for puffs of the cooler nighttime air. Each leaf cell builds up an overnight stockpile of carbon dioxide that it will use for the light-dependent reactions of photosynthesis the next day. It's a slower and less efficient method of growth, but it's a life-saving adjustment. There are hundreds of CAM plants in dozens of families, including a majority of succulents and desert plants. Some of the most advanced CAM users reside in many families we've already discussed, such as the living rocks, cacti, agaves, bromeliads, milkweeds, euphorbias, aloes, and stonecrops.

As a side note, desert plants have also cracked down on a particularly water-inefficient process that all plants perform during photosynthesis. It's a process called carbon fixation, wherein carbon is separated from carbon dioxide before being used to produce sugars and other carbohydrates. More than eight thousand plant species around the world have a special type of carbon fixation called C_4 fixation. This process efficiently uses water during drought and high temperatures, whereas the more popular alternative, C_3 fixation, uses around four times more water for the same task. Completely detailing these extremely complex chemical processes is beyond the scope of this book, but the major way C_4 reduces water usage is by eliminating waste. Both C_3 and C_4 fixation utilize a critical enzyme called RuBisCO, but C_4 cells make much more efficient use of it by creating an optimal low-oxygen and carbon dioxide–rich cellular environment for RuBisCO to operate in. C_3 fixation is more common because it is the evolutionarily more ancestral pathway for plants to fix carbon. C_4 is a more recent strategy that enables water conservation in more than nineteen plant families, including grasses (corn, sorghum, sugarcane), sedges, euphorbias, and amaranths.

Rooting for Moisture

As you might surmise, one of the most important parts of a drought survival system is a plant's roots. This collection of adaptations for moisture withdrawal is always buried below the desert sands. Xerophyte evolution has driven root systems toward two extremes: surface roots and taproots. To fight for an existence in some of the driest soils in the world, plants grow either deep-drilling taproots, web-like mats of surface roots, or a combination of both.

A spread of radial surface roots is specially adapted to take advantage of light, mist-like rains, fogs, and other brief precipitation events. These networks of extremely fine absorption roots spread like spiderwebs just barely underneath the soil's surface. Water collection morphology, such as funnel-shaped rosettes, spines, and trichomes, often channel water toward this type of root system. There isn't a point to burrow much farther down because most desert moisture disappears quite quickly. Unfortunately, the low humidity and high temperatures can evaporate precipitation away as soon as it falls. Surficial root systems are often accompanied by plentiful water storage tissues so that the plant has a chance to persist until the next light moistening. For that reason, you'll most commonly find radial surface roots in succulent plant families such as the cacti, euphorbias, agaves, aloes, stapeliads, stonecrops, and ice plants. It sounds like an excruciating existence, but fortunately this strategy works quite well for these remarkably adapted plants.

Plants that use extremely long roots to reach into water far below the soil surface are known as phreatophytes. Phreatophytes are often xerophytes, but not always. The goal of this type of root system is to hunt for groundwater and establish a plentiful, reliable "moisture mine." This underground jackpot, equivalent to a lifetime supply of water, consists of two subterranean areas called the water table and associated capillary fringe (the term given to the moist area right above the water table). For the roots to reach either, they must drill dozens of feet down. It's akin to winning the lottery in an arid habitat and can take a single plant years to accomplish—if it survives long enough. However, this dream is entirely within the realm of possibility for phreatophyte root systems. One of the most successful phreatophytes is the mesquite tree, holding the current world record with a more than eighty-foot-deep root network. Another unusual phreatophyte is *Welwitschia mirabilis*, the tree tumbo of the Namib Desert, whose strange, ribbon-like leaves sprout atop a massive taproot that serves as the plant's own private well. Some plants, such as creosote, are known for a hybrid root system consisting of both fine surface roots and formidable groundwater-tapping roots.

Hydraulic redistribution, a moisture-bartering system for plant roots, is a fascinating capability of xerophytes as well. Users of this trait can take the moisture they find from a successful, hydrated part of their root system and spread it to parts that may be drying out and dying from lack of moisture. This works because hydraulic redistributors have special vascular connections within their root systems that other plants lack. The most common situation for this behavior is in phreatophytes, whose deep taproots have successfully reached a groundwater supply. Prolonged hot and dry surface conditions can desiccate and kill surface roots. When a surface root is under severe moisture stress, it exerts a pull on specialized xylem channels to draw moisture from the taproot. This connection goes both ways—water collected from a strong surface rain can be funneled to help a struggling taproot. Hydraulic redistribution can sustain these drill-like taproots for many years, allowing them to persistently bore through endless feet of dry sand before finally winning the jackpot—a lifetime supply of groundwater!

Contractile roots give desert plants the ability to retreat underground during a drought. A contractile root system is thick and fleshy, positioned directly underneath a plant's crown, rosette, bulbs, tubers, or corms. Contractile roots are moisture sensitive, their fibers exerting vertical forces on plants as they moisten or dry. A precipitation event may cause the roots to swell, briefly popping the plant's leaves above the soil for photosynthesis. Contractile roots shrink as they dry, so they tug the plant underground during droughts. The tension force that contractile roots can bring to bear is great enough to shove small pebbles and soil particles out of the plant's path of retreat. This creates a snug underground channel for the plant to inhabit for a protracted length of time. Contractile-rooted species are gamblers, hoping to dodge just enough moisture loss to survive a drought. Plants that make prominent use of contractile roots include living rock genera (*Lithops, Fenestraria, Conophytum*); haworthias; rock cacti (Ariocarpus, Frailea); xerophytes in the lily, iris, and amaryllis families; and many more.

The Drought Dodgers

For some xerophytes, drought is too powerful to be reckoned with. Rather than give up the ghost completely, however, these plants simply find ways to put their growth on pause until better times. These types of survival adaptations are called drought escape or evasion. Let's meet some of the species that have adapted parts of their life cycles specifically to dodge drought.

Annual plants, which live for a single growing season, are among nature's best drought evaders. These species have naturally short life cycles and fast growth rates. To compensate, they produce a prodigious amount of seed by the end of their lives. By flinging out so many propagules, annuals are incredibly responsive to extreme environmental conditions such as drought. Unpredictable rain, dry soils, or heat waves? No problem! Annuals make the investment to ensure that a few seeds are always lying in wait for the optimal time for germination. Annuals can pop in and out between active growth and a dormant seed bank to dodge drought. As proof of concept that their strategy works, the number of annual plant species in a habitat is strongly correlated with the quantity and reliability of precipitation. Deserts and seasonally dry regions have relatively high annual species numbers and diversity. Comparatively, wet areas such as rainforests favor long-lived perennial species and stress seedlings with light competition from larger plants. Annuals have evolved into experts at identifying and capitalizing on niche microclimates. Their life cycles can be completed in an incredibly short period of time, which allows them to take advantage of any sudden shifts to favorable conditions. For example, a precious few weeks of surface moisture after a brief drizzle might be all an annual needs to germinate, grow, and set the next generation. Even an ephemeral puddle can be sufficient habitat.

While relying exclusively on their seeds confers enormous advantages, annuals are taking quite a risk if the seed bank is wiped out for any reason. Therefore, annuals have evolved clever seed dormancy tricks. A seed must successfully pass through a drought and then detect and germinate in favorable conditions. It is not uncommon for xerophyte seeds to remain viable for decades, dutifully biding their time for a perfect forecast. Xerophyte seeds are preloaded with germination inhibitors, which are chemicals that prevent the seeds from growing. These chemicals are often water soluble or affected by temperature. Annual seeds are, therefore, among the best botanical detectors of moisture, shade from nurse plants or natural features, and mild seasons. Inhibitor concentrations are calculated in advance such that only large supplies of moisture and cool temperatures can cause the seeds to spring into action. Seeds are on the lookout for several light mistings and tremendous flash floods, each event providing enough moisture to unlock the embryos within the seeds and sustain them through a brief but full growing cycle.

In a second drought avoidance strategy, some plants are fully prepared to sacrifice water-wasting tissues such as their leaves during dry spells. Such plants are said to be drought deciduous. Drought deciduous plants profit during wet seasons, funneling the food their productive leaves create into storage tissues such as trunks, caudexes, and taproots. A canopy that is green and active during spring can be completely shed in a hot, dry summer. Unusually, the trunks and branches of drought deciduous plants remain green and photosynthetic so that food production continues, albeit at a highly reduced pace that limits water loss. Examples of species with this strategy include baobab trees, palo verde, mesquite, the silk floss tree, brittlebush (*Encelia farinosa*), burro-weed (*Ambrosia dumosa*), cheesebush (*A. salsola*), acacias, ocotillo (*Fouquieria splendens*), and vegetation of southern Africa and the Madagascar spiny forests (genus *Pachypodium*, families Didiereaceae and Euphorbiaceae).

In this chapter, we've traveled through the parched world of drought survival. From their humble aquatic origins, plants have adapted to endure life in some of our planet's driest places. Xerophytes have unique ways to keep their cool, beat the heat, and extract every drop of water they can wring from the environment. Along the intense evolutionary journey this required, xerophytes assumed shapes and functions dramatically different from anything that has ever existed. In their unquenchable quest to stay alive, xerophytes have earned their place among Earth's most resilient plants.

CHAPTER 17

Ice, Snow, and Plantifreeze

How Plants Survive the Cold

Each year, winter's grasp settles upon much of the planet, causing dropping temperatures and the chance of ice or snow. How do you cope with these changing conditions? Perhaps you move to a warmer location, curl up by a fire, or don some extra layers of clothing. Would it be advisable to stand outside in the elements with no protection? Fortunately for humans, we are mobile creatures that can avoid these dangerous threats. For organisms anchored to the ground, there is no escape. Cold is an inevitable force of nature across much of the globe. When frost, ice, snow, and sleet are imminent, plants must prepare internal solutions to these challenges. The freezing of water may seem inconsequential to us, but ice is one of the greatest challenges plants have to endure.

Ice Isn't Nice

Ice puts a lot of pressure on plants—literally! They face this pressure internally and externally. Plants, like humans and other life-forms, contain mostly water. Water is critical for plants. It is one of the primary inputs of photosynthesis, their sugar generation technique. Thus, plants need a continual supply of water for this process. This high concentration of water poses a problem at low temperatures because when water freezes it expands. Plant cells are like little water balloons filled with the essence of life. If a cell gets too cold, it simply can't contain the expanding water and bursts. Oddly enough, plants also face the threat of dehydration when they're frozen. The water supply becomes essentially too solid to pull through the vascular system. Imagine if you couldn't self-regulate

Temperate trees shed snow loads with branch and leaf adaptations.

temperature and your inner workings froze. That would pose an immediate threat to your survival. Clearly, this is an issue plants have to contend with.

If we look externally, we can see that ice continues to pose problems for plants. Belowground, ice can create a significant amount of external pressure. When water freezes in the soil, it can exert enough force to crush roots. It can even cause frost heaving, which tugs plants out of the ground altogether. This happens when layers of the soil swell upward as the water contained within them begins to freeze and expand. Aboveground, freezing water can cause issues as well. The heavy weight of snow and ice building up on leaves and branches can often send them crashing to the ground. In extreme cases or with the help of wind, this force can cause an entire tree to topple. As we have seen, being inextricably linked to water becomes a serious liability for plant life when it gets cold.

The Floral Freeze Fighters

Yet across the planet, plants have colonized all manner of chilly habitats, from high elevations all the way to very northern latitudes. What are their secrets for internal cold protection? Plants that live in cold regions do a large amount of preplanning for plummeting temperatures. Some plants have evolved ways to manipulate the properties of chemistry. One way is by concentrating solutes such as salts, sugars, and special proteins inside their sap. This allows plants to lower their internal freezing point. Plant cells are less likely to freeze and rupture with this "plantifreeze" in place.

You're familiar with this type of plant defense if you've ever eaten pancakes or waffles, not the doughy treats themselves but rather that magically delicious maple syrup. Maples are a great example of a family of plants whose addiction to sugar is saving their lives. During the growing season, maple species produce sugars via photosynthesis and stash the sugars in their trunks and roots. When the maple sap rises in the spring, these concentrated sugars protect the trees from cold damage as they begin growing. This sweet sap hasn't gone unnoticed by humans. Maple tapping is an ancient practice that has a long history among the indigenous people of North America. European colonists continued to refine the production of maple syrup into one of the tastiest industries around.

By the end of the seventeenth century, European settlers and fur traders were heavily involved in the harvesting and production of maple syrup. This product was a convenient and less costly domestic substitute for cane sugar, which had to be imported from the West Indies. Rather than incise gouges into the trees, Europeans improved the safety and yield of the tapping process by boring holes in the trees using augers. Trees were far less likely to be girdled or killed because of this improvement. The sap was channeled into watertight buckets, then boiled in shallow metal pans to sterilize and reduce the watery sap to syrup. Maple tapping continues to evolve. Technological advancements brought to bear over the centuries since have included filtration methods that reduce contaminants, more efficient heating for boiling the sap, plastic tubing and vacuum pumps for higher collection rates, and pest management strategies. With modern technology and best management practices, a maple production farm, or sugarbush, can be expected to yield 3.2 gallons of sap per tree per day during the short tapping season between late winter and early spring. This amounts to merely 7 percent of a tree's total sap volume, making the tapping process quite sustainable. Trees typically begin to be harvested at thirty to forty years of age and continue producing sap annually until age one hundred.

The sap of maples (genus *Acer*) is rich in sugars, which act as a natural antifreeze during cold temperatures. This survival adaptation, along with further concentrating the syrup via a little boiling, is what renders maple syrup sweet.

While a good example, maples aren't the only species with such a sweet tooth. Birch sap is similarly tapped in the boreal regions of North America as well as in northern China. Humans turn this successful winter defense into birch beer. Walnuts, hickories, sycamores, box elders, and ironwood are also plants humans have historically kept "on tap." While these plants are more widely known for the richness of their sap, all plants concentrate sugars from photosynthesis. The ones that are better able to concentrate these solutes into their tissues are the ones that survive in colder environments. Though this process comes with inherent risks of increased pressure from pests looking for this sweet substance, it is worth the risk in a colder environment for the protection it provides. Depending

Tapping of birch sap.

on what's available to a plant in its immediate environment, some plants may also use the salty nutrients they extract from the ground for the same purpose. Desert plants such as cacti are particularly adept at this.

Another ingenious way plants avoid freeze damage is by moving their water outside their cells before temperatures drop. This is called extracellular freezing or freeze-induced dehydration. By dehydrating their cells just a little, plants avoid the internal pressure damage from ice. They do this by secreting a bit of water out of their stomata (leaf pores). The stomata then close tightly, preventing any intrusion from expanding ice. Plants can also pack the water into spaces between their cells to prevent damage. Once the water warms into a liquid, it's right where the cells can quickly absorb it again.

Great laurel (*Rhododendron maximum*) leaves in the spring, warm and happy.

Great laurel (*Rhododendron maximum*) leaves becoming "living thermometers" as they undergo freeze-induced dehydration.

Most temperate-zone plants have evolved extracellular freezing. Some of the most prominent examples of this strategy in North America have often been called "living thermometers." Leaves of such species change shape in direct response to outdoor temperatures. Examples are evergreens such as rhododendrons and conifers. Their leaves remain green and alive all year, braving winter temperatures. You can tell when a rhododendron is experiencing an extracellular freeze because the leaves will become frosty, curled, and droopy when the temperature drops below freezing. This defense operates day and night. You may begin to notice many plants around you performing this strategy on cold days. Overall, it's a risky strategy for plants to employ. Since sunny winter days can damage the dehydrated leaves, plants must balance how much water they excrete and absorb each day.

We've seen how plants deal with water inside their tissues, but what about the weight of ice and snow buildup outside on their leaves and branches? On average, a one-inch depth of snow weighs about 1.25 pounds per square foot. As this snow accumulates, it can become a lot more weight

It's all in the structure! The strongly pyramidal shape, open branching, and needle-shaped leaves of these lodgepole pines (*Pinus contorta*) are shedding a heavy snow before it causes breakage.

The natural flexibility of woods such as spruce and birch enables them to bear weight and slough off snow. It also enables the vibrations that provide resonating soundboards in musical instruments such as guitars, stringed instruments, pianos, and drums. Woods with these interesting acoustic properties are called tonewoods.

than the plant bargained for as it was growing. While walking in the woods after severe ice and snow, you'll often hear branches snapping. Even if snow loading doesn't cause breakage, it can cause permanent bends, uprooting, and enough stress to invite pest and disease pressure.

Temperate trees have become remarkable engineers in order to shed those unwanted pounds. Evergreen conifers retain up to 80 percent of their leaves in order to photosynthesize during winter—a gamble that risks collecting too much snow. Their leaves look quite different when compared to angiosperms. Conifers' thin, needle-like leaves have a smaller surface for ice to collect on than broad leaves. The snow that does collect slides off because the leaves are coated with a slick, waxy epidermis and are flexible enough to bend downward under weight. Pines, firs, spruces, yews, and hemlocks have this property.

Flowering plants in the willow and birch families are also notably elastic. The secret to this lies in the interlocking structure of the trees' natural grain, which enables it to uniformly distribute weight. In an interesting twist, flexibility renders a wood breakage-resistant while also imparting amazing acoustic properties. This springiness of the grain allows these trees to vibrate and resonate sound much more than other types of plants. Woods such as spruce and birch are integral to the manufacture of soundboards for musical instruments such as acoustic guitars, pianos, and drums. Humans have taken willow trees' pliability in another creative direction. The flexibility of their twigs renders them suitable for the production of wicker objects such as baskets and furniture.

Some gymnosperms, such as bald cypresses, larches, gingkoes, and dawn redwoods, opted against keeping their needles in the winter. They are deciduous, which means they lose their leaves completely in the fall. It works just as effectively at reducing winter injury, so much so that a vast number of flowering plant families are also deciduous. Deciduous plants dodge the risk of winter damage, but sacrificing leaves deprives them of the ability to photosynthesize in winter. Growing new leaves every year also has a higher nutrient cost. However, deciduous species do the best they can to extract all valuable resources from their leaves before signaling them to fall. The chemicals plants produce during the growing season are full of nutrients that plants can mobilize elsewhere in response to seasonal cues. Many of these chemicals have survival purposes, one example being pigments. In the fall, deciduous plants begin the deconstruction of the pigments they used to photosynthesize and protect their tissues from UV

As deciduous plants withdraw nutrient-rich pigments from their leaves to survive the cold, temperate regions are treated to the natural spectacle of fall color.

The branch and needle structure of this Sitka spruce (*Picea sitchensis*) is optimally engineered for shedding snow. The intricate feather-shaped placement of branchlets and waxy leaf surfaces slide snow right off. A natural flexibility causes spruce branches to droop under the weight of snow, enhancing its transfer to the ground. By reducing wide, flat surfaces for snow to pile up on, evergreen trees keep their leaves accessible to sunlight and their branches safe from damage.

radiation. As these pigments retreat from the leaves, they reveal a beautiful gradation of color changes that we call the natural spectacle of fall color. Fall color is produced by the complex interaction of pigments such as chlorophyll (green), tannins (brown), flavonoids (yellow, red), carotenoids (red, orange), and anthocyanins (purple). Environmental factors such as light, hydration, and temperature affect the rates at which plants can break down their leaf pigments. Therefore, fall color can differ uniquely from year to year. The next time you admire the leaves falling in autumn, think of plant survival.

Another thing to consider is that only the strongest plants are left standing after severe winter weather. The actual morphology, or shape, of trees affects their cold resistance. Winter landscapes, high mountains, and the iconic Christmas-tree look that many of us find charming are influenced by time-tested evolutionary defenses to the cold. Trees that live in cold environments have the best chance of winter survival when they follow a set of structural guidelines. Less snow collects on trees with symmetrical, narrow, and prominently tapering crowns resembling cone or pyramid shapes. A solitary, straight, vertical trunk is stronger than a split or widely branched one. The structure of cold-resistant branches is open, arranged in tiers, and designed to flex downward, all of which ensure that snow is dumped safely to the ground below. Some species, such as spruces, grow conspicuously drooping, downward-pointing secondary branches, which further reduce the landing surface for ice. It appears that getting in shape has helped plants shed quite a few pounds of snow!

This Engelmann spruce (*Picea engelmannii*) in Washington is well prepared for cold. Its needles droop downward from the main branches, and its straight, narrow, tapering crown prevents heavy snow loads. Most of its compatriots in the background are following suit.

Finally, one of the most fascinating microenvironments you'll find winter-adapted flora thriving in is the layer between snowpack and the surface of the ground. This special place is called the subnivean zone, after the Latin words *sub-* (under) and *nix* (snow). In the subnivean zone, where the snowpack creates a unique haven beneath its frozen surface, plant survival takes on a distinct strategy tailored to the challenges of winter. As snow accumulates on the ground, some of it catches on debris such as rocks and sticks or on frozen hardy vegetation, forming a protective shield against the elements. At six to eight inches of thickness, a layer of snowpack shelters the subnivean zone from wind and large temperature swings. Temperatures here can remain within a degree or two of the freezing point of water, regardless of exterior temperatures. Thus, the subnivean zone is a sheltered area full of an assortment of biodiversity. While small mammals such as mice, voles, and shrews construct intricate tunnel systems for protection against the elements and predators, certain plants also thrive in this concealed environment. Notable among them are hare's-tail cottongrass, marsh Labrador tea, lingonberry, and Arctic bell-heather. These resilient plants capitalize on the subnivean conditions, benefiting from elevated CO_2 levels generated by animals, subfreezing temperatures, and sufficient light for photosynthesis. The subnivean environment serves not only as a refuge for animals but also as a unique niche where select plant species strategically endure and even thrive amid the winter's challenges.

We've seen some amazing ways in which plants survive cooler weather. Every time you see a snowy landscape or feel the crisp chill of a winter morning, stop and think of all the fascinating ways plants have responded to the seasonality of higher latitudes. Whenever you're devouring maple syrup–coated pancakes, listening to musical instruments, watching leaves fall, or just admiring a stately conifer, you're experiencing millions of years of adaptations to the cold . . . and that's pretty cool.

CHAPTER 18

Wind-gineering

The Hidden World of Storm Survival

Being a plant isn't always a breeze. When fierce windstorms rage across the landscape, plants cannot evacuate. They remain as rooted witnesses to the devastating damage wind brings. However, in the catastrophic wake of hurricanes and tornadoes, among shredded forests and annihilated structures, you'll find some of the most amazingly adapted plants standing strong. Plants get quite a workout from wind as they twist and flex. However, these mechanical movements engage a hidden world of survival systems. As they dance in the wind, plants are perceiving, signaling, reacting, and generating stronger growth. When the next inevitable windstorm looms, a plant's advance preparation determines life and death. The defenses plants build against the wind just might *blow* your mind!

The survival systems of palm trees enable them to breeze through hurricane-force winds that annihilate their competitors. As trees bend and twist during storms, they are programming a hidden world of survival adaptations.

Let's find some of the windiest places around the world and see why they make survival challenging for plants. We'll be looking for a few basic geographic conditions. Plants experience the greatest wind pressure in areas that are flat, coastal, high elevation, or along the routes of major global wind currents such as the trade winds. Here, wind is strong and frequent enough to exert a selective pressure on plant communities. Resultingly, these areas hold species with some of the most impressive wind survival adaptations. Those with the strongest engineering survive long enough to reproduce, while the competition finds itself damaged, uprooted, or killed.

Wind-ing Up

Every breeze or storm you've ever experienced starts on a global scale and channels through the local influences of terrain. Wind is really a manifestation of solar energy, and it is created when the sun unevenly heats Earth. One reason for this variability is Earth's round shape, which causes the equator to heat more than its poles. Second, Earth is a rotating body, warming during the day and cooling at night. Finally, Earth's continents absorb more heat than its oceans. As these global imbalances occur, complex movements of air begin on a worldwide scale.

Once air begins moving, geographic features channel it to various locations and modify its strength. Wind gains strength flowing across expansive flat areas and loses strength when it encounters elevation features such as mountain ranges and large forests. Examples of flat terrain with wind-influenced plant evolution are the Great Plains in the midwestern United States, the Florida Everglades, the Eurasian steppe belt, Patagonia and the Pampas in South America, major African savannahs such as the Serengeti, and the tropical savannahs of northern Australia. Species complexes in these areas feature palms, grasses, and wind-tolerant forbs that are often exposed to damaging high winds, dust storms, hurricanes, and tornadoes. The ocean also lacks elevation changes, so wind strength typically amplifies across the ocean until it reaches coastal and island environments. For this reason, wind-adapted plants such as palms, live oaks, select conifers, and grasses proliferate in coastal plant communities. Hurricanes, cyclones, and typhoons are some of the dominant selection events in coastal maritime plant communities. There's also safety in numbers; plants

in the middle of forests experience lower wind speeds than those that grow on forest edges or all alone.

High-elevation areas also have regular strong winds, but they are caused by different mechanisms. Air crashing into a mountain range is squeezed uphill and temporarily accelerated. Most of its moisture is jettisoned behind it as it cools and condenses into humidity and precipitation. To make matters worse, wind begins moving faster at higher altitudes because it does not encounter friction with all of the obstacles on Earth's surface. Air molecules slow down as they bump into terrain features, structures, plants, and even other air molecules. The drag from this gaseous bumper car effect is highest at Earth's surface and sea level because air molecules are densest there. Higher altitudes are windier because the air is thinner and more mobile. At the top of mountains such as the Rockies, Sierra Nevada, Andes, Alps, Himalayas, Alleghenies, and Appalachians, wind-adapted trees predominate. Wind is one of the reasons why species of pines, junipers, spruces, firs, and larches are common on mountain slopes and at high elevations.

Wind strength typically amplifies across the ocean until it reaches coastal and island environments. For this reason, plant communities in those areas proliferate with wind-adapted plants such as palms. Hurricanes, cyclones, and typhoons are some of the dominant selection events in coastal maritime plant communities, such as this tropical coastal forest in Bako National Park, Sarawak, Borneo, Malaysia.

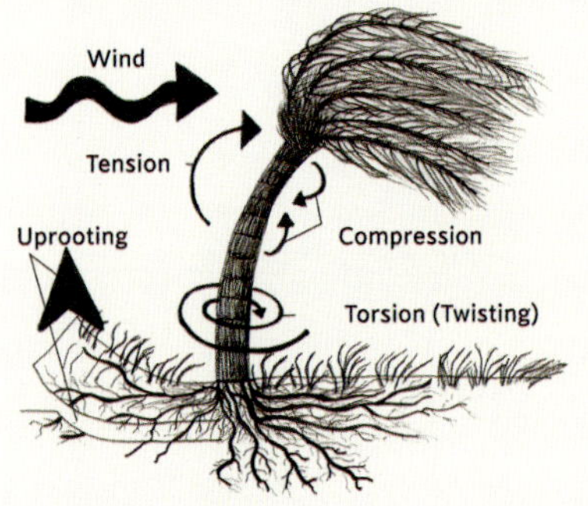

When wind meets a plant, it's a real drag—literally! As wind drags air molecules across plants, it exerts numerous damaging forces such as uprooting, tension, compression, bending, torque, sandblasting, and shearing.

When wind meets a plant, it's a real drag—literally! Plants act like giant, leafy sails. Plants experience wind as a friction force called drag, which is created by air molecules moving across their surface. Each plant's overall wind resistance is a product of its shape, called a sail area. The taller, wider, and leafier a plant is, the greater the forces that wind can exert upon it. Plants get quite a bit of exercise—they're tensed, compressed, bent, torqued, sanded, and sheared during every gust. Wind speed is measured with a scale such as the Beaufort (wind speed ratings: breeze force, gale force, storm force, and hurricane force) or the Fujita (the F in an F-5 tornado rating). Both of these scales incorporate speed measurements with observable effects on natural objects. Breeze-force winds (four to thirty-one miles per hour) are quite tolerable to most plants, causing their leaves and branches to move and their trunks to sway. However, these simple movements are where the interesting adaptations lie. The force of wind signals the plant to prepare survival systems.

A wild pinyon pine at the edge of the Grand Canyon in Arizona. Harsh winds have sculpted it into a natural and iconic bonsai. The shape of this plant is wind resistance engineering at its finest.

A pinyon pine in California. With less wind exposure, it has grown taller and denser than its counterpart at the Grand Canyon. If both plants experienced a severe wind event, which do you think would be more likely to survive?

Grow Like the Wind: A Plant's Response to Storms

Wind pressure causes most plants to grow stronger with slow, incremental, and additive developmental changes. This process of adaptive growth is called thigmomorphogenesis after three Greek words: *thigma* (touch), *morphos* (form), and *genesis* (create). The most common overall effect is the production of naturally dwarfed, short, and stocky plants with strong cells. Frequent wind exposure stimulates plants to reduce stem and shoot elongation, increase the anchoring strength of the root system, increase taper and stem diameter, and thicken cell walls. Even plants of the same species can assume strikingly different shapes if they've received contrasting wind exposure. The next time you find yourself in a high-wind environment, take a look around. You just might notice plants around you getting shorter.

Amazingly, plants can sense and calculate wind. During thigmomorphogenesis, wind stimulates plants to produce and regulate certain hormones such as auxin, jasmonates, and ethylene. These hormones control plant growth, messaging cells to grow sturdier, denser, or more flexible. Cells open additional signaling pathways as wind creates cellular stresses such as stretching, flexing, tension, compression, deformation, and pressure. Strained cells engage genes that produce signaling proteins such as

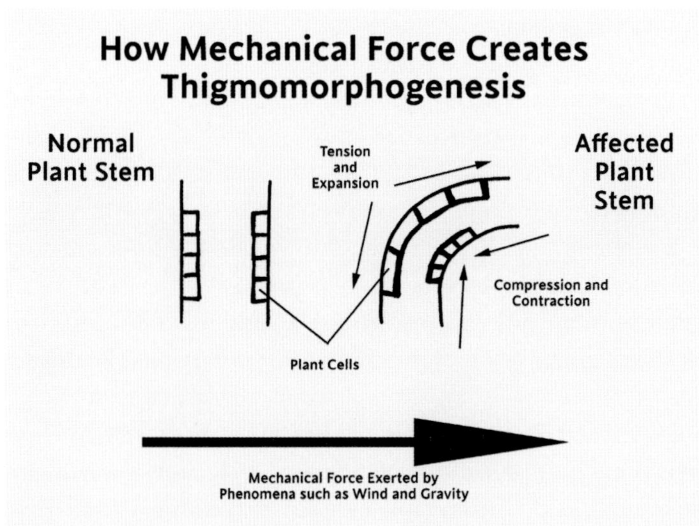

Mechanical strain on plant cells, such as bending, tension, elongation, and compression, induces adaptive growth called thigmomorphogenesis. The physical act of plant cells changing shape is enough for them to "sense" and calculate wind strength and direction. The response is to engage growth hormones, directing future growth to be stronger and more likely to survive windstorms.

Bonsai, a popular horticultural hobby, involves styling dwarf trees so that they resemble the natural sculpting effects of wind. Ideal features include thick trunks, reduced height, angled and contorted branches, and the reduction of foliage density. The frequent physical contact between a plant and its caretaker slowly induces thigmomorphogenic changes. Bonsai maintenance activities such as clipping, wiring, and bending branches influence a plant's growth through hormonal changes.

calmodulin and contain higher levels of signaling ions such as calcium. As the wind whips around plants, they calculate its force, direction, and frequency across every cell in every leaf, branch, and root. The next time you watch the trees wave during a storm, think of all that data perception occurring around you.

Thigmomorphogenesis might seem like a crazy concept—plants being able to perceive, quantify, and react to forces of nature. However, if you've ever seen a bonsai tree, you're already familiar with wind's natural sculpting forces. Bonsai is a Japanese art form involving the creation, maintenance, and sculpture of living dwarf trees. Bonsai's forerunner in China, called *penjing*, is similarly centered around growing and depicting wild-looking, naturally dwarfed landscapes. Both hobbies date back thousands of years and, at their core, receive inspiration from the occurrence of natural, wind-carved bonsai. These awe-inspiring trees even have a special name, *yamadori*, a Japanese word literally meaning "collecting plants from the mountains." Originally, horticulturists wild-collected yamadori, but over thousands of years, they developed horticultural techniques to replicate them. Some bonsai, such as the Yamaki Pine, which resides in the Smithsonian and survived the atomic bombing of Hiroshima, have been cared for on a daily basis for hundreds of years, passed down through generations of growers. While there are many styles of bonsai, general principles include reducing the tree's height and foliage density. Horticulturists encourage thick, sturdy, and tapered trunks. Deliberate mechanical forces from pruning, clamps, and wire slowly shape the bonsai's trunk and branches. In fact, the physical act of regular care and attention may be engaging thigmomorphogenic changes. In bonsai and penjing, the human hand rather than the wind is the force creating the living sculpture.

Plants change both outside and inside in response to wind. Plants can chemically alter the composition of their wood to produce reaction wood, which is a special shock-absorbing type of wood that plants form to acclimate to mechanical stresses such as those imposed by gravity (i.e., the weight of their branches) or wind. Reaction wood absorbs some of the energy from tension and compression, ideally preventing stems from bending and cracking under pressure. Reaction wood is individually tailored by each plant for its own trunks and branches depending on their level of wind exposure. The most highly stressed parts of the plant, such as the bases of branches and stems, exhibit the greatest changes. For example, a tree's growth ring might appear elliptical rather than circular if it contains reaction wood. This is because the tree likely experienced more wind across the wider axis of the ring, and the reaction wood cushioned the trunk as it flexed. In another example, a reaction wood cell might be programmed to contain a higher percentage of cellulose to increase strength and flexibility or more lignin to enhance stiffness. The cell's walls might be shorter, thicker, or longer depending on the forces it needs to resist or absorb. Reaction wood is denser than normal growth—essentially an army of specialized cells ready to counteract any strain imposed by the wind.

In extremely threatening windy conditions, such as the dry sandy salt spray of beaches or freezing alpine winds, entire trees can become sculpted into a natural phenomenon known as a flag tree, or krummholz, after the German *krumm* (crooked, bent, twisted) and *Holz* (wood). Flag trees are masterpieces of survival, living by the thinnest margin necessary to weather the elements. Flag trees form when buds on the top and windward sides of the crown are repeatedly frozen, desiccated, blown off, or killed by the wind. Since the only buds capable of living are all on the leeward side, the result is a canopy so one-sided that it literally resembles

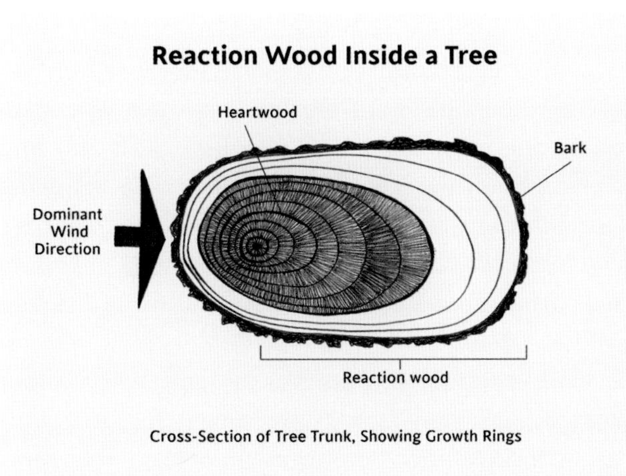

Reaction wood is a dense, cushioning tissue produced by most plants in response to natural forces such as wind. Shown here is a cross section of a hypothetical tree trunk. Reaction wood is displayed in its elliptical, directionally thickened growth rings. From this, we can tell that the plant experienced wind flow mostly from the right side of the photo, while the reaction wood on the left side helped absorb strain as the trunk bent. Until you look this deeply into the hidden world, you would never know that plants were this prepared for wind!

This red spruce (*Picea rubens*) atop the Dolly Sods Wilderness in West Virginia has become a krummholz, or flag tree, in response to extreme winds that prevent buds from staying alive on the tree's windward side.

a flag on a pole. Entire forests of flag trees might never grow taller than a human adult; some may never grow taller than your waist or knee! Flag trees are commonly found throughout the Northern Hemisphere in conifers such as the European spruce, mountain pine, balsam fir, red spruce, black spruce, subalpine fir, subalpine larch, Engelmann spruce, whitebark pine, limber pine, bristlecone pine, and lodgepole pine. You'll also find thigmomorphogenic changes on the leaves of these trees. To resist the drying, freezing, and sandblasting of the wind, leaf cells of these trees have evolved thicker cuticles and cell walls. Altogether, the changes are costly but are designed to keep moisture in, add insulation, and form a barrier against windblown sand and debris.

Even underground, plants react to wind stresses by adapting their root systems. In several experiments, researchers have shown that mechanically stressed plants grow thicker, stiffer, longer, and more numerous roots than unstressed plants. Amazingly, thigmomorphogenetic growth prepares every part of a plant for wind resistance. From root tip to shoot tip, plants are always ready.

Things turn grim when winds reach gale force (thirty-two to sixty-three miles per hour). Plants begin to face the potential for real damage. Leaves, twigs, and branches may be entirely stripped off or broken. Trunks begin bending and exerting upward forces that can loosen and break parts of the root system. The ultimate and most catastrophic winds are hurricane force (greater than seventy-three miles per hour). Entire trees can be bent, windsnapped, debarked, and windthrown (uprooted to such a degree that they fall over completely). Winds reaching over 250 miles per hour enact massive devastation. At their worst, they can level entire forests, demolish buildings, and fling large objects such as trees and vehicles long distances. After Florida's bout in 1992 with Hurricane Andrew, vegetation surveys by the University of Florida found that the storm's winds of 160-plus miles per hour annihilated almost 40 percent of urban trees throughout the state.

Yet plants cling to life in the aftermath of severe events like this, in the process exposing the intricate wind architecture inside. As we've seen, the best defense is to prepare for windstorms far in advance. For the true masters of wind-gineering, this preparation represents millions of years of unique evolution. It's no coincidence that species such as palms, live oaks, lodgepole pines, spruces, and grasses proliferate on windy beaches, mountainsides, and grasslands. Every part of these plants—leaves, trunks, and roots—is crafted to blow the competition away during high winds.

Palms have taken a remarkable evolutionary trajectory to emerge as some of the most stalwartly aerodynamic champions, dominating coastal habitats worldwide. The family includes iconic North American natives such as the cabbage palm (*Sabal palmetto*), as well as major ornamentals in the genera *Washingtonia*, *Roystonea*, *Trachycarpus*, and *Butia* and crops such as the coconut, date, açaí palm, carnauba wax palm, peach palm (known for hearts of palm), and African oil palm. As monocots, palms are more closely related to grasses, pineapples, ginger, and asparagus than to dicots such as oaks or gymnosperms such as pines. This means their distinctive and brilliant solutions to wind survival evolved independently. Palm trunks cannot grow the secondary xylem cells that give dicot trees their annual rings and reaction wood. A slice of palm wood is

Palms are one of the true masters of wind engineering, surviving hurricanes that annihilate structures and competing plants.

homogeneous throughout, lacking rings and resembling a severed piece of rope. This is because palms are constructed of long, dense, continuous, and tough fibrous strands encased within softer, flexible tissue. The entire stem is designed to resist severe bending and torsion forces as the wind whips around it. Palms can spring right back into place from forty-to-fifty-degree bends in hurricane-force winds. The sail area of a palm is significantly minimized at such steep angles. It's a stark comparison with trees whose woody and rigid construction keeps them standing upright. The foliage at the top of a palm grows as a branchless bunch from a single live bud called a meristem. This tuft of large leaves replaces branches and sits atop a strongly tapered single trunk—an amazingly aerodynamic shape. As long as the precious meristem survives, so does the palm, even if all the leaves are ripped away during a storm. Palm leaves are supported by a long, ropey midrib that is simultaneously strong yet flexible enough to bend and twist. Palm leaves are also corrugated into numerous folds and often pinnate. Under wind strain, these feathery leaves collapse like a paper fan and drastically reduce the sail area of the plant. The cuticles on palm leaves are also thick and smooth so that they repel sandblasting as well as stormwater, which creates detrimental weight and drag. Belowground, palms have perfected root systems that keep them anchored to the ground despite catastrophic gales and storm surges. The key difference between palms and other trees is the density of their root systems, which sacrifice size for quantity. Rather than utilize a few large and deep anchoring roots, palms generate a plethora of small, fibrous roots that spread uniformly in all directions, especially the upper layers of soil. With this root swarm, palms grip a larger volume of soil, which they use as a heavy and stable subterranean underpinning. Palms are so good at their jobs that architects use them as models for bioinspired building design. Palm fiber, often a by-product of crops such as coconuts and oil palms, is being explored as a valuable additive in construction materials such as concrete. This way, buildings become imbued with the powerful flexibility and wind resistance of plants. The next time you see the iconic palm in the storm, admire the hidden world that helps it stand majestically above the devastation.

There are other expertly adapted plants along the windy coastlines of the world. On the Caribbean islands and the coast of the southeastern United States, areas with an extensive hurricane season, you'll find a marvelous, muscular behemoth of a plant: the live oak. According to research conducted by the University of Florida, it is among the most wind-tolerant woody plants on the planet. *Quercus virginiana* is an elegant, intricately

On the Caribbean islands and southeastern U.S. coast, areas with an extensive hurricane season, you'll find a marvelous, muscular behemoth of a plant: the live oak. According to research conducted by the University of Florida, it is among the most wind-tolerant woody plants on the planet. *Quercus virginiana* is an elegant, intricately contorted, beautiful, and long-lived symbol of endurance. It is an enormous, spreading octopus of a plant. Its thick lateral branches and roots stretch almost three times as long as the tree's height, creating a squat and sturdy profile.

contorted, beautiful, and long-lived symbol of endurance. It is an enormous, spreading octopus of a plant. Its thick lateral branches and roots stretch almost three times as long as the tree's height, creating a squat and sturdy profile. The architecture of a live oak canopy is a powerfully aerodynamic design that reduces sail area and windthrow. Live oak leaves are prepared for battling windborne rain and sand with tough cuticles. They're also built to easily detach (called dehiscence) under extreme force, giving the oak a way to drastically reduce its sail area and avoid being toppled. The trees might be stripped bare in a hurricane, but they have the resilience to sprout anew rather than die completely. Within their trunks, you'll find a spiral wood grain with interlocking patterns that provide a coping mechanism against torsion stresses. Wood that grows in a spiral shape is less likely to be damaged by twisting forces. Live oaks also have the strongest, heaviest, densest, and most durable wood among all oak species. This keeps these giants anchored and steadfast over centuries of storms. A divergence from the flexible stems of palms, spiral grain has been a persistent evolutionary

The USS *Constitution*, commissioned by George Washington in 1797, is the oldest ship in the U.S. Navy. During the age of wooden ships, plant and national defense became one and the same. Cannonballs bounced off the *Constitution*'s durable live oak frame for over a century, notably in the War of 1812. Its triumphant successes earned it the famous nickname "Old Ironsides" despite the hull being made entirely of pine and live oak milled near St. Simons, Georgia.

feature for millions of years in dicots and gymnosperms. We know this by finding it in fossilized wood. Spiral grain occurs independently in numerous wind-tolerant species of pine, spruce, fir, and larch.

For centuries during the age of wood shipbuilding, naval defenses of entire nations and the defense systems of the live oak were one and the same. Live oak–clad warships like the USS *Hancock*, USS *Constitution*, and USS *Constellation* defended the fledgling United States against piracy, fought against the British during the Revolutionary War and the War of 1812, clashed on both sides of the Civil War, and helped end the slave trade by policing Atlantic waters. For the three centuries leading up to the invention of iron and steel ships, live oak triumphed as the toughest and most irreplaceable shipbuilding material. Warships became as invincible as a live oak, with windstorms, waves, and cannon fire reduced to mere nuisances. The USS *Constitution*, framed with live oak and planked with longleaf pine, earned the name "Old Ironsides" because of how much British cannon fire it repelled during the War of 1812. The oak's natural ability to sustain enormous pressures, especially on curves and angles, made it a key feature of a warship's inner hull. The L-shaped junction between the trunk and gargantuan lateral roots made dozens of invaluable curved and angular parts of a ship's frame, such as the keel, knees, futtocks, and bows. The U.S. government considered live oaks such a strategic national resource that by 1845 it owned a virtual monopoly on over 250,000 acres of live oak timberland throughout the American South. The Live Oak Tree Reservation Program began under President John Quincy Adams in 1828 and created the first national tree farm near Gulf Breeze, Florida, called Naval Live Oak Reservation. Deprived of American wood for shipbuilding, England was forced to build and maintain its navy with inferior European woods from Baltic, Scandinavian, and heavily deforested English forests. Wooden ship combat was a test of human strength, but it was also a clash between tree species. The strength of the live oak, forged through its evolutionary history with hurricane exposure, gave America an advantage that crushed and outlasted Europe's weaker shipbuilding woods. European native trees such as the English oak, Dutch elm, Scots pine, and common spruce, originating in forests prone to fewer windstorms, were no match for live oaks. Although obsolete and since returned to the public, U.S. live oak preserves throughout the Southeast are phenomenal legacies where human history and the hidden strength of plant evolution found a powerful symbiosis.

Though the live oak has a stately structure, plants don't have to be gargantuan to survive storms. Sea oats (*Uniola paniculata*) and salt cordgrass (*Spartina alterniflora*) are grasses native to coasts of the southeastern United States, Mexico, and the Caribbean. The strategy these tough pioneer species use to survive in the wind involves engineering the very ground they grow in. They are land builders, creating and stabilizing entire coastal dune ecosystems and supporting the growth of compatriot plants and animals. Humans esteem these diligent plants for the ecosystem services they provide, such as breaking the force of coastal hurricanes, which protects us from storm surges, beach erosion, and property damage. They're such a valuable natural asset that picking or disturbing sea oats is punishable by law in Georgia, South Carolina, North Carolina, and Florida.

It all starts when the grasses begin a sand collection. Sea oats and salt cordgrass have strong, fibrous stems that can reach six feet high. They also grow horizontally from underground stems called rhizomes. This pattern creates a dense thicket of leaves that's perfect for trapping sand, which simply piles up beneath the grass colony. Underground, tenacious and fibrous roots are at work, stockpiling and binding the loose sand grains. The

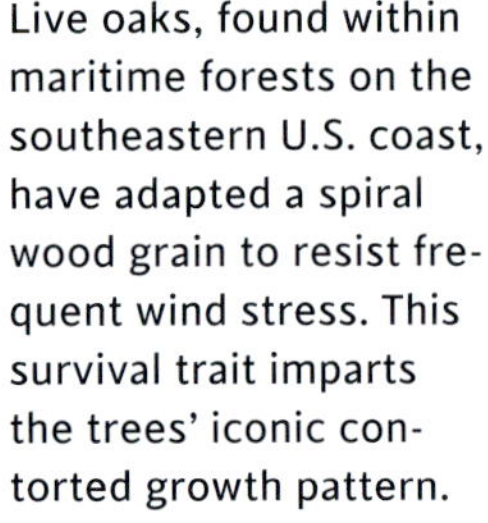
Live oaks, found within maritime forests on the southeastern U.S. coast, have adapted a spiral wood grain to resist frequent wind stress. This survival trait imparts the trees' iconic contorted growth pattern.

This dead limb shows a prominent spiral grain once its bark has been removed. This pattern of growth helps trees resist the twisting (torsion) forces of wind.

Plants don't have to be gargantuan to survive storms. The sea oats (Uniola paniculata) seen here build and anchor themselves into enormous sand dunes. Dune grass root systems are massive, stretching dozens of feet underground.

grasses aren't building up this sand for a beach castle, however. Rather, their sand collection helps stimulate growth by anchoring them against the wind, holding moisture, and releasing adsorbed nutrients. Sea oats' root systems are massive but remain hidden underground. In cases where entire dunes have been cleaved apart by hurricanes, sea oats' roots and rhizomes have been discovered weaving the dunes together. The roots may sink anywhere from twelve to forty feet below the peak of the dune, not only providing the plant's life insurance but also tapping into the water table in the process. In some cases, the remains of entire layers of grass colonies exist underground. This is because sea oats have adapted to being regularly buried alive by growing vertical rhizomes to claw back to the surface. Essentially, the sea oat plant *is* the dune—if a storm wants to move

Table 9 A Selection of Wind-gineered Plant Species

Species	*Common name*	*Habitat (generalized)*
Sabal palmetto	Cabbage palm	Coastal
Butia odorata	Jelly palm	Flatlands
Coconut	*Cocos nucifera*	Coastal
Panicum virgatum	Switchgrass	Flatlands
Paspalum vaginatum	Seashore paspalum	Coastal
Uniola paniculata	Sea oats	Coastal/flatlands
Spartina alterniflora	Salt cordgrass	Coastal
Quercus virginiana	Live oak	Coastal
Quercus geminata	Sand live oak	Coastal
Bursera simaruba	Gumbo limbo	Coastal
Magnolia grandiflora	Southern magnolia	Coastal
Pinus contorta	Lodgepole pine	High elevation
Pinus ponderosa	Ponderosa pine	High elevation
Pinus banksiana	Jack pine	High elevation
Solidago spp.	Goldenrod	Flatlands
Asclepias spp.	Milkweeds	Flatlands
Species within the amaranth, amaryllis, aloe, mustard, aster, borage, pink, bean, mint, spikemoss, and grass families	Tumbleweeds	Flatlands
Pinus pungens	Table mountain pine	High elevation
Pinus flexilis	Limber pine	High elevation
Pinus monticola	Western white pine	High elevation
Picea rubens	Red spruce	High elevation
Picea abies	Norway spruce	High elevation
Larix lyallii	Alpine larch	High elevation
Abies lasiocarpa	Subalpine fir	High elevation
Juniperus virginiana	Eastern red cedar	Coastal/flatlands
Myrica cerifera, *M. californica*, *M. pensylvanica*	Wax myrtle, California wax myrtle, bayberry	Coastal

the plant, it has to haul all the connected sand along with it. The final adaptation of sea oats is the investment they make in their offspring. Sea oat seedheads are iconically beautiful, oat-like clusters that top the stems in the fall. As these seedheads whip in the wind, they scrape shallow furrows in the sand in crescent-shaped arcs around the plant. These small channels catch the plant's seeds as they fall and ensure they become anchored and protected with sand. In time, the germinating seeds are fertilized by the new sand and sheltered from the wind, heat, and desiccation.

It hasn't been easy for plants to succeed in windy habitats around the world. While many of us find it relaxing to vacation on beaches and mountains, the flora there has a different experience. These plants are clinging to their very existence. Palms, live oaks, montane conifers, dune grasses, and many others have evolved every cell, structure, and strategy possible to brave savage storms. In the process, they have interwoven with human history, our sense of place, and the aesthetics of the landscapes around us. The next time you watch wind rustling through the leaves, you're catching a glimpse of this world coming alive.

CHAPTER 19

Sink or Swim

The Hidden World of Flood Survival

Water. Serene, powerful, beautiful, and one of the most vital building blocks of all life. It is almost unimaginable to think that during floods, water's life-giving essence transforms into destruction and death. In fact, over three billion years ago, water was the scene for the evolution of primordial plant life. Algae, some of the first photosynthesizing cells that gave rise to plant life, began here. In only a mere blink of geological time—the past five hundred million years—plants have been waging the massive and successful colonization of dry land. No matter how much evolutionary distance land plants place between their aqueous beginnings, they are still inextricably linked to water for survival. It is water that supports them, carries nutrients to their tissues, and enables the photosynthesis of sugars from solar energy. But when floodwaters seize the land, everything changes. Plants are once again at the mercy of a force of nature. Floods limit plant growth and development by starving them of oxygen. Worldwide, floods are responsible for billions of dollars annually in crop losses and damages. However, in wetlands around the world, the ebb and flow of evolution have created a unique world that enables plants to keep their heads above the water. In a boomerang-like evolutionary trajectory, wetland plants such as water lilies, rice, cranberries, bald cypresses, and mangroves broke off their quest for drier places. Instead, the niches they specialized in were the shallow, watery domains that harbored their ancient forerunners.

Chronic or sporadic flooding characterizes several fascinating and biodiverse habitats around the world. The general challenge for plants in these environments is that more water enters and remains than can exit via gravity. In wetlands, flooding pressures the flora's evolutionary

Water lilies are specialized colonizers of flooded habitats like this pond in the southeastern United States. Floating leaves and flowers, air transportation tissues, and other hidden adaptations allow aquatic plants to swim where other species sink.

trajectory. There is an endless diversity of wetland types, each with its own challenges. Perhaps the water's source differs; moisture can originate via rain, surface flow, groundwater, or tidal shifts. Or maybe the water itself is distinct. Wetlands may be fresh, saline, brackish, acidic, alkaline, stagnant, or flowing. Floods may occur chronically or seasonally, or standing water may be permanent. The structure and compactness of the underlying soil also affect how much water a wetland can absorb and how slowly that water drains away. Highly organic, compacted, or clay-rich soil can "perch" an entire water table just below the soil. You'll see flood survival adaptations float to the surface in bogs, mires, marshes, fens, swamps, peatlands, bottomlands, lakes, ponds, and riparian zones around the world. Some of these are the most spectacular wetlands in the world: the Florida Everglades, the Okefenokee Swamp in Georgia, the Pantanal in Brazil, Llanos de Moxos in Bolivia, the West Siberian Plain, the Sundarbans and Kerala backwaters in India and Bangladesh, iSimangaliso Wetland Park in South Africa, and Kakadu National Park in Australia. Others exist in river systems, such as the Mississippi, Nile, and Amazon River basins, the Mekong Delta in Vietnam, and the Okavango Delta in Botswana. Let's examine what plant life faces in these challenging environments.

A Flood of Problems

Water by itself is a harmless and absolutely necessary little molecule for plants. During floods, however, the buildup of water creates a menacing chain reaction. Flooding affects plants by denying them oxygen and sunlight. As these problems impair basic biological functions, toxins and diseases simultaneously proliferate in the flooded soil. Plant roots, like animals and microbes, require oxygen for the process of respiration. They all compete for this precious resource in the soil, leading to a mad rush for the last gasp of air whenever floods occur. Respiration allows plants to break their sugars from photosynthesis into energy, so a failure of the process begins starving cells to death. Roots absorb oxygen from air pores between soil particles. During waterlogging, these pores become deficient in (hypoxic) or devoid of (anoxic) oxygen. Without access to air, the roots begin to suffocate and die. Tragically, a plant with dead roots will wilt and desiccate to death even though it is surrounded by water. As plants become successively submerged, the sunlight reaching their leaves is greatly diminished. Water molecules, muck, and floating vegetation further impair plants' ability to generate energy. As you can see, flooding's effects are particularly vicious.

Even if the roots manage to survive without light or oxygen, a secondary series of reactions causes the situation to further deteriorate by causing a buildup of toxins in the soil. Anoxic conditions prevent plant cells from using aerobic respiration and performing critical life functions such as waste removal. Plant cells must respire to stay alive, and they prefer aerobic respiration for doing so. Plant cells might, in a last-ditch attempt at life, switch to anaerobic respiration during a flood. Anaerobic respiration uses molecules other than oxygen, such as sulfur, iron, nitrogen, fumarate, and carbon. However, anerobic respiration comes with significant risks and costs attached. It is inefficient and yields much less overall energy. Anaerobic is also riskier than aerobic respiration because it produces toxic waste products such as acids, metals, and gases. An especially harmful by-product is called a reactive oxygen species (ROS) or oxygen free radicals. These oxygen-containing molecules are so highly unstable and reactive that they can damage vital components of cells such as DNA, RNA, and proteins. When free radicals build to this level, it is called oxidative stress and starts killing cells. Animal cells like ours produce free

Panama disease (*Fusarium oxysporum* f. sp. *cubense*), a semiaquatic root rot, infects a banana plant in waterlogged soils. The healthy white cells are large, spongy, and full of aerenchyma, and they conduct a great deal of gas flow through the plant. The brown cells are portions of the plant's vascular system that have been necrotized by the devastating Panama disease.

radicals as well, and they cause long-term damage such as cancer, diabetes, and heart disease. Plant and animal cells detoxify free radicals with antidotes called antioxidants, but cells are unable to produce these without oxygen. Plant wastes begin accumulating to such excesses that cells begin dumping them overboard into the soil in a desperate bid to stay alive. Plant cells are not the only ones engaged in this process. Soil organisms such as anaerobic bacteria and yeasts spill their waste right alongside, only intensifying the problem. Chemicals such as ROS, acetaldehyde, ethylene, hydrogen sulfide, nitrates, methane, iron, and carbon dioxide can build to deadly levels. You may have smelled this potent blend of chemicals before as the phenomenon we know as swamp gas. Inside cells, the release of acids begins toxifying the cell's cytoplasm and vacuoles and causes failure of the outer membrane. Flooded soil is one frightening chemical soup.

Like sharks drawn to blood in the water, plant diseases love to invade during floods. Waterlogged soils are the breeding ground for virulent pathogens. Among these is *Fusarium*, a genus that includes Panama disease, a killer that has caused the monocropping banana industry a lot of strife. The

For tens of thousands of years, humans have grown rice in “paddies”: flooded fields that suppress weeds and favor the growth of this flood-tolerant plant. Rice holds genes that enable it to build air-transport tissues, sense flooding, and stretch its leaves in response to rising floodwaters. Aquaculture, in which fish and shrimp are cultivated for food in the paddies, also fertilizes the rice. Here, rice farmers transplant small plugs of rice directly into the flooded paddy’s mucky soil. A finished planting is on the right paddy, while the farmers are just beginning to plant on the left paddy.

Mangroves are famous for their snorkeling semiaquatic stilt roots, which can be seen marking the high-water line in the photo. They provide anchorage and air exchange. Mangroves are critical keystone plants for a variety of coastal, marsh, and wetland ecosystems. Mangrove thickets buffer against storms and provide dense cover for immense animal biodiversity.

historic banana cultivar 'Gros Michel' was plagued with this disease so severely that it has nearly been wiped out. Its replacement is the banana cultivar we're most familiar with—the 'Cavendish,' which is also severely threatened by this fungicide-resistant rot. Additional examples of pathogens that love waterlogged soils include *Phytophthora* (the cause of the Irish Potato Famine), *Pythium*, *Cylindrocladium*, and *Rhizoctonia*. While these pathogens might live on or inside plant cells, they are armed with sporulating structures that release microscopic spores in wet soil. The little spores actually swim, propelled by hair-like, whipping tails called flagella. This type of propulsion is called flagellation. Once root cells are dead or injured, the dying plant's open wounds become the site of a microbial buffet. Entire crops and plant populations might fall to disease during a prolonged flood.

Dealing with Deluges

Thankfully, these grim situations are not much of a problem for aquatic plants. They've been working on solutions for these issues over millions of years of evolutionary time. For starters, most aquatic plants have learned how to perceive signals in their environment so they can prepare for floods in advance. Plants use the adjustment of chemical quantities in their environment to realize that a flood is imminent. When water bottles up their gas exchange, flood-tolerant plants sense low soil oxygen and high CO_2 and ethylene gas levels. They're also specially equipped to detect and respond to the harmful levels of the free radicals and toxins discussed earlier.

Once plants detect flooding, they don't start grabbing sandbags, though, and they can't evacuate. Instead, the first response aquatic plants make to an incoming flood is to slow their growth and respiration and stockpile their oxygen and energy. Within six hours of a soil becoming hypoxic (oxygen-starved), aquatic plants begin these metabolic changes, while flood-intolerant plants continue normally, wasting resources with no clue of the impending crisis. One major controller of flood response is ethylene, a gaseous hormone produced by plant cells. In submerged plants, ethylene levels can skyrocket because the hormone's diffusion rate in water is ten thousand times slower than in air. When flood-tolerant plants sense this buildup, it signals a host of beneficial morphological changes that help them survive. For example, woody plants such as mangroves and water tupelos begin enlarging breathing pores in their stems (lenticels). Most species, including water lilies and elephant ears, ratchet up production of air-holding tissues called aerenchyma. This process gives them the ability to float their tissues toward the surface to reach precious sunlight and air. Emergent plants without the ability to float, such as rice, cattails, papyrus, and rushes, begin growing toward and above the water's rising surface as fast as possible so that they can also maintain access to air and light. Ethylene also initiates the production of mass quantities of adventitious roots—mangrove stilt roots, the fibrous roots of willows, and bald cypress "knees" are just a few prominent examples. As a result, aquatic plant roots perform some of the most peculiar stunts to seek oxygen anywhere they can find it, including growing against gravity, reaching out horizontally across the water's surface, and shooting straight upward if the roots are underwater. If you've ever overwatered a houseplant, you might just see some of these "weirdo" roots grasping for air. All of these genetically encoded survival responses are just part of the successful lives of wetland plants.

Aquatic plants also took a few chemistry lessons that allowed them to detoxify the nasty chemical soup created by a flood. They simply use their HIPs—their hypoxically induced proteins, of course! Aquatic plants evolved specialized genes that signal the perception of and response to a flood. Some of these genes turn aquatic plants into factories for HIPs. In general, there are many different types of HIPs, but they all target critical parts of flood survival. One type assists the plant with energy metabolism. Another counteracts cell acidification. A third lends a hand building aerenchyma tissues. A fourth scavenges and deactivates reactive oxygen species. A fifth type is critical for sensing and sending signals within the plant, especially those relating to flooding. The faster a plant can correctly identify low soil oxygen or high levels of harmful toxins, the better its chances to prepare and survive.

We've mentioned aerenchyma a couple of times, and this amazing adaptation is the botanical world's version of the snorkel or scuba gear. Aerenchyma is a specialized tissue, spongy and full of pores. Plants use it to conduct and store large quantities of air. Flood-tolerant plants such as water lilies and water hyacinth use networks of aerenchyma to pipe

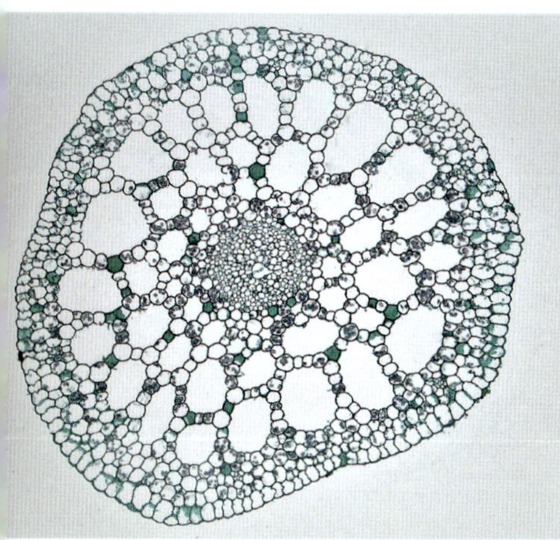

Aerenchyma is the botanical world's version of the snorkel or scuba gear. Aerenchyma is plant tissue that holds or conducts air. The large, tube-like aerenchyma cells in this *Elodea* stem pipe oxygen throughout the plant, enabling it to survive anoxic conditions. The leaf stem of water lilies can snorkel air from the water's surface to the rhizome several feet underwater.

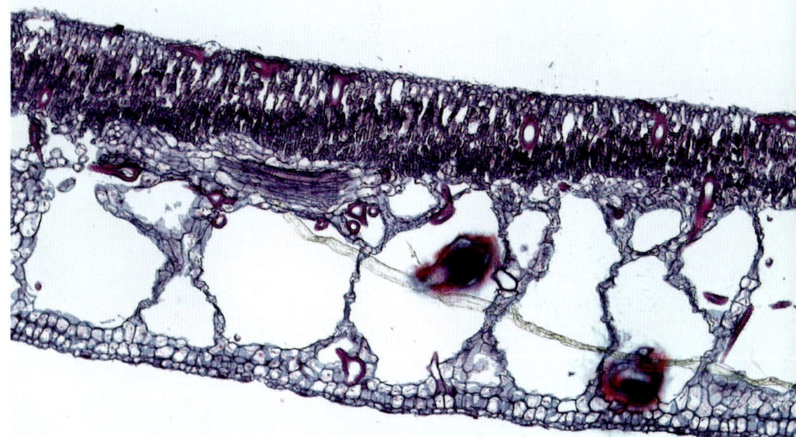

Aerenchyma allows plants to float and adjust their growth to the rise and fall of floodwaters. This cut-through of a water lily pad has evolved to be naturally buoyant because of its large air cells. Also helping the leaf breathe, stomata are concentrated on the top of the leaf. The outer epidermis has evolved the extremely water-repellant "lotus effect" to prevent water from clogging the stomata during gas exchange.

air from the water's surface to their stems and roots. Aerenchyma occurs throughout all of a plant's tissues, usually in long, connected tubes from the tips of leaves, through the stems and flowers, and down into the root tips. To create aerenchyma tissue, a plant grows huge open cells that are basically air balloons. Once they've reached the appropriate size, each aerenchyma cell receives a suicide signal and dies, leaving a large sacrificial air pore inside the plant that channels air to surrounding cells. It is a brilliant evolutionary risk that pays off tremendously. With this single adaptation, aquatic plants solved most of the problems inherent to living in flooded soils.

By inventing snorkeling, plants such as the water lily and lotus can generate and transport vast quantities of oxygen underwater. Diverging from all other plants that primarily concentrate gas exchange pores (stomata) underneath their leaves, aquatic plants only make them on top of their

Lotus roots (genus *Nelumbo*) are edible, dug from mucky soils that may lie several feet underwater below the floating leaves. The intricate holes inside are not damage but rather massive aerenchyma pipes for air conductance.

Victoria amazonica, the giant Amazon water lily, is the largest water lily on the planet. Each twenty-foot-diameter leaf has enough natural buoyancy to support a human being. Using aerenchyma to float confers vast advantages for harvesting sunlight, dumping wastes, and supplying gases to the plant.

leaves. Lily pads, particularly in species such as the giant Amazon water lily (*Victoria amazonica*), are in one respect enormous oxygen vacuums. These leaves can be up to ten feet in diameter and so powerful that they can pipe air more than twenty-five feet underwater along the leaf's long, flexible, aerenchyma-rich, and tube-like stem.

Oxygenating their roots benefits aquatic plants in multiple ways. The most obvious way is that their tissues can aerobically respire and operate normally. More subtly, aquatic plants create an oxygen-rich, life-sustaining "bubble" around their roots that hosts colonies of *aerobic* microbes. In return for this precious oxygen, aerobic bacteria detoxify and eliminate many of the harmful anoxic compounds from the surrounding sediments, such as iron, sulfides, and nitrates. It is one example of symbiosis that keeps all of these underwater life-forms thriving and breathing.

If you've ever noticed plants floating placidly in a pond or wetland, you're also looking at the power of aerenchyma. This tissue imbues plants with the ability to float but not because they're having a hot, relaxing summer. Floating is powerful because it allows placement of the leaves right at the water's surface, the prime location for absorbing sunlight. It is also a prime location for venting built-up gases such as carbon dioxide and ethylene as well as grabbing oxygen to spread around for respiration. Most aquatic plants also utilize aerenchyma to place their flowers above the water, effectively luring pollinators for a tropical vacation. These beautiful, scented botanical islands offer a feast of nectar and pollen, opportunities for mating, and safety from predators. Curiously, several aquatic species—the lotus, skunk cabbage, and elephant ear—also heat their flowers. This effect, called thermoregulation, occurs because these plants burn some of their food reserves to generate almost a full watt of energy for each flower at any given time. Presumably, the additional heat offers a helping hand to their cold-blooded insect pollinators, whose improved energy level enhances pollination. A healthy friskiness level also ensures flourishing populations of pollinators. In addition to the benefits floating provides during flowering, it also allows plants to adjust effortlessly when water levels rise and fall during flooding events. Aquatic plants usually have long elastic stems that easily stretch back and forth with the water level and currents. Floating plants come in all shapes and sizes. Water lilies, lotuses, duckweed, mosquitofern (*Azolla*), and sphagnum moss all rise and fall with their habitats, the seasons, and the weather. Floating is also so powerful that it has enabled the spread of pernicious invasive species such as water hyacinth, salvinia, hydrilla, parrot's feather, and water lettuce.

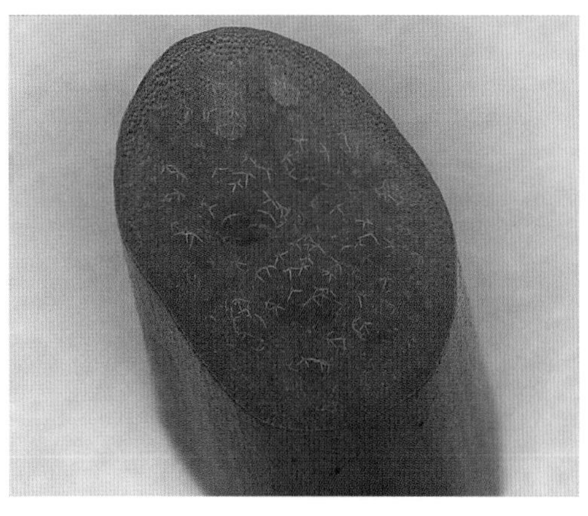

You're looking at a cross section of a water lily leaf stem—a natural pipe. Large aerenchyma channels run like gas lines through the water lily's long underwater leaf stems. Water lily leaves operate like giant oxygen vacuums. The largest species can pipe air more than twenty-five feet underwater using these snorkel-like stems. The star-shaped devices are called astrosclereids—sharp, crystalline defense cells full of silica and calcium oxalate.

To protect their investment in floating leaves, aquatic plants such as the lotus evolved some of the most water-repellent epidermises in the plant kingdom. This amazing phenomenon is called the lotus effect. Because aquatic plants transferred their stomata to the upper surfaces of their leaves, the stomata risked being clogged by splashing water and rain. The lotus effect causes water droplets to bead and roll off the leaves. Droplets

The North American native water lily (*Nymphaea odorata*) has floating leaves and flowers because it evolved aerenchyma. The three-foot-diameter leaves can vacuum oxygen to roots more than eight feet below the surface. For pollinators, aquatic flowers are a safe, edible paradise. The regular patrols of an alligator might also keep common insect predators at bay.

In the beautiful tranquility of bottomland swamps, two flood-tolerant plants with different strategies coexist. The first is duckweed (*Lemna minor*), a floating mat–forming species made of millions of tiny clonal units. The bald cypresses (*Taxodium distichum*) in the background utilize their iconic tapered trunks and knees (not visible) to aid in underwater gas exchange. Taken at Duckweed Pond in Riverbend Wildlife Management Area on the Oconee River in Laurens County, Georgia.

The lotus effect, seen here as water beads off elephant ear (*Colocasia esculenta*) leaves, is a survival adaptation for aquatic life. The epidermis of aquatic plants is ultrahydrophobic, giving it incredible water repellency. The primary goal of the adaptation is to clear water from the stomata on the surface of the leaf. Any debris, pest eggs, and microbes on the leaves also slide right off.

hitting this type of surface bounce away like basketballs. This way, water never interferes with the stomata or wets the leaf surface. Any debris and insect eggs are also washed away, so the lotus effect simultaneously plays a role in defense against disease and herbivory. Let's take a closer look at how plants figured out how to manufacture this magically self-cleaning surface.

Despite humans having known about the lotus effect since at least the writing of the Bhagavad Gita in 200 BCE, we only discovered the hidden mechanism for the lotus effect in the 1970s with the invention of the scanning electron microscope. The lotus effect has evolved not only in aquatic plants such as the lotus, water lily, and golden club but also in numerous nonaquatics such as nasturtiums, columbine, and maidenhair fern. The prickly pear cactus uses the lotus effect for fog collection. The microscopic structure of these plants' epidermises renders them ultrahydrophobic or extremely water repellent and unable to get wet. The lotus effect is

Lotus survival adaptations have inspired water-repellent and self-cleaning materials and coatings. Here, an ultrahydrophobic coating (clear) coats a colored surface. As water beads and rolls off the surface to uncoated areas, it removes any debris and microorganisms. Self-cleaning medical tubing, windshields, clothes, deicing paints, fog collectors, and desalinization are just some of the applications for this valuable nanotechnology.

so extremely effective that if you dunk a lily, lotus, or elephant ear below the water, it will emerge completely dry! Spaced across the epidermis are thousands of tiny, bumpy, and waxy trichomes. When droplets land on the leaf, they "crowdsurf" on these trichomes, which prevent the droplets from making contact with the leaf surface. It might seem strange for a plant to repel the water it usually loves. However, when the leaf remains dry, stomata remain clear, and plants photosynthesize happily. Plants find this epidermis so valuable that they continuously heal its integrity after any kind of damage.

Aquatic plant cuticles have unparalleled water repellency and are so advanced that we use them as models for nanotechnology and engineering. Through the use of this technology in paints, coatings, and materials, virtually anything can be rendered immune to water, ice, and germs. Let's imagine a world where clothes, buildings, and windshields can never get wet or dirty. Restroom and hospital surfaces can be made completely self-cleaning with a simple coating. The same technology is already helping ships, satellite dishes, and aircraft self-shed water, ice, and corrosion. Ultrahydrophobic materials are enhancing steam and fog collection systems and bringing greater efficiency to power generation, agriculture, and desalinization. Waterproof electrical circuits and microchips are increasing the water resistance of our gadgetry. Medical tubing and surgical tools can be engineered so slick that blood, bacteria, and viruses cannot adhere to them. This is the type of future being made possible through bioinspired design from aquatic plants.

Seed dispersal is another area where aerenchyma comes in handy. Many wetland species have evolved seeds or fruits that float, which is perfect for finding a starting point that's neither too flooded nor too dry for survival. This is basically like plants equipping their offspring with life preservers. There's no better example of this than the American cranberry (*Vaccinium macrocarpon*), whose sour red berries have air pockets that enable them to float and disperse before they rot underwater with their cargo of precious seeds. The cranberry is native to the northeastern United States and Canada, where it grows in rain-fed wetlands called bogs. Bogs were carved

The cranberry (*Vaccinium macrocarpon*) grows in waterlogged bogs, using air tissues to disperse its sour fruits during seasonal flooding patterns, which just happen to coincide with the Thanksgiving harvest! Humans utilize this plant's natural flood survival adaptation to float and harvest the seeds en masse.

out of impermeable bedrock during the ice age, essentially forming giant water-filled bowls of swampy soil and highly adapted plants. The unique, acidic, and mucky soil of bogs consists of the living and decayed remains of another flood-tolerant bog plant, sphagnum moss, whose air tissues float and sink in response to the water level. When cool, rainy weather sets in in the fall, bogs become even more flooded. Enter the cranberry, whose marvelous fruits ripen and disperse at exactly this time.

The fall cranberry harvest is the result of human interaction with the natural survival system of a plant. Harvesting the cranberry hasn't always been easy—in fact, only since the 1800s have cranberries been harvested by flooding their fields. Before that, the cranberry was cultivated for food by Native Americans / First Nations peoples throughout North America. The fruit became an important and recurring ingredient in a jerky-like food called pemmican, which was produced using a seasonally variable mixture of wild fruits and meats. For centuries, pemmican has been a portable, decay-resistant, and calorie- and nutrient-dense survival food. To make it, cranberries and other wild fruits are mashed and dried along with game meat and fats from bear, bison, elk, deer, moose, waterfowl, and even salmon. Cranberry-infused pemmican sustained Native and European settlers and explorers through the harsh winters of the northeastern United States and Canada. Pemmican made the perfect "on-the-go" food and was a staple ration for itinerant explorers and armies during wartime. Cranberry-laced pemmican powered the exploits of some of the first Europeans to navigate to the North and South Poles, including Roald Amundsen, Ernest Shackleford, Richard E. Byrd, and Fridtjof Nansen.

However, these little red bubbles of floating tastiness had yet to participate in the Thanksgiving dinners they do today. What changed things was human understanding of the plant's dispersal mechanism in floods. We developed a technique called wet harvesting specifically for the cranberry in order to take advantage of its floating fruits. Up until the 1800s, cranberries were picked by hand, one by one, in the soggy, peaty bogs. It was a thankless, freezing, and laborious task. In the 1900s the cranberry grew in prominence because of the efficiency of wet harvesting. Humans learned that because of the berry's internal air pockets, we could flood the fields and harvest floating cranberries all at once. The flooding also offers freeze protection for the plants and suppresses the growth of weeds. For

Just in time before Thanksgiving, cranberry bogs are artificially flooded in October. Special machinery shakes the berries off the plants, and the hordes of floating fruit turn the bog red.

a cranberry harvest, bogs are deliberately flooded to approximately six to eight inches. Then, giant machines called water reels—essentially massive egg beaters—vigorously thrash the plants to dislodge the cranberries. The hordes of floating fruit turn the bog red. Farmers corral and collect the berry-laden water using booms, pumps, and conveyors. The process is so efficient that farmers produce up to four hundred million pounds a year of this bog fruit. Cranberries remain popular fresh, canned, juiced, dried, and as a constituent of energy-rich foods such as trail mix, granola, and energy bars. Nowadays, by co-opting the cranberry's natural flotation, companies have taken the cranberry from a historic survival food to the mass market and holiday dinner plate.

In this chapter, we've taken a dive into the depths of the hidden world of flood tolerance. There are plants that snorkel, plants that strap their offspring into fruity flotation devices, and plants prepared to float above the competition. These amazing survival systems resulted from a tremendous journey through natural history when plants jumped out of the water onto dry land and then splashed back again. The next time you see a thriving wetland, take a deep breath of air before you swim, or devour a humble cranberry, remember how strong plant survival had to be to survive that journey.

Table 10 Selected Species with Notable Flood Tolerance

Scientific name	*Common name*
Acorus spp.	Acorus
Azolla caroliniana	Azolla
Carex intumescens	Bladder sedge
Ceratophyllum submersum	Hornwort
Colocasia spp.	Elephant ear
Cyperus papyrus	Papyrus
Eichhornia crassipes	Water hyacinth
Elodea spp.	Waterweed
Equisetum spp.	Horsetails
Isoetes spp.	Quillworts
Juncus spp.	Rushes
Lemna spp.	Duckweed
Lysichiton americanus	Western skunk cabbage
Musa spp.	Banana
Myriophyllum aquaticum	Parrot's feather
Nelumbo spp.	Lotus
Nymphaea spp.	Water lily
Nymphoides spp.	Floatingheart
Nyssa aquatica	Water tupelo
Orontium aquaticum	Golden club
Oryza sativa	Rice
Peltandra virginica	Tuckahoe
Pistia stratiotes	Water lettuce
Pontederia cordata	Pickerelweed
Rhizophora spp.	Mangroves
Sagittaria spp.	Arrowhead
Salvinia molesta	Salvinia
Spartina alternifolia	Salt cordgrass
Sphagnum spp.	Sphagnum moss
Taxodium distichum	Bald cypress
Typha latifolia	Cattails
Vaccinium macrocarpon	Cranberry
Wolffia spp.	Wolffia
Zea mays	Corn

CHAPTER 20

Life in the Barrens

How Plants Survive Inhospitable Soils

We've spent quite some time discussing the visible parts of plants. However, exploring what goes on below the surface can also divulge some hidden plant secrets. Soil can present plants with a slew of difficulties. One of the biggest challenges for plants transitioning to terrestrial life was how to access the nutrients they need from the soil and air. Prior to that, aquatic photosynthesizers got what they needed through diffusion of materials out of a nutrient-rich soup known as the ocean. Soils provide limited consistency and vary greatly. We will spend this chapter looking at some of the most challenging soils: low nutrient, saline (or salty), and serpentine. Each of these types of soil has led to interesting adaptations by the plants that call them home. Only the most specialized survival systems can keep our leafy green friends alive as these environments starve, desiccate, and poison them. Shortages of some minerals and excesses of others can be severely limiting for plant growth or even lethal. Plants, though, have come up with clever solutions. Low-nutrient soils, for example, drove some plants to form symbiotic relationships with fertilizer-producing microbes in order to obtain nutrients. In other low-nutrient, acid systems, we see the evolution of carnivorous plants—plants that access the much-needed nitrogen, phosphorus, and potassium through the breakdown of animals. Saline soils have their own suite of specialists called halophytes, which can literally stay alive while utilizing ocean water. Finally, we'll look at hyperaccumulation, a set of adaptations that allow plants called hyperaccumulators to withstand lethal concentrations of heavy metals in serpentine soils and industrially contaminated sites.

The Eternal Famine: Low-Nutrient Soils

While plants may make their own sugar through photosynthesis, they need a lot more nutrients to keep themselves going. The process of survival for plants isn't quite as simple as collecting water, carbon dioxide, and sunlight. To grow and develop properly, they need to supplement nutrients from the environment that they cannot manufacture for themselves. Most plants are able to harvest the nutrients they need from the soil using their roots. However, there are soils across the world that are so low in nutrients that the plant communities growing in them might slowly stunt and starve to death unless the plants find work-arounds.

To survive and obtain a competitive edge in these low-nutrient environments, plants evolved two amazing solutions that rely on other organisms. The first, hidden below the soil's surface, involved plants transforming their roots into fertilizer factories with the helpful colonization of symbiotic bacteria and fungi. By developing mutualistic relationships, these plants are able to trade for the resources they need. A second, more sinister solution was carnivory. Yes, you heard right—in order to supplement their diet, some plants break down animals into a big multivitamin. Instead of relying on inhospitable soils to provide the much-needed nitrogen, phosphorus, potassium, and other macro- and micronutrients plants need to survive, these plants decided to utilize the little packets of nutrients flying, crawling, and hopping around them. Carnivorous plants, specialized to survive in low-nutrient habitats around the world, are capable of attracting, capturing, killing, and absorbing the nutrients within live animals such as insects. Now that we've "captured" your attention, let's follow plant survival systems along their eternal quest to seek nutrients.

Rooting for Each Other: Mutualistic Associations

One of a plant's great loves is the element nitrogen. Plants crave this molecule because it is a critical building block for other molecules that support most of a plant's basic life functions: amino acids, proteins, nucleic acids (such as DNA and RNA), chlorophyll (by which they absorb sunlight), and defensive chemicals. Without nitrogen, plant life could not exist—there would be no more photosynthesis, no more rejuvenation and growth, and no replication of genetic code. Plants can only use nitrogen in certain chemical forms, such as the molecules ammonia and urea. These forms of soilborne nitrogen are readily leached by water and often limiting. Ironically, a virtually limitless amount of nitrogen gas exists in our atmosphere, floating past plants all of the time. Seventy-eight percent of our atmosphere is, in fact, nitrogen gas. Nitrogen gas molecules contain two nitrogen atoms. Unfortunately, the nitrogens are so strongly bonded to each other that they become antisocial to reactions with most other molecules—this is called being chemically "inert." Nitrogen gas's inertness deprives plant cells of the ability to directly utilize it in chemical reactions.

Cycads like this sago palm (*Cycas revoluta*) have evolved branching, specialized coralloid roots to add some additional nitrogen to their diet. Cycads have forged an ancient symbiotic relationship with nitrogen-fixing cyanobacteria in order to survive low-nutrient conditions. The cycads allow these cyanobacteria to colonize their coralloid roots, providing them carbon and a protected environment. In return, the cyanobacteria fix nitrogen from the air into a usable form by the cycad. Note that the cycad still forms normal roots for anchorage and the transport of water and nutrients. These can also be seen in the photo.

Cutting into a cycad's coralloid root will reveal blue rings of cyanobacteria flourishing inside. Seen here is a cross section of a coralloid root. The blue bands are millions of cyanobacterial cells living in harmony within the protected roots of the cycad, where they pump out extra nitrogen for their plant partner.

Microbes have known about this problem for billions of years longer than land plants. As such, microbes evolved a solution, a process called nitrogen fixation. Using fixation, microbes transform nitrogen gas into a nitrogen-rich, usable molecule called ammonia. On a daily basis, billions of years before plants and animals even existed, teeny blue-green bacteria began producing ammonia from our atmosphere for their own needs. Before humans came along with our ability to manufacture nitrogen-rich fertilizers from animal waste and fossil fuels, plants had extreme pressure to seek out novel sources of ammonia. Many natural areas still experience this pressure.

At a subsequent critical point in the natural history of life, the roots of land plants made interspecies contact with nitrogen-fixing microbes in the soil. The helpful partnership they forged has lasted through eons. While plants were seeking nitrogen, the microbes were seeking resources *they* couldn't manufacture, such as carbon. Mutually beneficial relationships like this are a type of symbiosis called mutualism.

Cyanobacteria are a type of blue-green bacteria that fix nitrogen, and they are among the oldest living photosynthetic organisms on our planet. There are many different kinds of them. The ones that form symbiotic relationships with plants are primarily in the *Nostoc* genus. Cycads, a group of ancient gymnosperms that predominated during the time of the dinosaurs, have truly rolled out the red carpet for cyanobacteria. Cycads make two different types of roots. One of them, the branching coralloid root, is unique to cycads. These lie close to the soil surface and house a city of thriving cyanobacteria. Their other roots are more familiar, growing deep into the soil to provide anchorage as well as water and mineral uptake. The areas on plants where cyanobacteria dwell are usually at or above the soil surface for adequate access to sunlight. Both plants and cyanobacteria photosynthesize by absorbing light. If you take a peek inside any of these tissues, especially the coralloid roots of cycads, you'll find numerous dark bluish-green bands. These are home to millions of cyanobacteria, supplying both themselves and their plant hosts like little nitrogen fertilizer packets. All *Nostoc* requests from these plants in return is a place to live and a bit of carbon. Cycads aren't the only plants that get a nitrogenous boost from cyanobacteria. The stems of the giant-leaved gunneras, the floating foliage of aquatic *Azolla* ferns, and the tissues of diminutive hornworts are all places that cyanobacteria inhabit.

There are other types of bacteria utilized by plants. *Rhizobium* bacteria and plants in the bean (or legume) family are historically and intimately entwined. Most of the bean family utilizes *Rhizobium* bacteria to supplement nitrogen, including peas, chickpeas, beans, clovers, alfalfa, soybeans, lupines, kudzu, peanuts, acacias, redbuds, locusts, senna, tamarinds, mimosas, and many others. The high amount of nitrogen in the seeds and foliage of these plants' tissues is valuable not only in our diets but also as forage for our livestock and as leftover biomass tilled and incorporated into fields. We often see legumes used as a cover crop during the offseason in agricultural settings. When used in this way, we often see these plants referred to as nitrogen-fixing plants. However, it really should be their microbes getting the credit. These cover crops, such as clover or alfalfa, are often tilled into the soil to help restore nutrients lost through the growing and harvesting of crops. This process of incorporating these plants into our soil has earned the name "green manure." Legumes offer supplies of carbon to rhizobia and even build them little homes to live in called root nodules. Billions of little nitrogen-pumping rhizobia live happily in each nodule, and each leguminous plant can have thousands of nodules. Rhizobia live freely in the soil, but upon sensing exudates from the roots of leguminous hosts, they "infect" them beneficially. When the rhizobia and root hairs of legumes

Rhizobium root nodules can be seen here as small globules of tissue on the roots of this peanut plant (*Arachis hypogaea*). The nodules are a survival partnership between *Rhizobium* bacteria and plants in the bean family. Each single nodule holds an entire civilization of nitrogen-fixing *Rhizobium* bacteria. The plant and bacteria chemically signal each other back and forth during colonization, and the result is this special, spherical root growth that allows the bacteria a safe place to manufacture nitrogen.

chemically recognize each other, the rhizobia attach to the root hairs and send threads of infection into further root cells. To create the characteristic nodules, the plant signals developmental changes in its own root cells. Root hairs begin curling protectively around the little rhizobia, while other cells proliferate into a large, nodule-like mass. Legumes double-check the work of the rhizobia, monitoring how much nitrogen they're receiving from each nodule. If the microbes' work isn't up to par, legumes such as soybeans and lupines actually punish them with sanctions. The plants restrict the oxygen and carbon flowing to underperforming nodules, slowing the reproduction of the bacteria until their nitrogen output is satisfactory. Who knew plants could be such taskmasters?

Legumes and cycads don't hold a monopoly on microbial symbiosis, however. Bacteria in the genus *Frankia*, called actinobacteria, can also fix nitrogen. Eight families of plants, including the birch, wax myrtle, oleaster, rose, buckthorn, and she-oak families, prefer *Frankia* as the source of their nitrogen. Similar to legumes, these plant family members construct elaborate root nodules for little *Frankia*s to frolic happily in. As long as the bacteria deliver their nitrogen on schedule, they're set for life in their subterranean, nitrogen-rich utopia. Alders, casuarinas, wax myrtle, bayberry, *Ceanothus*, mountain mahogany, *Eleagnus*, sea buckthorn, and the unusual anchor plants (genus *Colletia*) are all nonleguminous nitrogen fixers, thanks to the helpful *Frankia*s.

Phosphorus is another ingredient critical to plants, but naturally rich sources of phosphorus, such as rock phosphate, are a globally rare occurrence. Not surprisingly, many soils around the world are highly limited in phosphorus. This is especially true in areas such as Australia and Africa, each of which contains specialist groups of phosphorus-extracting plants such as the protea family and carnivorous plants. To enhance and maintain access to this nutrient, many plants around the world hire the help of some fungal friends.

Chemically, obtaining phosphorus is a whole different ball game for plants compared to obtaining nitrogen. Unlike the gaseous nitrogen, phosphorus is not found in our atmosphere. Therefore, microbes can't fix phosphorus from the air. Instead, we gain phosphorus into the natural world via the weathering of rock, a slow process. Thus, plants must obtain phosphorus by mining it from the soil with their roots. This can be difficult work, so plants often enlist the help of another mutualistic relationship. After all, the best phosphorus miners of all time aren't even plants—they're fungi.

Plant roots recruit special fungi called mycorrhizae (or mycorrhizal fungi) to help gather and share water and nutrients such as phosphorus and sulfur. In return, the fungi receive carbon, something plants readily manufacture. Mycorrhizal fungi are an amazing group of organisms that have lived symbiotically with plants for millions of years. More than 90 percent of all land plants rely on these fungi for survival. Plants can actually sense their own phosphorus levels. When they notice supplies running low, they respond by increasing the length of their roots, producing more root hairs, and sending signals to beneficial fungi. Many of the mushrooms you see during a walk in the forest and the complex web of life below the soil are a secret network where a vast trade of nutrients helps both plants and fungi survive.

Fungi have special mining abilities that plants lack. For example, unlike plants, fungi are fantastic decomposers. Fungi wield the power of enzymes and acids that can weather an unavailable source of phosphorus such as bedrock and break down organic matter. Both of these activities dissolve phosphorus into a solution that plant roots can absorb. You've heard of

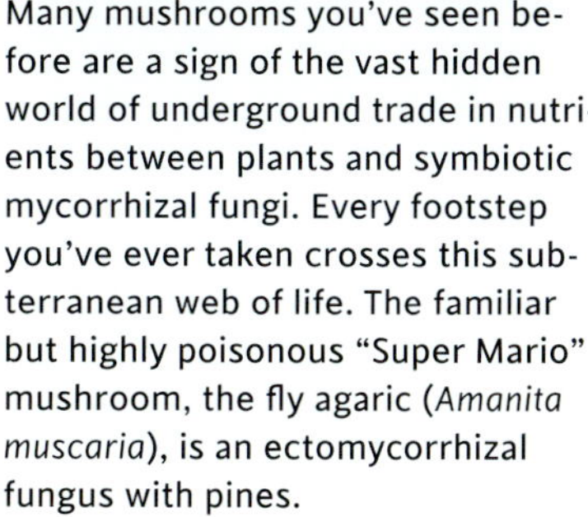

Many mushrooms you've seen before are a sign of the vast hidden world of underground trade in nutrients between plants and symbiotic mycorrhizal fungi. Every footstep you've ever taken crosses this subterranean web of life. The familiar but highly poisonous "Super Mario" mushroom, the fly agaric (*Amanita muscaria*), is an ectomycorrhizal fungus with pines.

The vegetative part of a fungus, the mycelium, is a white, expansive, branching network of tiny, root-like threads called hyphae. You can easily encounter these threads at work by turning over rotting leaves and logs or digging in the soil. Hyphae are microscopically small, so much so that they can mine water and nutrients from pores in the soil that even plant roots are too bulky and clumsy to penetrate.

many of these acids: citric (the tang of lemons), acetic (vinegar), oxalic (a defense molecule plants also make), malic (the tartness of fruits such as apples), and ascorbic (vitamin C). The vegetative part of a fungus, the mycelium, is a white, expansive, branching network of tiny root-like threads called hyphae. You can easily encounter these threads at work by turning over rotting leaves and logs or digging in the soil. Hyphae are microscopically small, so much so that they can mine water and nutrients from pores in the soil that even plant roots are too bulky and clumsy to penetrate. The result is a dramatic increase in the surface area from which plants can access water and nutrients. Working in tandem, plant roots and fungal hyphae connect and bind soil particles, improving their structure and drainage while fighting erosion. These shared networks can even link the roots of genetically related plants through a forest floor, allowing the sharing of resources between plants.

When fungi are ready to reproduce, they make a ubiquitous spore-bearing structure we know as the mushroom. That's right: that little mushroom you see poking out of the ground is just the tip of the iceberg when it comes to the whole organism. Many fungi branch out in large networks just below the soil, rising a reproductive structure—a mushroom—out of the soil only when conditions are ideal for propagation. Not all mycorrhizal fungi reproduce via mushrooms, nor are all mushroom-producing fungi beneficial or mycorrhizal in nature. However, mycorrhizae we're familiar with include puffballs, earth stars, stinkhorns, chanterelles, coral mushrooms, boletes, truffles, and poisonous species such as the recognizable fly agaric and death cap mushrooms. In exchange for their valuable services, plants share carbohydrates and B vitamins with mycorrhizae.

The delectable black truffle (*Tuber melanosporum*) is a type of mycorrhizal fungus that helps plants obtain nutrients. Truffles are spore-bearing structures—similar in purpose to mushrooms, except they're completely underground. Their wonderful taste, earthy smell, and difficulty of cultivation make them a culinary luxury. Truffles colonize the roots of trees such as hazelnuts and are located via smell and then dug up with the help of specially trained pigs and dogs. The use of truffle-hunting pigs dates back to the Roman Empire; a pig's extraordinary sense of smell can locate truffles up to three feet underground!

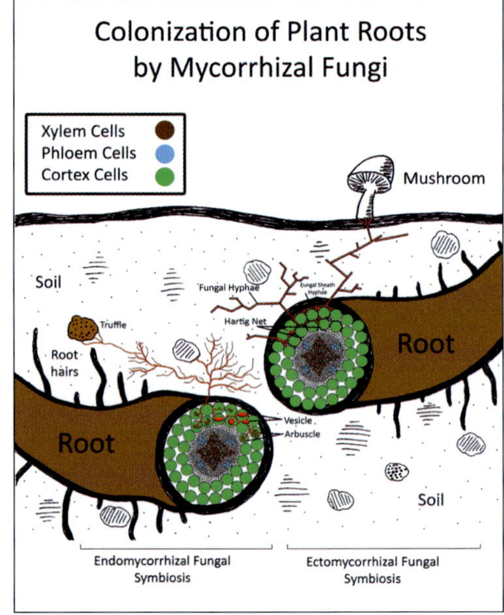

Seen here are different types of mycorrhizal fungi colonizing plant roots.

Ectomycorrhizae, which form a spiderweb-like sheet of protection over the roots of their plant associates, are seen here colonizing the brown-colored roots of white spruce (*Picea glauca*). Ectomycorrhizae form sheaths of mycelium that coat plant roots, supplementing them with water, nutrients, and secreting chemicals that fight a wide range of plant diseases, including pathogenic fungi, bacteria, protozoans, and nematodes.

Coral mushrooms (genus *Artomyces*) are a mycorrhizal fungus.

Mycorrhizal fungi colonize plant roots by using highly evolved tools of infection. Their spores germinate on the soil, sending hyphae underground and homing in on chemicals produced by the plant roots they hope to infect. A fundamental difference between pathogenic fungi and mycorrhizae is that, rather than forming defenses against these fungi, plant roots actually emit chemical advertisements to signal the mycorrhizae to move in. Fungi have highly developed adaptations for entering and spreading through plant cells, such as enzymes that dissolve plant cell walls. These tools are used by pathogenic fungi to parasitize and extract nutrients from their hosts, but mycorrhizae decided to use their skills for good rather than evil. When cell barriers collapse between the two organisms, which is quite a risk for living things to take, the symbiotic trade network comes alive.

Mycorrhizae fungi, a surprisingly diverse group, have just two general plant colonization strategies. Endomycorrhizal fungi (from the Greek *endon*, meaning "within") utilize their fungal wiles to penetrate the cells of their host plants, cramming an interface of branched hyphae (called arbuscles) or balloon-shaped hyphae (called vesicles) directly into plant root cells. These intricate interfaces are the mycorrhizal marketplace where water, nutrients, and carbohydrates are exchanged between the partners. Endomycorrhizae are evolutionarily older and much more widespread across Earth. In contrast, ectomycorrhizal fungi (from the Greek *ektos*, meaning "outside") rarely progress into plant cells, instead surrounding and protecting plant roots like a layer of biological armor. They are evolutionarily more recent and occur on a smaller percentage of plant families, but what they are capable of is simply amazing. Ectomycorrhizae form a dense exchange site called the Hartig net, which is a sophisticated network of fungal cells spread over and connected with cells of plant roots. Altogether, this net forms a protective and absorptive sheath that surrounds these roots while wider-ranging hyphae shoot far into the surrounding soil to mine. Not only do ectomycorrhizae increase the uptake of water and nutrients such as phosphorus, but they also secrete antimicrobial chemicals that kill many soilborne plant diseases. Obviously, these fungi have quite a stake in seeing that their plant partners survive assaults from parasitizing fungi, protozoa, nematodes, and bacteria.

While plants have their own equipment for defense and resource acquisition, the edge provided by mycorrhizal fungi is another critical part of their survival systems. Quite often, the network of belowground exchange connects an entire community of fungal species and plant species. A single mycorrhiza may have dozens of species of plant partners ranging across an entire forest; plants cooperate simultaneously with multiple fungi as well. Each contributes its specialty to the survival of the whole—there is even resource partitioning and shifting allocation to plants or fungi that advertise certain needs. This process has been ongoing underneath every footstep we've ever taken. It can even benefit us in agriculture. For decades, we saw fungi as either unimportant or detrimental. However, we now know that we can save money on pesticides if these mutualistic fungal communities are in our fields bolstering our plants.

Ferocious Foliage: The Carnivorous Plants

As the hidden world opens up more and more before us, it may seem bizarre and far-fetched to us to learn that for some plants, killing and breaking down animals are the only ways to survive in soils that are extremely low in accessible nutrients. Carnivorous plants survive by reverse-engineering live animals into nutrient-rich fluid that the plants absorb as fertilizer. It is

This phenomenal pitcher plant (*Nepenthes veitchii*) uses pitcher-shaped leaves to attract, capture, and digest insect prey. To carnivorous plants like this, bugs might as well be little vitamin pills with legs and wings. Carnivorous adaptations in plants are survival strategies that work exceptionally well in habitats with nutrient-poor soils. The hunting grounds of the pitcher plant seen here are the tropical peat-swamp rainforests of the Maliau Basin on the island of Borneo, Malaysia. A ravenous appetite allows carnivorous plants to thrive where other species starve.

The evolutionary crucible for plant carnivory was the muck of mossy, waterlogged swamps around the world. Since the dawn of land plants 450 million years ago, one plant species has engineered, maintained, and stewarded some of the most nutritionally sterile places in the world: sphagnum moss. This particular one is building up nutrient-poor pitcher plant habitat on the tropical rainforested slopes of Mount Kinabalu, Malaysia.

a sinister and desperate way of life driven by the starvation plants have faced for millions of years in nutrient-poor wetlands around the world. Carnivorous plants require three phases to complete this task: they must attract, capture, and break down living insects and small animals. At each step of this process, carnivorous plants circumvent insects' ability to perceive and avoid danger. Carnivorous plants have adaptations that outsmart their prey's vision, sense of smell, perception of danger, ability to crawl or fly away, and ability to grapple and stick to most known surfaces. To predatory plants, bugs might as well be little vitamin pills with legs and wings. The cells of insects and other invertebrates are rich in exactly what plants seek: nitrogen, phosphorus, potassium, calcium, and other trace minerals. Are you *hungry* to learn more about how these plants work?

The evolutionary crucible for plant carnivory was the muck of mossy, waterlogged swamps around the world. Hot spots for carnivorous plants

include swampy areas of the southeastern US Coastal Plain, Indonesia, Australia, South Africa, and Central and South America as well as temperate bogs throughout North America, Europe, and Asia. Since the dawn of land plants 450 million years ago, one plant species has engineered, maintained, and stewarded some of the most nutritionally sterile places in the world: sphagnum moss. Sphagnum is a denizen of the damp, able to adjust to rising and falling water levels using aerenchyma tissues. The special thing this moss does differently, however, is pump out loads of acidic ions into its surroundings. Over years, the soil pH acidifies, dropping so low that nutrients become chemically altered, rendering them unavailable to absorption by plants. As bigger, taller, and more advanced plants began evolving, masses of sphagnum competed with them, using acres of pH adjustment to starve them to death. Few woody shrubs or trees grow alongside sphagnum; those that try become stunted, miniature dwarfs, unable to collect the nutritional raw materials for taller growth. Until the evolution of carnivorous plants, which evolved prey capture as a workaround, sphagnum remained a champion of the bogs and swamps across the world. Sphagnum's powerful acidity and antimicrobial secretions as well as the flooded, stagnant, and anaerobic soils of these swamps render the ground inhospitable for decomposing bacteria and fungi as well as beneficial rhizobia, *Frankia*, and mycorrhizae. As a result, without decomposition, the decay of vegetation becomes as static as the swamps' flooded soil. Between the lack of rot and the binding from low pH, nutrients remain locked within the depths of bog vegetation for thousands of years. Even perfectly preserved human remains, called bog bodies, have lain frozen in time for centuries in this abyssal muck.

To become fully carnivorous, plants compiled the forces of a wide-ranging series of defensive and nondefensive botanical adaptations. For example, carnivorous plants repurposed common pollination attractants from flowers to lure in unsuspecting insects. These attractants include nectar baits (wickedly laced with paralytic chemicals such as coniine), ultraviolet signaling, colorful pigmentation, and scent attractants. The only thing that changed was the location: carnivorous plants place these elements on their trap-like leaves instead of their flowers. Insects land to feed, only to discover other repurposed defensive adaptations on the plant's leaf surfaces. Carnivorous plant traps teem with some of the slickest epidermal

The slope of Mount Kinabalu towers in the distance on the island of Borneo, Malaysia. Along its face are expansive peat-swamp rainforests where the soil is so poor that it stunts the trees shown in this photo. The short stature and scarcely leaved branches are indicative of how stressful this environment is for noncarnivorous plants. Clambering over these naturally diminutive thickets are *Nepenthes* pitcher plants, enriched with the nutrient-rich bug carcasses they expertly glean using slippery pitfall traps.

cells in nature as well as waxes, glues, and trichomes all specially designed to imprison and kill prey. The final step of digestion relies on other seemingly common defensive adaptations: the production of disinfectants and digestive enzymes and the affinity of plant cells for absorbing nutrients. The antimicrobial and prey-dissolving compounds in carnivorous plants originally evolved as defenses, having been used for millions of years to erode the cells of plant pathogens and herbivores. Carnivorous plants, combining all of these adaptations in such frighteningly novel ways, have learned to conquer the riches of the insect kingdom for their own survival.

The Venus flytrap (*Dionaea muscipula*) is a rare but fascinating carnivorous plant native only to North and South Carolina. Its jaw-like leaves produce red pigmentation and sugary nectar

Venus flytraps (*Dionaea muscipula*), rare endemic plants from North and South Carolina, form a deadly obstacle course for insects. Each jaw-like leaf is baited with pigmentation and nectar. When trigger hairs inside the leaves sense multiple movements from live prey, the flytrap snaps shut, forming a botanical stomach that fills with digestive enzymes and liquefies prey. If one leaf fails, each successive leaf shall receive its chance.

as attractants. Each plant forms a colony of traps along the soil surface and hoisted a few inches into the air. Flytraps form treacherous ground for insects to traverse—if one trap fails to catch prey, each successive leaf shall receive its chance. The inner trap surface possesses touch-sensitive trichomes. When stimulated by the movements of prey, flytraps *move*, clamping shut and initiating the process of digestion. A flytrap's power of movement is a big commitment of the plant's resources, so its efforts are efficiently calculated and targeted. To avoid wasting energy closing on an inanimate object or a phenomenon such as a raindrop, a brush from a nearby plant in the wind, or a falling leaf, the flytrap requires multiple trichomes to be triggered by live, moving prey. At this point, it closes most of the way, leaving the bug in a tiny green jail cell created by the folded leaf and barred by the trichomes that line the leaf's margins. Each trap only gets a few shots at sealing completely, so the plant must be sure of the presence of edible prey. If the prey is too small, it will climb out, and the trap can reopen. If the insect's size is large enough, the trapped creature will continue spinning around, repeatedly hitting the trigger hairs and securing its fate. Flies, wasps, ants, crickets, beetles, spiders, snails, and slugs all meet their demise within these traps on a regular basis. After shutting airtight and suffocating their prey, flytraps pump these foliar "stomachs" full of digestive enzymes. After liquefying the bug's guts (and leaving the exoskeletons to be cleansed away in the next storm), flytraps slurp up the valuable nutrients using special glands lining their leaves. The entire process takes up to a week. During that time, if the light is just right, you can make out the silenced silhouette of the plant's meal.

Another type of carnivorous plant, the pitcher plant, sends bugs "down the tubes" using specialized leaves called pitfall traps. Pitfall traps are hollow, tube-like leaves. Multiple carnivorous genera have evolved pitchers convergently. These include the North American *Sarracenia* and *Darlingtonia*; the South American *Heliamphora*, *Brocchinia*, and *Catopsis*; the Indonesian *Nepenthes*; and the Australian *Cephalotus*. The leaves of pitcher plants form elegant, slippery tubes filled with pools of digestive juice. To attract prey, these pitchers are brightly colored, patterned with ultraviolet advertisements, and often perfumed with floral fragrances. Most

This yellow trumpet pitcher plant (*Sarracenia flava*) population from the coastal plain of Georgia is a spectacular sight. But these aren't flowers—they're vibrant, tube-shaped leaves! North American pitcher plants use a combination of visual and scent attractants, drugged nectar baits, slippery surfaces, and pointy trichomes to lure insects into the digestive fluid inside their pitchers.

dubious is that the outer surfaces of these plants' pitchers are loaded with extrafloral nectaries. Each trap is baited with beads of nectar that become richer in sugar the closer they are located toward the trap's treacherous opening. The top of a pitcher's long tube has an edge coated in slippery epidermal cells and flaky wax layers that offer no grip to the sticky pads and hooks on insect feet. This helps make it more likely that the insect will fall inside. Once in, the insect will find it quite difficult to escape, as the interior is lined with stiff, downward-pointing hairs that make it easy to go in but challenging to climb up. Some pitchers, in order to secure flying prey that might otherwise be able to fly out, lace their nectar with a paralytic that makes it difficult for an insect to fly successfully. Others, such as the hooded pitcher plant (*Sarracenia minor*) or cobra plant (*Darlingtonia californica*), put light windows—thin areas of tissue—on the hood of the plant stretching over the opening. If you've ever seen a fly at a window, then you have a good idea of how this works. The insect eventually wears itself out beating against these false openings, falling backward down into the pitcher.

Without needing to move at all, these marvels of evolution often fill to the brim with insects every season. Glands inside the pitchers secrete digestive juice to decompose prey. Botanical surfactants in the fluid break its surface tension such that prey sink immediately to the bottom. There is no hope of swimming out of the grasp of pitcher plants. A host of commensal invertebrates such as slime mites, maggots, mosquito larvae, and microorganisms also capitalize on the bug buffet, rending and shredding prey into smaller and more absorbable chunks—a disgusting but insanely effective practice.

Sundews (genus *Drosera*) and butterworts (genus *Pinguicula*), glistening terrors of the bogs, have leaves that are intricately festooned with glandular trichomes. Each trichome appears to be topped with sparkling droplets of

The mountain purple pitcher plant (*Sarracenia purpurea* var. *montana*) maintains pools of digestive liquid in its unique leaves. Botanical surfactants in the fluid break its surface tension such that prey sink immediately to the bottom. There is no hope of swimming out of the grasp of pitcher plants. A host of commensal invertebrates such as slime mites, maggots, mosquito larvae, and microorganisms also capitalize on the bug buffet, rending and shredding prey into smaller and more absorbable chunks.

Gloopy death awaits insects venturing too close to these sundew plants (genus *Drosera*) in the University of Georgia Plant Biology Teaching Collection. Within each pot of sundews grows the acidic sphagnum mosses that helped create the nutrient-poor bog soils that drove carnivorous plant evolution.

Each of this threadleaf sundew (*Drosera filiformis*) plant's glandular trichomes secretes a dewdrop full of nanofibers, glues, surfactants, disinfectants, and digestive enzymes. Sundews sense touch and then respond with slow and steady leaf movements. Each leaf blade and all nearby trichomes curl over prey, forming a slimy "bug sandwich."

"dew" or nectar freely available to any wayward insect. However, these droplets are among the most complex and threatening fluids exuded by any plant. Sundews and butterworts create a sticky situation for prey using these drops of powerful, gluey mucus. They've basically made themselves into large sticky traps to any small creatures who may be attracted in. If you rub your finger across these surfaces, you will come away with a long trail of sticky slime. Contained in these adhesives is some of the best prey-capturing biological nanotechnology, including nanofibers and nanoparticles that lend this seemingly simple fluid such viscoelasticity that it can be stretched into threads up to three feet in length! The fluid not only stretches but also contains other chemicals that have a high affinity for attaching to living cells of other organisms—an organic glue. Bugs have quite a lot of moving parts that can fall victim to glue, unfortunately. Their wings, antennae, and legs now become massive liabilities, as one brush with a sundew may be fatal. Surfactants, which reduce the surface tension of the dew, assist the fluid's ability to spread and coat insect exoskeletons. This not only helps secure prey but also has the potential to clog the insect's breathing pores (called spiracles), suffocating it. After all, prey that lives and struggles is prey capable of escape. Bugs can be germy little creatures, so the next set of chemicals found in the fluid are disinfectants. It is as if the plant is washing its hands and food before its meal! The last thing sundews and butterworts want is for their rotting prey to be consumed by bacteria and fungi or for dirty prey to unleash a pathogen into the plant. Digestive enzymes, the coup de grâce, liquefy prey right on the leaves, whose surfaces hold absorptive glands to capture the nutrient-rich slurry. On top of everything else, sundews and butterworts are capable of slow but determined movement. Each touch of their glandular trichomes activates action potentials that curl entire leaf blades into nice bug tacos and sandwiches. Every available trichome near the prey slowly bends to pinch it in a gloopy death grip. After all, catching prey is such a spectacular feat that these plants must protect their food from escape, rain, wind, and thieving insect predators. We have grown carnivorous plants for over a decade, and we still think that these plants have developed one of the most wicked ways to end insects.

Adrift in the shallow, stagnant waters of swamps lurk further dangers, this time for waterborne insects. Aquatic carnivorous plants known as the

Bladderworts bear thousands of underwater traps such as the ones seen here on a yellow bladderwort (*Utricularia australis*). The traps' ability to catch and digest small aquatic animals helps carnivorous plants survive and compete in nutritionally impoverished waters.

On bladderwort traps, prey is baited to the inward-opening door by sugary secretions from the doorframe cells as well as the bubble of fresh air inside the trap. Touching the feathery sensory hairs outside the trap sucks a column of water into the trap, along with the prey.

Adrift in the shallow, stagnant waters of swamps lurk dangers for waterborne insects. Aquatic carnivorous plants known as the bladderworts (genus *Utricularia*) hunt these waters using thousands of diminutive, powerful, and lightning-fast bladder traps. The delightful golden mass flowering of this inflated bladderwort (*Utricularia inflata*) population does not betray the presence of the deadly bladders lying in wait underwater.

bladderworts hunt these waters using thousands of diminutive, powerful, and lightning-fast bladder traps. Bladder traps grow preloaded with air, which not only helps bladderworts float but also baits prey looking for oxygen. Bladderwort prey includes tiny aquatic animals such as water fleas, protozoans, insect larvae (such as mosquitoes), nematodes, and, for large bladderwort species, even tiny young fish and tadpoles. Bladder traps feature a hinged, inward-opening door whose entrance is adorned with intricate, protruding trigger hairs. Cells around the trap's door secrete mucilage laced with sugar. Once an animal swims within proximity of the trigger hairs, the door opens and the surrounding water, ideally containing prey, is sucked into the trap. The mechanism works in as little as one-hundredth of a second, so fast that even slow-motion or high-speed cameras can barely capture bladderworts in action. While secreting enzymes to dissolve their prey, bladderworts pump all of the water back out of their traps. Bladderworts are capable of reloading their traps entirely in just fifteen to thirty minutes, making them one of the most ravenous and sophisticated carnivorous plants.

Plant nutrient acquisition goes down two evolutionary trajectories. The friendly route recruited the help of civilizations of beneficial microbes. The lethal route, taken by carnivorous plants, subjugates animals into liquid fertilizer. As you're sitting down for your next meal, think of how challenging it can be for plants to obtain nutrition.

A Salt Assault

We've looked at plant survival when there is a scarcity of resources, but plants can just as easily struggle in excessive quantities of minerals. One particularly difficult mineral for plants to contend with is salt at excessive levels. Our planet is so chock-full of salt that you can see it from space. Seventy percent of our planet's surface is covered with salty oceans, the place where photosynthetic life first evolved 4.5 billion years ago. As a result, almost 7 percent of the land area of Earth exposes plants to water and soils with excessive levels of salt. There are many kinds of salts, but the most commonly occurring one is sodium chloride (NaCl). You know it better as table salt.

That's not snow you're looking at! This is a habitat called a salt flat. The Great Basin of North America is a vast, dry area with numerous salt flats. The Bonneville Salt Flats near the Great Salt Lake in Utah are some of the largest.

Mangroves proliferate in saline coastal environments. Mangroves have succulent leaves with salt-excreting glands so powerful that they often leave salt crystals on the leaf surfaces. Their thick stilt roots, seen here, are coated with a spongy layer that acts as a salt filter, enabling mangroves to extract fresh water even from seawater!

Salt marshes, such as the one seen here on Tybee Island, Georgia, are teeming with salt-tolerant species such as saltmarsh cordgrass (*Spartina alterniflora*). *Spartina* species can pump excess salt away through special leaf glands.

While you're enjoying the sunshine and waves along the beach, the salt-adapted plants around you continue a struggle for survival dating back millions of years. Called halophytes (plants adapted for salt tolerance), these plants have amazing abilities that allow them to survive salt's assault on their tissues. Their briny cauldrons of evolution are predominantly coastal beaches and dunes. Farther inland, where the intense salinity of pure ocean water is mediated with fresh water, you'll also find these plants dominating brackish-water habitats such as marshes, mangrove swamps, and estuaries. The dry, saline soils of deserts also feature halophytes. This is because these areas experience low precipitation combined with high temperatures and evaporation rates, which concentrates salts in the soil. Over the past few thousand years, human agricultural activities have led to the emergence of farming with increasingly saline soils. More than 11 percent of the world's irrigated areas now contain salt-affected soils, and this trend continues to rise. There are numerous reasons for agricultural salinization, including climate change and poor agricultural practices. As temperatures around the world climb, rainfall patterns are changing. Droughts are becoming more frequent, prolonged, and intense, leaving less precipitation to dilute the salts in soils. As sea levels rise, coastal soils and the groundwater supplies utilized for irrigation face a greater saltwater intrusion than modern societies have ever faced. In addition to these pressures, improper agricultural practices continue to salinize arable land. We salt the earth when we clear deep-rooted forests for pasture and crops, irrigate with poor-quality saline groundwater and wastewater, and profligately abuse salt-laden fertilizers. Salinization is an expensive and often irreversible problem, too, costing billions of dollars per year worldwide in crop losses and reduced yields. Entire landscapes of flora can be obliterated when a soil has become too saline. The unusual use of salt-tolerant genetics and strategies inside halophytes may offer us a variety of ways to abate the agricultural and environmental damages of salinization.

Salinity is a tremendously difficult stress for plants, including many vital agricultural crops. For billions of years, marine algae have had no problem adapting to oceanic conditions. Oddly enough, this tolerance seems to have waned as their terrestrial descendants evolved to life on dry land. Today, only 1–2 percent of all higher land plants can survive in near-seawater salt concentrations. Scientists have uncovered more than seventy evolutionary lineages of plants that each independently evolved its own solutions to succeed in excessively saline substrates. Overall, salt tolerance in plants is an incredibly rare phenomenon.

As the Tetris-like salt molecules accumulate around them, halophytes must find ways to move, store, or excrete them before they reach threatening levels. These fascinating solutions are what make halophytes truly distinct from other drought-tolerant plants.

With salt molecules, plants face a problem that's a lot like playing the video game Tetris. In Tetris, players race to manage continuously falling blocks and lose when too many accumulate. Similarly, plants face a serious overaccumulation and management problem with salt in their environments. As salt accumulates ad infinitum, plants must successfully find places to quickly stash the accumulation of molecules lest the plants dry out and die. Salt is so ubiquitous in the water, soil, and ocean spray that it builds up inside plant cells at every opportunity. There are two main issues with salt that make it toxic. First, salt molecules are highly attracted to the same water molecules vital to living cells, so they're capable of depriving plants of moisture and desiccating them to death. Plants, as the multicellular turgid green water balloons they are, need fresh moisture to survive. Plants' tissues quest for water in a game of tug of war with ambient salt molecules. Halophytes are more successful in that game because they've evolved ways to harvest fresh water from incredibly saline environments. Second, if the ions (charged elements) of salt get inside the cell, they wreak havoc on cellular processes. Salt is composed of the ions of sodium (Na^+) and chlorine (Cl^-), and each of these differently charged ions can disrupt the delicate balances within cells. In excess, salt ions interfere with vital processes such as photosynthesis and the transportation of other ions such as potassium (K^+), and they induce oxidative stress. As we'll recall from the flood chapter, oxidative stress is caused by the proliferation of reactive oxygen species, also known as free radicals. Reactive oxygen species are a type of unstable molecule that contains oxygen and that easily and adversely reacts with other molecules in a cell. Halophytes evolved survival systems that protect the water and reactions within their cells from salt intrusion. To do so, they safely micromanage how salt moves into, through, and out of their cells. As the Tetris-like salt molecules accumulate around them, halophytes have evolved ways to move, store, or excrete them before they reach threatening levels. These fascinating solutions are what make halophytes truly distinct from other drought-tolerant plants. Gather water, exclude salt—that's the key to halophyte survival.

To ensure they have a steady supply of fresh water at all times, halophytes have evolved tissues much like drought-tolerant desert plants. Thick, waxy cuticles, protected and recessed stomata, and succulent leaves come standard on most halophytes. Most halophytes have also evolved the C_4 photosynthesis pathway, which is a highly water-efficient type of photosynthesis often employed by drought-tolerant plants. Once halophytes have taken up fresh water, these adaptations both protect the supply and regulate its use at maximum efficiency.

Salt-intolerant plants absorb sodium chloride ions passively until they accumulate to toxic and drying levels. Because salt is so attracted to dissolving in water, plants that can't screen it out when they drink are in trouble. Halophytes, however, have a genetic arsenal of ion regulation that strictly controls every movement of salt ions. The membranes of halophyte cells and vacuoles have specialized gates that regulate and restrict the passage of sodium chloride ions, much like mandatory security checkpoints or bouncers outside public establishments. This makes halophytes effective at many levels of salt regulation, as salt is prevented from entering cells in the first place. It is guided through the cytoplasm safely without interfering with any vital chemical reactions. Finally, it is locked away or dumped in a final destination. This might be the cell's vacuole, a salt bladder, or a salt-spitting gland. We'll demonstrate how all of these work shortly. Salt only has a one-way ticket through the cell. The salt moves completely through the plant, ideally, without ever being able to interact with it in a damaging manner.

Halophytes essentially invented salt imprisonment and expulsion. One of these salt prisons is the vacuole. All plant cells have vacuoles, which serve as a multipurpose area for toxin storage, waste product dumping, and excess water storage. Halophytes lock salt in theirs, so the vacuole becomes incredibly saline while the rest of the cell and its organelles safely operate in fresh, pure water. Basically, the cell creates its own botanical desalinization. This works because halophytes have evolved specialized salt-restriction gates and transporter molecules. The concentrated salt in these vacuoles still has an intense atomic attraction to water molecules, a

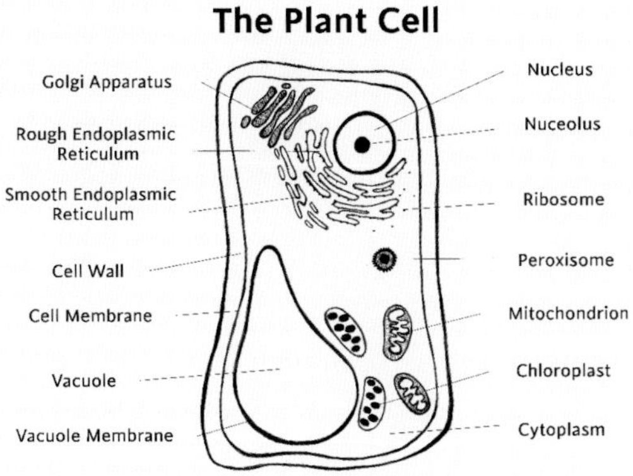

The Plant Cell

Golgi Apparatus
Rough Endoplasmic Reticulum
Smooth Endoplasmic Reticulum
Cell Wall
Cell Membrane
Vacuole
Vacuole Membrane
Nucleus
Nuceolus
Ribosome
Peroxisome
Mitochondrion
Chloroplast
Cytoplasm

Halophytes essentially invented salt imprisonment and expulsion. One of these salt prisons is the vacuole, seen above. All plant cells have vacuoles, which serve as multipurpose storage areas. Halophytes lock salt in theirs, desalinating the water that the rest of the cell operates safely in. The massive levels of salt building inside these vacuoles attract even more fresh water into the cell. When salt levels in their tissues reach critical mass, halophytes simply drop their leaves.

fact not lost on halophytes. Using all of this freely available salt, halophytes turn their vacuoles into electromagnetic lures that increase the amount of freshwater molecules entering their cells. Halophyte roots are especially adept at using this technique. Even if they grow in seawater, extreme halophytes electromagnetically lure fresh water to themselves, and their ion regulation takes care of the rest by acting as a bouncer to incoming salt ions. When its foliage's lifespan is complete, the plant simply sheds it, along with the excess of salt it has collected, back into the environment. Halophyte leaves, such as those in mangroves, are often referred to as sacrificial. The whole halophyte is always in transition as new leaves emerge for salt storage and old ones sequester and shed salt away from the plant.

Other types of salinity tolerance have evolved across the world. Another type of salt prison is the salt bladder, which is a balloon-like cell that evolved from either epidermal cells or repurposed trichomes. As seen in the ice plant photo, salt bladders make plants look like they're covered in shiny, crystalline drops of dew. Within these bladders you'll find highly concentrated salt water, which creates a salty atomic attraction to draw clean water into the plant. Many halophytes possess salt bladders of various sizes, but they are especially evident within the ice plant and amaranth families. Ice plants, quinoa, and saltbushes have stretchy salt bladders on all their leaves, flowers, and stems that are large enough to be seen with the naked eye. The light-scattering, glistening bladders have an additional

Salt bladders, a feature of halophytes like this ice plant (*Mesembryanthemum crystallinum*), offer extra storage space for fresh water as well as excess salt.

benefit of helping cool the plant's surface by reducing light and heat from the sun. Salt bladders simultaneously reduce water use while increasing water uptake. The bladders are so good at absorbing salt that older ones eventually explode, releasing a salty, crusty film onto the plant's leaves that is eventually removed by rain or tidewaters. While it remains, however, this white, crystalline crust is highly reflective to both light and heat and is also defensive against herbivores. Salt can be intensely drying to hungry mammals and bugs, so even while it isn't a plant-produced chemical poison, it works just as effectively. Scientists have found that salt bladders play a great role at reducing herbivory of halophytes. Given a choice between leaves with salt bladders and leaves with bladders removed, insects greatly preferred their food unsalted.

Salt-spitting glands are another halophyte invention. These specialized adaptive structures occur primarily on the plant's leaves and stems. They are efficient desalinization tools that remove salt from plant tissues. The way they perform this feat is by secreting the salt to the outside environment, but the process must be powered by some of the plant's energy supplies. The glands have evolved in eleven families of plants, including the amaranth family (including the saltbushes and quinoa), the grass family (saltmarsh cordgrass, saltgrass, seashore *Paspalum*), the verbena family (mangroves), and more. Like the salt crust from exploded salt bladders, expelled salt is useful for its sun-reflective properties as well as its defensive function against herbivores. Mangroves are so adept at pumping salt through these glands that you'll often find salt crystals on both surfaces of their leaves. Salt glands help clear room within the plant for when more salt inevitably arrives from the surrounding environment.

Salt survival adaptations in plants have been responsible for critical points in human history. In ancient Mesopotamia and Egypt, early civilizations

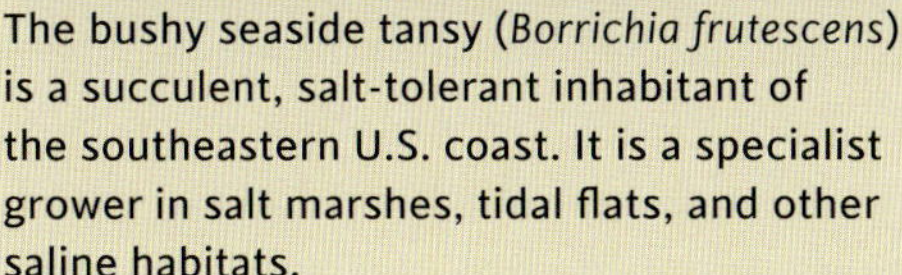

The bushy seaside tansy (*Borrichia frutescens*) is a succulent, salt-tolerant inhabitant of the southeastern U.S. coast. It is a specialist grower in salt marshes, tidal flats, and other saline habitats.

Sea beans (genus *Salicornia*) are widespread, edible halophytes. They possess succulent tissues for water storage, reduced leaves for water conservation, and the ability to sequester sodium chloride in their vacuoles. The salty flavor of sea beans makes them a brilliant pairing with seafood.

Salt-tolerant plants have been responsible for critical points in human history, like the development of pottery glazing, glassmaking, paper, and soap. The ash of halophytes such as the saltwort (*Salsola soda*) is extremely rich in sodium carbonate (soda ash), which is useful for the chemical reactions required by these processes. This Egyptian scarab, dating from circa 1070–945 BCE, was made out of a glass-like glaze called faience around the same time as the invention of glassmaking.

accidentally discovered that the ash of burned halophytes has important chemical properties. When halophytes were burned by these societies, their salt-rich tissues generated ashes that were especially rich in sodium. This is distinctly different from normal plants, whose ash is predominantly rich in carbon. Thus, these halophytes produce one of the highest natural sources of soda ash (sodium carbonate). Soda ash made new chemical reactions possible. While this may not sound like much, plant-derived soda ash contributed to the invention and development of civilization's first pottery glazes, glass, paper, and soap. Even the words "sodium" and "soda" originate from the name for the Mediterranean-native halophyte *Salsola soda*, the humble saltwort. We have a lot of salt-tolerant dead and burned plants to thank for the origins of many products we use today.

Several halophytes around the world also make great food. Imagine cooking with a vegetable that was already perfectly seasoned because of its ability to take up salt from the environment. For thousands of years, salt-tolerant plants such as pearl millet, *Salsola soda*, sea beans, New Zealand spinach, and quinoa have been mainstays in the diets of human cultures. Many other obscure and undomesticated halophytes are also useful and being scientifically explored as crops. Wild, salt-tolerant relatives of crops such as sunflowers and wheat may hold genetics that allow plant breeders to improve salt tolerance in crop plants. This would allow us to grow these crops in damaged or compromised fields that would otherwise be unusable. Saltwater cress (*Thellungiella halophila*) is a diminutive, weedy-looking plant in the mustard family that grows in extremely salty habitats. It happens to be a halophytic relative of the scientifically important model plant *Arabidopsis*, and the genomes of both organisms have been sequenced for better understanding of their adaptations. Because of this apt genetic comparison, these two plants have given us amazing insight into understanding the differences between and evolutionary history of halophytes and nonhalophytes. Environmentally friendly, salt-tolerant crops produced from this research and breeding could someday tolerate more marginal environments and utilize fewer inputs (such as fresh water) than they do now.

Halophytes aren't just being explored as food. Using their ability to sequester and mitigate salt, they can be used to revegetate and purify water and soils ruined by salinization or industrial pollution. Imagine a chance for green to emerge once more even in disastrous barrens such as the former Aral Sea. Once the fourth-largest lake in the world, the Aral began salinizing and drying up in the 1960s as a result of massive Soviet diversion of its waters for irrigation. By the 1980s the sea had divided in two (North and South), with further divisions occurring in the 2000s. Most of the eastern side of the sea had disappeared by the end of the 2000s. This unprecedented and rapid loss of an entire ecosystem created salt and toxin contamination of the soil, generating dust storms around the region. This area has now come to be called the Aralkum Desert. The survival systems of wild plants offer ways we might, in part, repair or prevent these types of events. These destroyed agricultural areas could once again have life if salt-tolerant species could help cleanse their soils.

As ornamental crops, many halophytes can be irrigated with seawater rather than fresh water, and the salt even kills the weeds! *Paspalum vaginatum*, the seashore *Paspalum*, is a tough, salt-tolerant coastal dune grass that is gaining attention for lawns and golf courses. These artificial environments are traditionally quite heavy on freshwater use, but *Paspalum* looks perfect even when watered with pure seawater. However, seawater irrigation is an extreme solution with the downside that it tends to contaminate soils for future use. Still, this offers a solution in areas where soils are already damaged or quality fresh water is otherwise limited. Other halophytes, such as *Salicornia bigelovii*, are emerging oilseed crops, biofuels, and biofilters for salt-contaminated effluent.

It's a salty world here on our well-seasoned planet. We hope that just a few more of the commonplace events in your life might connect you to the hidden world of plant salt survival. As you sprinkle salt crystals to flavor your next meal, pause to contemplate how plant life is fighting against the assault of salt.

Heavy Metal Rocks: Metallophytes and Hyperaccumulation

Snaking through the geology of Earth's continents are bands of rock called serpentinite, uplifted from their metamorphic origins in the extreme depths of Earth's mantle. Serpentinite rocks receive their name from their eerie green hue, scale-like patterning, and greasy feel to the touch. Serpentine soils are weathered from this rock by erosion as well

Seen here is a serpentine rock, naturally laden with levels of heavy metals that have contributed to the evolution of metallophytes and hyperaccumulators. Serpentine's name comes from its greenish color, scale-like patterning, and slippery texture.

as mining. Problematically, serpentine rock contains botanically deadly levels of heavy metals. Plants utilize many metals as micronutrients, but serpentine formations simply have too much of the wrong ones. Serpentine soils are extremely unique and varied, but their unifying trait is their contamination with one or more metals such as lead, copper, molybdenum, nickel, mercury, selenium, cadmium, chromium, zinc, magnesium, manganese, arsenic, iron, and cobalt. Additionally, "good" nutrients such as nitrogen, phosphorus, potassium, and calcium are usually scarce in serpentine soils. Plants thus face a nutritional double whammy on serpentine, and it's miraculous any of them survive at all. Serpentine flora is generally uncommon but can be found in the Calaminarian grasslands in the Scottish highlands, the island territory of New Caledonia, the Klamath Mountains of Oregon and California, Soldiers Delight Natural Environment Area in Maryland, and outcroppings in the Alps, Appalachians, and Balkan peninsula.

Enter the metallophytes, metal-loving superhero plants that have evolved to meet the challenges of these barren lands. The survival systems of metallophytes use three adaptations that are quite different from those of other plants. First, they deliberately extract heavy metals from the soil. Second, they move them much faster through their tissues. Third, they have a greater ability to detoxify and sequester the metals inside their cell walls, leaves, and roots. As a result of these abilities, many metallophytes are also called hyperaccumulators. The term "hyperaccumulator" simply refers to these extraordinary plants, which may concentrate metals in their living tissues at levels hundreds or thousands of times higher than is typical for most plants. Metal levels can climb so high that they are normally considered toxic or even lethal to nonhyperaccumulating plants. Hyperaccumulation not only allows growth in some of the world's worst soils, including the slag left behind from mines and industrial disasters, but also fights herbivores. Hyperaccumulators become so metallicized that they gain the ability to poison things that eat them. Imagine if your next salad were rich in arsenic instead of vitamins! That's what metallophyte herbivores face. If you were hiking around a serpentine outcrop, you might just think the diminutive plants around you were harmless wildflowers. However, concealed discreetly within their tissues is a chemical armor that keeps them alive.

So how do metallophytes accomplish this feat? They actually use a combination of strategies similar to what we've already seen in plants adapted to low nutrition and salt. Because there are so many different unique serpentine soils, equally plentiful are plants' solutions to dealing with heavy metals. Metallophyte violets, such as *Viola lutea* from the Calaminarian grasslands and the Alps in Europe, employ arbuscular mycorrhizal fungi to capture and filter heavy metals before they can enter the plant's cells. *Armeria maritima* subsp. *helleri*, also from the Calaminarian grasslands, simply squirts out excess lead and zinc using metal glands (much like

The bladder campion (*Silene vulgaris*) is a hyperaccumulating metallophyte. It can tolerate soils polluted by heavy metals, including zinc and cadmium, and is being evaluated for its potential to remediate old mine spoil tips and areas surrounding sheet metal galvanizing factories.

the way some halophytes excrete salt). *Silene vulgaris*, *Minuartia verna*, *Quercus durata*, and *Pteris vittata* create metal-loaded sacrificial leaves full of copper, zinc, cadmium, and arsenic. The plants thereby become toxic to herbivores. When the metal content of these tissues reaches dangerous levels, the plants simply dump the threat by dropping their leaves. Metallophytes in the mustard family such as *Arabidopsis halleri* and *Thlaspi caerulescens* are some of our best scientific models for how plants deal with metal stress. Most metallophytes have genes that produce protective chemicals (such as proline) that shield their cells and biological processes from damage by metals. Metallophytes also carefully guide heavy metals safely through their cells using specialized transporter molecules. This is quite similar to the strategy halophytes use for salt. The metals are bound into the construction of new cell walls, locked away in vacuoles, or pumped completely out of glands in a metallophyte's leaf.

Heavy metal pollution is a significant global issue. Chemical spills, industrial accidents, and mining slag have deposited toxic levels of heavy metals into the environment. We've unleashed arsenic from pesticides and wood preservatives. Cadmium contaminants originate in the waste from electroplating, chemical pigments, and spent batteries. Chromium is a frequent pollutant from tanneries and steelworks. Lead originates from battery waste, the spraying of herbicides, and the petroleum industry. Mining slag is a common source of copper, nickel, and zinc. Mercury emanates from lighting, scientific instruments, and medical waste.

Hyperaccumulating plants are being increasingly utilized to rescue these wastelands, however, in a technique called phytoremediation. Phytoremediation technologies use living plants to extract and remove hazardous contaminants from soil, air, and water. In this process, we deliberately inoculate some of the nastiest sites with hyperaccumulators capable of cleaning up heavy metals. Such plants are incredibly efficient, although an unsolved problem is that the metal-laden vegetation still has to be moved, processed, and disposed of somewhere else. Usually, the vegetation is harvested and sent to hazardous material landfills, which have specialized techniques to isolate the contaminants from the surrounding environment. Phytoremediation is a slow but sure and cost-effective process. Only a fraction of the contaminants in a site are removed each year, but progress is definitely in the right direction. Depending on the severity

New Caledonia's extremely serpentine geology is home to almost 10 percent of the world's nickel reserves. On the iron-, cobalt-, and nickel-rich soils shown here, metallophytes take on surreal, blue-green colors because of their hyperaccumulation of heavy metals. Even their exuded sap can be blue or turquoise at high levels of hyperaccumulation. New Caledonia, one of the world's metallophyte hot spots, contains more than twenty-one hundred plants that survive metallic soils, and 80 percent of these plants are found nowhere else in the world.

of the contamination, it may take decades or longer before sites are completely restored.

There is also interest in phytomining. This is the use of hyperaccumulators to sequester desirable metals such as copper, zinc, and nickel. In this situation, the plant biomass becomes a rich "bio-ore" whose tissues hold a strikingly high percentage of valuable metal. Nickel phytomines in Indonesia, for example, have reported yields of up to 25 percent nickel per unit of dry biomass. The trees in these forests can be crushed into an eerie, blue-green, nickel-colored juice whose nickel makeup is more concentrated than most of the traditional nickel ores mined from the earth. Yields of one to three hundred pounds of metal per acre every six to twelve months isn't unheard of on metal farms like these. Scientists have even found ways to trick hyperaccumulators into uptaking metals they normally wouldn't, such as gold, silver, palladium, and platinum. Once the metal crops are harvested, metal can be extracted from the resulting biomass via burning (pyrometallurgy), liquid solutions (hydrometallurgy), or other industrial processes. All of these rare and valuable metals are incredibly useful for a variety of products, including catalytic converters, jewelry, electronics, stainless steel, plumbing, scientific instruments, medical equipment, and coinage.

No matter how the very land itself poisons and deprives them, plants have triumphed. Without the ability to choose where they begin to grow, plants either eke out a living or perish. The grass may be greener in more verdant lands, but soil extremophile plants have the strength to be green amid some of the bleakest and most challenging places on Earth. We have seen plants at their most friendly, stewarding entire civilizations of microbes. We have seen them at their most murderous, wringing nutrients from the innards of hapless bugs. We've also seen them at their most stressed as salts and metals push them to the brink of existence. We hope the hidden stories of their struggles will be inspiration the next time life gives you lemons.

Epilogue

We hope this work has illustrated the unusual and often unnoticed world of plants. Their continual struggle for survival is as fascinating as it is complex. *Hidden World* has attempted to highlight a few of the massive evolutionary milestones that have made plant defenses the formidable systems we see today. We hope this book has given you a better understanding of the complex, hidden network operating around you every day. Deep and astounding struggles in the natural world parallel and overlap our lives.

Whether we realize it or not, plants intertwine with the entire natural world. It is a relationship we benefit from. Plant survival systems are not only a captivating subject but also of colossal importance in ecology and human societies. We wind up becoming major benefactors of plants' success. When plants thrive, we also thrive. The battles raging throughout this secret world hold great consequences for the ecological web, including all of humankind. Earth's countless species need plants for food, shelter, moisture, and reproduction. If plants fail to survive, so do all the animals that depend upon them. Biodiversity is one of life's strongest insurance policies against large-scale catastrophe. Plant life has emerged, though severely scathed, through multiple mass extinctions—our planet's darkest times. Through the crucible of natural selection, plant survival systems continually grow stronger, more inventive, and more resilient. The extraordinary pressures plants face on a daily basis, as unpleasant as it sounds, prepare them for continued survival. Plant defense evolution is an ongoing story. In many of the ways we interact with plants, we actively participate in and shape the future of this story. When we set aside natural preserves, biodiversity has the space it needs to continue to thrive and evolve as it has done for many millions of years. Artificial selection, the selection and breeding of species with valuable properties for agriculture, ornament, industry, and science, has also influenced the path taken by many plants' survival systems. Environmental resilience, pest resistance, beauty, and the yield of useful chemicals or food are all reasons we propagate the lineages of some plants as opposed to others.

Our survival is dependent upon our flora's successful and continued survival. Plants feed us, clothe us, and shelter us from the environment. Plant tissues hold up our buildings, strengthen our tools, and fuel our heat sources and machinery. The chemicals plants manufacture for their own defense yield medicines, flavors, scents, and components of many important products, industrial processes, and technologies. The collapse of a plant species inevitably ricochets through our economy and well-being. Enormous ramifications are at stake when plant survival is thrown out of balance or fails to succeed. This is especially true during this global industrial age. Humans have converted land for a variety of uses, altered the climate, introduced a variety of chemicals, and moved animal and plant species around the world. With these changes, we have seen examples of plant defenses overwhelmed, from disease outbreaks in our inbred or monocultured crops to newfound pests attacking our native plants and everything in between. These changes can be costly for us and can threaten our food security, our air and water quality, and our natural resources. Those of us who have never picnicked in the shade of an American elm, roasted American chestnuts, or savored a 'Gros Michel' banana are living in a world where plant defenses occasionally and completely fail. Complete defeats of plant survival systems are sporadic but devastating. In the disease realm, even the name of each pathogen evokes powerful sentiments about its wrath: chestnut blight, Dutch elm disease, the Irish Potato Famine, rice blast, citrus greening, Panama disease, and sudden oak death. This is also true of destructive herbivores: the hemlock woolly adelgid, boll weevil, emerald ash borer, plant-parasitic nematodes, gypsy moth, and Japanese beetle. Greater consciousness of the magnitude of these ecological issues can help us find ways to prevent and manage them.

Frequently, mundane activities within our lives bring us in contact with the hidden world. Pruned branches and mown landscapes mysteriously heal and regenerate. A peeled citrus fruit may squirt us in the eye with the fungicidal repellent oil that protects its offspring. The pigmented palette of our flora surrounds us with visual delight, carrying cryptic warnings all the while. The air around us and the ground below us flow with strange communications and bartering systems we scarcely perceive. The smells and tastes of our food enrich our lives—and, secretly, our health. But in the process, onions may sting our eyes, chilies may singe our tongues, and bitter mustard family vegetables will remain the bane of picky eaters everywhere. We may find some plant survival strategies unpleasant, but we are supposed to. Perhaps we might come to understand the necessity behind such wickednesses as spines, trichomes, poisons, bitter repellents, and barriers of silica and wood. It is likely we would fight just as hard for our own survival.

The natural world we belong to is precious. This work is dedicated to inspiring a greater awareness of and appreciation for our planet's wonders. We hope the insights we have provided here may enrich your quality of life, heighten your sense of belonging to this planet, and spur your sense of stewardship toward biodiversity. What an amazing "pale blue dot" we are a part of! Plants are *fascinating*. The story of natural history they tell is incredible. The systems keeping them alive are as intricate as they are extraordinary. We sincerely hope this work succeeds in bringing nature alive by opening up entirely new perspectives from which you can view the world around you. The vantage point we have as inhabitants of this spectacular planet is a privilege worth savoring forever.

Appendix of Featured Plants

Plants are listed in alphabetical order by family scientific name.

Common name	*Family common name*	*Family scientific name*	*Genus*	*Species*
Elderberry	Moschatel	Adoxaceae	*Sambucus*	Various
Cone plants	Fig-marigold	Aizoaceae	*Conophytum*	Various
Ice plants	Fig-marigold	Aizoaceae	*Delosperma*	Various
Baby's toes	Fig-marigold	Aizoaceae	*Fenestraria*	Various
Living rocks	Fig-marigold	Aizoaceae	*Lithops*	Various
Split rock plant	Fig-marigold	Aizoaceae	*Pleiospilos*	*nellii*
New Zealand spinach	Fig-marigold	Aizoaceae	*Tetragonia*	*tetragonioides*
Titanopsis	Fig-marigold	Aizoaceae	*Titanopsis*	Various
Palmer amaranth	Amaranth	Amaranthaceae	*Amaranthus*	*palmeri*
Saltbushes	Amaranth	Amaranthaceae	*Atriplex*	Various
Beet	Amaranth	Amaranthaceae	*Beta*	*vulgaris*
Quinoa	Amaranth	Amaranthaceae	*Chenopodium*	*quinoa*
Chenopods	Amaranth	Amaranthaceae	*Chenopodium*	Various
Glassworts	Amaranth	Amaranthaceae	*Salicornia*	Various
Saltwort	Amaranth	Amaranthaceae	*Salsola*	*soda*
Spinach	Amaranth	Amaranthaceae	*Spinacia*	*oleracea*
Leeks	Amaryllis	Amaryllidaceae	*Allium*	*ampeloprasum*
Wild onion	Amaryllis	Amaryllidaceae	*Allium*	*canadense*
Onion, shallots	Amaryllis	Amaryllidaceae	*Allium*	*cepa*
Star of Persia	Amaryllis	Amaryllidaceae	*Allium*	*cristophii*

Common name	*Family common name*	*Family scientific name*	*Genus*	*Species*
Giant allium	Amaryllis	Amaryllidaceae	*Allium*	*giganteum*
Ornamental onion	Amaryllis	Amaryllidaceae	*Allium*	*hollandicum*
Garlic	Amaryllis	Amaryllidaceae	*Allium*	*sativum*
Chives	Amaryllis	Amaryllidaceae	*Allium*	*schoenoprasum*
Ramps	Amaryllis	Amaryllidaceae	*Allium*	*tricoccum*
Scallions	Amaryllis	Amaryllidaceae	*Allium*	Various
Fire lilies	Amaryllis	Amaryllidaceae	*Cyrtanthus*	Various
Daffodil	Amaryllis	Amaryllidaceae	*Narcissus*	Various
Mango	Sumac	Anacardiaceae	*Mangifera*	*indica*
Pistachio	Sumac	Anacardiaceae	*Pistacia*	*vera*
Sumacs	Sumac	Anacardiaceae	*Rhus*	Various
Quebracho	Sumac	Anacardiaceae	*Schinopsis*	Various
Cilantro/coriander	Carrot	Apiaceae	*Coriandrum*	*sativum*
Cumin	Carrot	Apiaceae	*Cuminum*	*cyminum*
Carrot	Carrot	Apiaceae	*Daucus*	*carota* subsp. *sativus*
Fennels	Carrot	Apiaceae	*Ferula*	Various
Parsley	Carrot	Apiaceae	*Petroselinum*	*crispum*
Anise	Carrot	Apiaceae	*Pimpinella*	*anisum*
Hedge parsley	Carrot	Apiaceae	*Torilis*	*arvensis*
Milkweeds	Dogbane	Apocynaceae	*Asclepias*	Various
Carrionflowers	Dogbane	Apocynaceae	*Hoodia*	Various
Carrionflowers	Dogbane	Apocynaceae	*Huernia*	Various
Oleander	Dogbane	Apocynaceae	*Nerium*	*oleander*
Madagascar palm	Dogbane	Apocynaceae	*Pachypodium*	Various
Stick plants	Dogbane	Apocynaceae	*Rhytidocaulon*	Various
Carrionflowers	Dogbane	Apocynaceae	*Stapelia*	Various
Gallberry	Holly	Aquifoliaceae	*Ilex*	*glabra*
Guayusa	Holly	Aquifoliaceae	*Ilex*	*guayusa*
Yerba maté	Holly	Aquifoliaceae	*Ilex*	*paraguariensis*
Yaupon	Holly	Aquifoliaceae	*Ilex*	*vomitoria*

Common name	Family common name	Family scientific name	Genus	Species
Hollies	Holly	Aquifoliaceae	Various	Various
Corpse flowers	Arum	Araceae	*Amorphophallus*	Various
Jack-in-the-pulpit	Arum	Araceae	*Arisaema*	*triphyllum*
Elephant ear	Arum	Araceae	*Caladium*	Various
Calla lilies	Arum	Araceae	*Calla*	*palustris*
Elephant ear, Taro	Arum	Araceae	*Colocasia*	Various
Dumb canes	Arum	Araceae	*Dieffenbachia*	Various
Duckweed	Arum	Araceae	*Lemna*	Various
Golden club	Arum	Araceae	*Orontium*	*aquaticum*
Philodendrons	Arum	Araceae	*Philodendron*	Various
Water lettuce	Arum	Araceae	*Pistia*	*stratiotes*
Skunk cabbage	Arum	Araceae	*Symplocarpus*	*foetidus*
Devil's walkingstick	Ginseng	Araliaceae	*Aralia*	*spinosa*
Kauris	Araucarians	Araucariaceae	*Agathis*	Various
Jelly palms	Palm	Arecaceae	*Butia*	Various
Coconut	Palm	Arecaceae	*Cocos*	*nucifera*
African oil palm	Palm	Arecaceae	*Elaeis*	*guineensis*
Açaí palm	Palm	Arecaceae	*Euterpe*	*oleracea*
Date palm	Palm	Arecaceae	*Phoenix*	*dactylifera*
Royal palms	Palm	Arecaceae	*Roystonea*	Various
Cabbage palms	Palm	Arecaceae	*Sabal*	Various
Saw palmetto	Palm	Arecaceae	*Serenoa*	*repens*
Windmill palms	Palm	Arecaceae	*Trachycarpus*	Various
Palms	Palm	Arecaceae	Various	Various
Washingtonia	Palm	Arecaceae	*Washingtonia*	Various
Ant plant	Milkweed	Asclepiaceae	*Dischidia*	Various
Agaves	Asparagus	Asparagaceae	*Agave*	Various
Asparagus	Asparagus	Asparagaceae	*Asparagus*	*officinalis*
Ponytail palm	Asparagus	Asparagaceae	*Beaucarnea*	*recurvata*
Camas	Asparagus	Asparagaceae	*Camassia*	*quamash*

Common name	Family common name	Family scientific name	Genus	Species
Soaproot	Asparagus	Asparagaceae	*Chlorogalum*	Various
Dragon trees	Asparagus	Asparagaceae	*Dracaena*	Various
Garden hyacinth	Asparagus	Asparagaceae	*Hyacinthus*	*orientalis*
Snake plant	Asparagus	Asparagaceae	*Sansevieria*	Various
Yuccas	Asparagus	Asparagaceae	*Yucca*	Various
Aloes	Asphodel	Asphodelaceae	*Aloe*	Various
Bellyflowers	Asphodel	Asphodelaceae	*Gasteria*	Various
Haworthias	Asphodel	Asphodelaceae	*Haworthia*	Various
Australian grass trees	Asphodel	Asphodelaceae	*Xanthorrhoea*	Various
Bird's nest fern	Spleenwort	Aspleniaceae	*Asplenium*	*nidus*
Burro-weed	Aster	Asteraceae	*Ambrosia*	*Dumosa*
Cheesebush	Aster	Asteraceae	*Ambrosia*	*Salsola*
Burdock	Aster	Asteraceae	*Arctium*	Various
Tarragon	Aster	Asteraceae	*Artemisia*	*dranunculus*
Sagebrush	Aster	Asteraceae	*Artemisia*	Various
Asters	Aster	Asteraceae	*Aster*	Various
Beggarticks	Aster	Asteraceae	*Bidens*	Various
Horseweed	Aster	Asteraceae	*Conyza*	*canadensis*
Brittlebush	Aster	Asteraceae	*Encelia*	*farinose*
Sunflowers	Aster	Asteraceae	*Helianthus*	Various
Georgia aster	Aster	Asteraceae	*Symphytotrichum*	*georgianum*
Everlasting snow	Aster	Asteraceae	*Syncarpha*	*vestita*
Dandelion	Aster	Asteraceae	*Taraxacum*	Various
Cocklebur	Aster	Asteraceae	*Xanthium*	Various
Jewelweed	Balsam	Balsaminaceae	*Impatiens*	*capensis*
Begonias	Begonia	Begoniaceae	*Begonia*	Various
Barberry	Barberry	Berberidaceae	*Berberis*	Various
Oregon grape	Barberry	Berberidaceae	*Mahonia*	*aquifolium*
Alders	Birch	Betulaceae	*Alnus*	Various
Birches	Birch	Betulaceae	*Betula*	Various

Common name	Family common name	Family scientific name	Genus	Species
Hound's-tongue	Borage	Boraginaceae	*Cynoglossum*	*officinale*
Whispering bells	Borage	Boraginaceae	*Emmenanthe*	*penduliflora*
Rockcress	Crucifers	Brassicaceae	*Arabidopsis*	*halleri*
Rockcress	Crucifers	Brassicaceae	*Arabidopsis*	Various
Horseradish	Crucifers	Brassicaceae	*Armoracia*	*rusticana*
Cabbage	Crucifers	Brassicaceae	*Brassica*	*oleracea*
Kale	Crucifers	Brassicaceae	*Brassica*	*oleracea*
Broccoli	Crucifers	Brassicaceae	*Brassica*	*oleracea* var. *botrytis*
Cauliflower	Crucifers	Brassicaceae	*Brassica*	*oleracea* var. *botrytis*
Brussels sprouts	Crucifers	Brassicaceae	*Brassica*	*oleracea* var. *gemmifera*
Bok choi	Crucifers	Brassicaceae	*Brassica*	*rapa* subsp. *chinensis*
Rapeseed, canola	Crucifers	Brassicaceae	*Brassica*	*rapa* subsp. *oleifera*
Turnip	Crucifers	Brassicaceae	*Brassica*	*rapa* subsp. *rapa*
Cruciferous vegetables	Mustard	Brassicaceae	*Brassica*	Various
Mustard	Crucifers	Brassicaceae	*Brassica* and *Sinapis*	Various
Camelina	Crucifers	Brassicaceae	*Camelina*	*sativa* var. *linicola*
Arugula	Crucifers	Brassicaceae	*Eruca*	*vesicaria*
Wasabi	Crucifers	Brassicaceae	*Eutrema*	*japonicum*
Watercress	Crucifers	Brassicaceae	*Nasturtium*	*officinale*
Radish	Crucifers	Brassicaceae	*Raphanus*	*raphanistrum* subsp. *sativus*
Indian hedgemustard	Crucifers	Brassicaceae	*Sisymbrium*	*orientale*
Saltwater cress	Crucifers	Brassicaceae	*Thellungiella*	*halophila*
Pennycress	Crucifers	Brassicaceae	*Thlaspi*	*caerulescens*
Pineapple	Bromeliad	Bromeliaceae	*Ananas*	*comosus*
Carnivorous bromeliad	Bromeliad	Bromeliaceae	*Brocchinia*	*reducta*
Lantern of the forest	Bromeliad	Bromeliaceae	*Catopsis*	*berteroniana*
Air plants	Bromeliad	Bromeliaceae	*Tillandsia*	Various
Bromeliads	Bromeliad	Bromeliaceae	Various	Various

Common name	*Family common name*	*Family scientific name*	*Genus*	*Species*
Frankincense	Torchwood	Burseraceae	*Boswellia*	*sacra*
Elemi	Torchwood	Burseraceae	*Canarium*	*luzonicum*
Copal tree	Torchwood	Burseraceae	*Protium*	*copal*
Rainbow plant	Rainbow plants	Byblidaceae	*Byblis*	Various
Rock cactus	Cactus	Cactaceae	*Ariocarpus*	Various
Silver torch cactus	Cactus	Cactaceae	*Cleistocactus*	*strausii*
Copiapoa	Cactus	Cactaceae	*Copiapoa*	Various
Frailea	Cactus	Cactaceae	*Frailea*	Various
Feather cactus	Cactus	Cactaceae	*Mammillaria*	*plumosa*
Old man of the Andes	Cactus	Cactaceae	*Oreocereus*	*celsianus*
Parodia	Cactus	Cactaceae	*Parodia*	Various
Paperspine cactus	Cactus	Cactaceae	*Sclerocactus*	*papyracanthus*
Cannabis	Hemp	Cannabaceae	*Cannabis*	Various
Hops	Hemp	Cannabaceae	*Humulus*	*lupulus*
Papaya	Papaya	Caricaceae	*Carica*	*papaya*
Sandwort	Pink	Caryophyllaceae	*Minuartia*	*verna*
Soapwort	Pink	Caryophyllaceae	*Saponaria*	*officinalis*
Bladder campion	Pink	Caryophyllaceae	*Silene*	*vulgaris*
Giant chickweed	Pink	Caryophyllaceae	*Stellaria*	*aquaticum*
She-oaks	She-oaks	Casuarinaceae	*Allocasuarina*	Various
Australian pine	She-oaks	Casuarinaceae	*Casuarina*	Various
Australian pitcher plant	Cephalotus	Cephalotaceae	*Cephalotus*	*follicularis*
Rockrose	Rockrose	Cistaceae	*Cistus*	Various
Spiderworts	Spiderwort	Commelinaceae	*Tradescantia*	Various
Dodder	Morning glory	Convolvulaceae	*Cuscuta*	Various
Dogwoods	Dogwood	Cornaceae	*Cornus*	Various
Tupelos	Dogwood	Cornaceae	*Nyssa*	Various
Stonecrop	Stonecrop	Crassulaceae	Various	Various
Javan cucumber	Pumpkin	Cucurbitaceae	*Alsomitra*	*macrocarpa*

Common name	Family common name	Family scientific name	Genus	Species
Cantaloupe	Pumpkin	Cucurbitaceae	*Cucumis*	*melo*
Cucumber	Pumpkin	Cucurbitaceae	*Cucumis*	*sativus*
Squashes	Pumpkin	Cucurbitaceae	*Cucurbita*	Various
Exploding cucumbers	Pumpkin	Cucurbitaceae	*Ecballium*	*elaterium*
Cypresses	Cypress	Cupressaceae	*Cupressus*	Various
Junipers	Cypress	Cupressaceae	*Juniperus*	Various
Dawn redwood	Cypress	Cupressaceae	*Metasequoia*	*glyptostroboides*
Giant sequoia	Cypress	Cupressaceae	*Sequoia*	*giganteum*
Coast redwood	Cypress	Cupressaceae	*Sequoia*	*sempervirens*
Baldcypress	Cypress	Cupressaceae	*Taxodium*	*distichum*
Sandarac	Cypress	Cupressaceae	*Tetraclinis*	*articulata*
Western redcedar	Cypress	Cupressaceae	*Thuja*	*plicata*
Arborvitae	Cypress	Cupressaceae	*Thuja*	Various
Tree ferns	Tree ferns	Cyathaceae	*Cyathea*	Various
Sago cycad	Cycad	Cycadaceae	*Cycas*	*revoluta*
Sedges	Sedge	Cyperaceae	Various	Various
Bracken fern	Bracken	Dennstaedtiaceae	*Pteridium*	Various
Tree ferns	Tree ferns	Dicksoniaceae	*Dicksonia*	Various
Elephant food plant	Didierea	Didiereaceae	*Portulacaria*	*afra*
	Didierea	Didiereaceae	Various	Various
Diptocarps	Diptocarp	Diptocarpaceae	Various	Various
Venus flytrap	Sundew	Droseraceae	*Dionaea*	*muscipula*
Sundews	Sundew	Droseraceae	*Drosera*	Various
Persimmons, ebonies	Ebony	Ebenaceae	*Diospyros*	Various
Russian olive	Oleaster	Elaeagnaceae	*Eleagnus*	*angustifolia*
Autumn olive	Oleaster	Elaeagnaceae	*Eleagnus*	*umbellata*
Sea buckthorn	Oleaster	Elaeagnaceae	*Hippophae*	Various
Canada buffaloberry	Oleaster	Elaeagnaceae	*Shepherdia*	*canadensis*
Giant horsetail trees (extinct)	Horsetail	Equisetaceae	*Calamites*	Various

Common name	*Family common name*	*Family scientific name*	*Genus*	*Species*
Horsetails	Horsetail	Equisetaceae	*Equisetum*	Various
Manzanita	Heath	Ericaceae	*Arctostaphylos*	Various
Arctic bell-heather	Heath	Ericaceae	*Cassiope*	*tetragona*
Sourwood	Heath	Ericaceae	*Oxydendrum*	*arboretum*
Marsh Labrador tea	Heath	Ericaceae	*Rhododendron*	*tomentosum*
Rhododendrons	Heath	Ericaceae	*Rhododendron*	Various
Cranberry	Heath	Ericaceae	*Vaccinium*	*macrocarpon*
Lingonberry	Heath	Ericaceae	*Vaccinium*	*vitis-idaea*
Blueberry	Heath	Ericaceae	Various	Various
Fire croton	Euphorbia	Euphorbiaceae	*Codiaeum*	*variegatum*
Baseball plant	Euphorbia	Euphorbiaceae	*Euphorbia*	*obesa*
Dead wood plant	Euphorbia	Euphorbiaceae	*Euphorbia*	*platyclada*
Euphorbia	Euphorbia	Euphorbiaceae	*Euphorbia*	Various
Buddha belly plant	Euphorbia	Euphorbiaceae	*Jatropha*	*podagrica*
Parasol-leaf tree	Euphorbia	Euphorbiaceae	*Macaranga*	*tanarius*
Cassava	Euphorbia	Euphorbiaceae	*Manihot*	*esculenta*
Castor bean	Euphorbia	Euphorbiaceae	*Ricinus*	*communis*
Acacias	Bean	Fabaceae	*Acacia*	Various
Hairy rattleweed	Bean	Fabaceae	*Baptisia*	*arachnifera*
Redbuds	Bean	Fabaceae	*Cercis*	Various
Chickpea	Bean	Fabaceae	*Cicer*	*arietinum*
Scotch broom	Bean	Fabaceae	*Cytisus*	*scoparius*
Stick-tights	Bean	Fabaceae	*Desmodium*	Various
Honeylocust	Bean	Fabaceae	*Gleditsia*	*triacanthos*
Soybean	Bean	Fabaceae	*Glycine*	*max*
Dwarf lupine	Bean	Fabaceae	*Lupinus*	*pusillus*
Lupines	Bean	Fabaceae	*Lupinus*	Various
Alfalfa	Bean	Fabaceae	*Medicago*	*sativa*
Sensitive plant	Bean	Fabaceae	*Mimosa*	*pudica*
Mimosa	Bean	Fabaceae	*Mimosa*	Various

Common name	Family common name	Family scientific name	Genus	Species
Palo verde	Bean	Fabaceae	*Parkinsonia*	*florida*
Beans	Bean	Fabaceae	*Phaseolus*	Various
Mesquite	Bean	Fabaceae	*Prosopis*	Various
Black locust	Bean	Fabaceae	*Robinia*	*pseudoacacia*
Senna	Bean	Fabaceae	*Senna*	Various
Tamarind	Bean	Fabaceae	*Tamarindus*	*indica*
Red clover	Bean	Fabaceae	*Trifolium*	*pratense*
Lentil	Bean	Fabaceae	*Vicia*	*lens*
Common vetch	Bean	Fabaceae	*Vicia*	*sativa*
American chestnut	Beech	Fagaceae	*Castanea*	*dentata*
Leather oak	Beech	Fagaceae	*Quercus*	*durata*
English oak	Beech	Fagaceae	*Quercus*	*robur*
Oaks	Beech	Fagaceae	*Quercus*	Various
Live oak	Beech	Fagaceae	*Quercus*	*virginiana*
Ocotillo	Ocotillo	Fouquieriaceae	*Fouquieria*	*splendens*
Common stork's bill	Geranium	Geraniaceae	*Erodium*	*cicutarium*
Ginkgo	Ginkgo	Ginkgoaceae	*Ginkgo*	*biloba*
Parrot's feather	Watermilfoil	Haloragaceae	*Myriophyllum*	*aquaticum*
Witch-hazel	Witch-hazel	Hamamelidaceae	*Hamamelis*	Various
Waterthyme	Frog's-bit	Hydrocharitaceae	*Hydrilla*	*verticillata*
Saffron	Iris	Iridaceae	*Crocus*	*sativus*
Irises	Iris	Iridaceae	*Iris*	Various
Pecan	Walnut	Juglandaceae	*Carya*	*illinoinensis*
Hickories	Walnut	Juglandaceae	*Carya*	Various
Walnuts	Walnut	Juglandaceae	*Juglans*	Various
Lavender	Mint	Lamiaceae	*Lavandula*	Various
Mints	Mint	Lamiaceae	*Mentha*	Various
Catnip	Mint	Lamiaceae	*Nepeta*	*cataria*
Basil	Mint	Lamiaceae	*Ocimium*	*basilicum*
Oregano	Mint	Lamiaceae	*Origanum*	*vulgare*

Common name	*Family common name*	*Family scientific name*	*Genus*	*Species*
Rosemary	Mint	Lamiaceae	*Rosmarinus*	*officinalis*
Chia	Mint	Lamiaceae	*Salvia*	*hispanica*
Sage	Mint	Lamiaceae	*Salvia*	*officinalis*
Thyme	Mint	Lamiaceae	*Thymus*	Various
Love vine	Laurel	Lauraceae	*Cassytha*	Various
Cinnamon	Laurel	Lauraceae	*Cinnamomum*	Various
Bay leaf	Laurel	Lauraceae	*Laurus*	*nobilis*
Corkscrew plant	Bladderwort	Lentibulariaceae	*Genlisea*	Various
Butterworts	Bladderwort	Lentibulariaceae	*Pinguicula*	Various
Bladderworts	Bladderwort	Lentibulariaceae	*Utricularia*	Various
Scale trees (extinct)	Scale trees	Lepidodendraceae	*Lepidodendron*	Various
Flax	Flax	Linaceae	*Linum*	*usitatissimum*
Electric shock plant	Loasa	Loasaceae	*Blumenbachia*	*insignis*
Stickleaf	Loasa	Loasaceae	*Caiophora*	Various
Loasa	Loasa	Loasaceae	*Loasa*	Various
Stickleaf	Loasa	Loasaceae	*Mentzelia*	Various
Blazing star	Loasa	Loasaceae	*Nasa*	Various
She-oak mistletoe	Showy mistletoes	Loranthaceae	*Amyema*	*cambagei*
Mangrove mistletoe	Showy mistletoes	Loranthaceae	*Amyema*	*mackayensis*
Australian mistletoe	Showy mistletoes	Loranthaceae	*Amyema*	*thalassia*
Mistletoe shrub	Showy mistletoes	Loranthaceae	*Dendrophthoe*	*homoplastica*
Mangrove mistletoe	Showy mistletoes	Loranthaceae	*Lysiana*	*maritima*
Showy mistletoes	Showy mistletoes	Loranthaceae	Various	Various
Clubmosses	Clubmosses	Lycopodiaceae	*Lycopodium*	Various
Pomegranate	Loosestrife	Lythraceae	*Punica*	*granatum*
Tulip poplar	Magnolia	Magnoliaceae	*Liriodendron*	*tulipifera*
Velvetleaf	Mallow	Malvaceae	*Abutilon*	*theophrasti*
Baobab trees	Mallow	Malvaceae	*Adansonia*	Various
Silk floss tree	Mallow	Malvaceae	*Ceiba*	*speciosa*
Kola nut	Mallow	Malvaceae	*Cola*	Various

Common name	*Family common name*	*Family scientific name*	*Genus*	*Species*
Cotton	Mallow	Malvaceae	*Gossypium*	Various
Longsepal globemallow	Mallow	Malvaceae	*Iliamna*	*longisepala*
Chocolate	Mallow	Malvaceae	*Theobroma*	*cacao*
Devil's claw	Devil's claw	Martyniaceae	*Proboscidea*	Various
Turkeybeard	Bunchflower	Melanthiaceae	*Xerophyllum*	*asphodeloides*
South American mistletoes	South American mistletoes	Misodendraceae	*Misodendrum*	Various
	Fig	Moraceae	*Ficus*	*politoria*
Osage orange	Fig	Moraceae	*Maclura*	*pomifera*
Figs	Fig	Moraceae	*Morus*	Various
Banana	Banana	Musaceae	*Musa*	*paradisiaca*
Nutmeg	Nutmeg	Myristicaceae	*Myristica*	*fragrans*
Bottlebrushes	Myrtle	Myrtaceae	*Callistemon*	Various
Eucalyptus	Myrtle	Myrtaceae	*Eucalyptus*	Various
Tea tree	Myrtle	Myrtaceae	*Melaleuca*	Various
Wax myrtle	Myrtle	Myrtaceae	*Morella*	*cyrifera*
Allspice	Myrtle	Myrtaceae	*Pimente*	*dioica*
Cloves	Myrtle	Myrtaceae	Syzygium	aromaticum
Lotus	Lotus	Nelumbonaceae	Nelumbo	nucifera
Tropical pitcher plants, monkey cups	Pitcher plant	Nepenthaceae	Nepenthes	Various
Sand verbena	Four o' clock	Nyctaginaceae	Abronia	Various
Bougainvillea	Four o' clock	Nyctaginaceae	Bougainvillea	Various
Water lily	Water lily	Nymphaeaceae	Nymphaea	Various
Giant Amazon water lily	Water lily	Nymphaeaceae	Victoria	amazonica
Water tupelo	Tupelo	Nyssaceae	Nyssa	aquatica
Oleaster	Olive	Oleaceae	Olea	oleaster
Tricolored vanda	Orchid	Orchidaceae	Vanda	tricolor
Vanilla	Orchid	Orchidaceae	Vanilla	planifolia
American bluehearts	Broomrapes	Orobanchaceae	Buchnera	Various

Common name	*Family common name*	*Family scientific name*	*Genus*	*Species*
Indian paintbrush	Broomrapes	Orobanchaceae	Castilleja	Various
Beechdrops	Broomrapes	Orobanchaceae	Epifagus	virginiana
Broomrape	Broomrapes	Orobanchaceae	Orobranche	Various
Witchweed	Broomrapes	Orobanchaceae	Striga	Various
Wood sorrel	Wood sorrel	Oxalidaceae	Oxalis	Various
Screw pines	Screw pine	Pandanaceae	Pandanus	Various
Fire poppies	Poppy	Papaveraceae	Papaver	californicum
Poppies	Poppy	Papaveraceae	Papaver	Various
Passionflowers	Passionflower	Passifloraceae	Passiflora	Various
Monkeyflower	Lopseed	Phrymaceae	Erythranthe	Various
Pokeweed	Pokeweed	Phytolaccaceae	Phytolacca	americana
Balsam fir	Pine	Pinaceae	Abies	balsamea
Subalpine fir	Pine	Pinaceae	Abies	lasiocarpa
Firs	Pine	Pinaceae	Abies	Various
Cedars	Pine	Pinaceae	Cedrus	Various
Subalpine larch	Pine	Pinaceae	Larix	lyallii
Larches	Pine	Pinaceae	Larix	Various
Norway spruce, European spruce	Pine	Pinaceae	Picea	abies
Engelmann spruce	Pine	Pinaceae	Picea	engelmannii
Black spruce	Pine	Pinaceae	Picea	mariana
Red spruce	Pine	Pinaceae	Picea	rubens
Spruces	Pine	Pinaceae	Picea	Various
Whitebark pine	Pine	Pinaceae	Pinus	albicaulis
Jack pine	Pine	Pinaceae	Pinus	banksiana
Sand pine	Pine	Pinaceae	Pinus	clausa
Lodgepole pine	Pine	Pinaceae	Pinus	contorta
Limber pine	Pine	Pinaceae	Pinus	flexilis
Bristlecone pine	Pine	Pinaceae	*Pinus*	*longaeva, aristata, balfouriana*
Mountain pine	Pine	Pinaceae	Pinus	mugo

Common name	*Family common name*	*Family scientific name*	*Genus*	*Species*
Longleaf pine	Pine	Pinaceae	Pinus	palustris
Table mountain pine	Pine	Pinaceae	Pinus	pungens
Scots pine	Pine	Pinaceae	Pinus	sylvestris
Loblolly pine	Pine	Pinaceae	Pinus	taeda
Pines	Pine	Pinaceae	Pinus	Various
Douglas fir	Pine	Pinaceae	Pseudotsuga	menziesii
Eastern hemlock	Pine	Pinaceae	Tsuga	canadensis
Western hemlock	Pine	Pinaceae	Tsuga	heterophylla
Hemlocks	Pine	Pinaceae	Tsuga	Various
Radiator plants	Black pepper	Piperaceae	Peperomia	Various
Black pepper	Black pepper	Piperaceae	Piper	nigrum
Snapdragons	Plantain	Plantaginaceae	Antirrhinum	Various
Sycamore	Plane-trees	Platanaceae	*Platanus*	Various
Sea thrift	Leadwort	Plumbaginaceae	*Armeria*	*maritima* subsp. *helleri*
Oats	Grass	Poaceae	—	Various
Blackgrass	Grass	Poaceae	*Alopecurus*	*myosuroides*
Wiregrass	Grass	Poaceae	*Aristida*	Various
Bamboo	Grass	Poaceae	*Bambusoideae* subfamily	Various
Grama grasses	Grass	Poaceae	*Bouteloua*	Various
Pinegrass	Grass	Poaceae	*Calamagrostis*	*rubescens*
Sandspur	Grass	Poaceae	*Cenchrus*	Various
Sawgrass	Grass	Poaceae	*Cladium*	Various
Pampas grass	Grass	Poaceae	*Cortaderia*	*selloana*
Lemongrass	Grass	Poaceae	*Cymbopogon*	*citratus*
Hairy crabgrass	Grass	Poaceae	*Digitaria*	*sanguinalis*
Barnyard grass	Grass	Poaceae	*Echinochloa*	*crus-galli*
Early barnyard grass	Grass	Poaceae	*Echinochloa*	*oryzoides*
Hare's-tail cottongrass	Grass	Poaceae	*Eriophorum*	*vaginatum*

Common name	*Family common name*	*Family scientific name*	*Genus*	*Species*
Needle and thread grass	Grass	Poaceae	*Hesperostipa*	Various
Barley	Grass	Poaceae	*Hordeum*	*vulgare*
Cogongrass	Grass	Poaceae	*Imperata*	*cylindrica*
Perennial ryegrass	Grass	Poaceae	*Lolium*	*perenne*
Needlegrasses	Grass	Poaceae	*Nasella*	Various
Rice	Grass	Poaceae	*Oryza*	*sativa*
Seashore paspalum	Grass	Poaceae	*Paspalum*	*vaginatum*
Black needle grass	Grass	Poaceae	*Piptochaetium*	*avenaceum*
Annual meadow grass	Grass	Poaceae	*Poa*	*annua*
Grasses	Grass	Poaceae	*Poaceae*	Various
Sugarcanes	Grass	Poaceae	*Saccharum*	Various
Rye	Grass	Poaceae	*Secale*	*cereale*
Wild rye	Grass	Poaceae	*Secale*	*montanum*
Sorghums	Grass	Poaceae	*Sorghum*	Various
Salt cordgrass	Grass	Poaceae	*Spartina*	*alterniflora*
Needlegrasses	Grass	Poaceae	*Stipa*	Various
Spinifex	Grass	Poaceae	*Triodia*	Various
Wheat	Grass	Poaceae	*Triticum*	Various
Sea oats	Grass	Poaceae	*Uniola*	*paniculata*
Corn	Grass	Poaceae	*Zea*	*mays*
Olotón corn	Grass	Poaceae	*Zea*	*mays* 'Sierra Mixe'
Honey-scented pincushion	Phlox	Polemoniaceae	*Navarretia*	*mellita*
Rhubarb	Knotweed	Polygonaceae	*Rheum*	× *hybridum*
Basket ferns	Polypod	Polypodiaceae	*Aglaomorpha*	Various
Oakleaf ferns	Polypod	Polypodiaceae	*Drynaria*	Various
Ant fern	Polypod	Polypodiaceae	*Lecanopteris*	Various
Ant fern	Polypod	Polypodiaceae	*Microgramma*	Various
Staghorn ferns	Polypod	Polypodiaceae	*Platycerium*	Various

Common name	*Family common name*	*Family scientific name*	*Genus*	*Species*
Resurrection ferns	Polypod	Polypodiaceae	*Pleopeltis*	Various
Water hyacinth	Water hyacinth	Pontederiaceae	*Pontederia*	*crassipes*
Banksias	Protea	Proteaceae	*Banksia*	Various
Leucadendrons	Protea	Proteaceae	*Leucadendron*	Various
Pincushions	Protea	Proteaceae	*Leucospermum*	Various
Sugarbushes	Protea	Proteaceae	*Protea*	Various
Maidenhair fern	Brake	Pteridaceae	*Adiantum*	Various
Ladder brake	Brake	Pteridaceae	*Pteris*	*vittata*
Soapbark tree	Soapbark	Quillajaceae	*Quillaja*	*Saponaria*
Rafflesia	Rafflesia	Rafflesiaceae	*Rafflesia*	Various
Columbines	Buttercup	Ranunculaceae	*Aquilegia*	Various
Restios	Restiads	Restionaceae	*Retio*	Various
Snowbrush	Buckthorn	Rhamnaceae	*Ceanothus*	*velutinus*
Anchor plants	Buckthorn	Rhamnaceae	*Colletia*	Various
Coffeeberry	Buckthorn	Rhamnaceae	*Frangula*	*californica*
Redberry	Buckthorn	Rhamnaceae	*Rhamnus*	*crocea*
Buckthorn	Buckthorn	Rhamnaceae	*Rhamnus*	Various
Red mangrove	Mangrove	Rhizophoraceae	*Rhizophora*	*mangle*
Australian mangrove	True mangroves	Rhizophoraceae	*Rhizophora*	Various
Mountain mahogany	Rose	Rosaceae	*Cercocarpus*	Various
Hawthorn	Rose	Rosaceae	*Crataegus*	Various
Strawberry	Rose	Rosaceae	*Fragaria*	Various
Avens	Rose	Rosaceae	*Geum*	Various
Apple	Rose	Rosaceae	*Malus*	*domestica*
Almond	Rose	Rosaceae	*Prunus*	*amygdalus*
Apricot	Rose	Rosaceae	*Prunus*	*armeniaca*
Cherry	Rose	Rosaceae	*Prunus*	*avium* and *cerasus*
Stone fruits	Rose	Rosaceae	*Prunus*	Various
Firethorn	Rose	Rosaceae	*Pyracantha*	Various

Common name	*Family common name*	*Family scientific name*	*Genus*	*Species*
Callery pear	Rose	Rosaceae	*Pyrus*	*calleryana*
Roses	Rose	Rosaceae	*Rosa*	Various
Thimbleberry	Rose	Rosaceae	*Rubus*	*parviflorus*
Raspberries, blackberries	Rose	Rosaceae	*Rubus*	Various
Coffee	Coffee	Rubiaceae	*Coffea*	*arabica*
Bedstraw	Coffee	Rubiaceae	*Galium*	*aparine*
Ant plant	Coffee	Rubiaceae	*Hydnophytum*	Various
Ant plant	Coffee	Rubiaceae	*Myrmecodia*	Various
Ant plant	Coffee	Rubiaceae	*Pachycentria*	*glauca*
Grapefruit	Citrus	Rutaceae	*Citrus*	*aurantium* f. *aurantium*
Citrus	Citrus	Rutaceae	*Citrus*	Various
Lime	Citrus	Rutaceae	*Citrus*	× *latifolia*
Lemon	Citrus	Rutaceae	*Citrus*	× *limon*
Orange	Citrus	Rutaceae	*Citrus*	× *sinensis*
Gas plant	Citrus	Rutaceae	*Dictamnus*	*albus*
Prickly ash	Citrus	Rutaceae	*Zanthoxylum*	Various
Aspens	Willow	Salicaceae	*Populus*	Various
Willows	Willow	Salicaceae	*Salix*	Various
Mosquito fern	Salvinia	Salviniaceae	*Azolla*	*caroliniana*
Watermoss	Salvinia	Salviniaceae	*Salvinia*	Various
Sandalwoods	Sandalwoods	Santalaceae	Various	Various
Sugar maple	Soapberry	Sapindaceae	*Acer*	*saccharum*
Maples	Soapberry	Sapindaceae	*Acer*	Various
Buckeyes	Soapberry	Sapindaceae	*Aesculus*	Various
Guarana berry	Soapberry	Sapindaceae	*Paullinia*	*cupana*
Soapberry	Soapberry	Sapindaceae	*Sapindus*	Various
Greasewood	Greasewood	Sarcobataceae	*Sarcobatus*	Various
Cobra plant	Trumpet pitcher plant	Sarraceniaceae	*Darlingtonia*	Various
Sun pitchers	Trumpet pitcher plant	Sarraceniaceae	*Heliamphora*	Various

Common name	*Family common name*	*Family scientific name*	*Genus*	*Species*
Trumpet pitcher plants	Trumpet pitcher plant	Sarraceniaceae	*Sarracenia*	Various
Sigillaria (extinct)	Arborescent lycophytes	Sigillariaceae	*Sigillaria*	Various
Chili pepper	Nightshade	Solanaceae	*Capsicum*	*annuum*
Chiltepin pepper	Nightshade	Solanaceae	*Capsicum*	*anuum* var. *glabriusculum*
Aji pepper	Nightshade	Solanaceae	*Capsicum*	*baccatum*
Habanero, ghost and Scotch bonnet peppers	Nightshade	Solanaceae	*Capsicum*	*chinense*
Tabasco, Thai, and piri piri peppers	Nightshade	Solanaceae	*Capsicum*	*frutescens*
Rocoto pepper	Nightshade	Solanaceae	*Capsicum*	*pubescens*
Coyote tobacco	Nightshade	Solanaceae	*Nicotiana*	*attenuata*
Tobacco	Nightshade	Solanaceae	*Nicotiana*	*tabacum*
Petunia	Nightshade	Solanaceae	*Petunia*	Various
Horse nettle	Nightshade	Solanaceae	*Solanum*	*carolinense*
Tomato	Nightshade	Solanaceae	*Solanum*	*lycopersicum*
Eggplant	Nightshade	Solanaceae	*Solanum*	*melongena*
Naranjilla	Nightshade	Solanaceae	*Solanum*	*quitoense*
Potatoes	Nightshade	Solanaceae	*Solanum*	*tuberosum*
Sphagnum moss	Sphagnum	Sphagnaceae	*Sphagnum*	Various
Triggerplants	Triggerplant	Stylidiaceae	*Stylidium*	Various
Sweetgum	Snowbell	Styracaceae	*Styrax*	Various
Tamarisk	Tamarisk	Tamaricaceae	*Tamarix*	Various
Tea-oil camellia	Tea	Theaceae	*Camellia*	*oleifera*
Tea	Tea	Theaceae	*Camellia*	*sinensis*
Nasturtiums	Nasturtium	Tropaeolaceae	*Tropaeolum*	Various
Cattails	Cattail	Typhaceae	*Typha*	Various
Elms	Elm	Ulmaceae	*Ulmus*	Various
Dutch elm	Elm	Ulmaceae	*Ulmus*	× *hollandica*

Common name	*Family common name*	*Family scientific name*	*Genus*	*Species*
Trumpet tree	Nettle	Urticaceae	*Cecropia*	Various
Stinging tree	Nettle	Urticaceae	*Dendrocnide*	*moroides*
Stinging nettle	Nettle	Urticaceae	*Urtica*	*dioica*
Tumbleweed	Various	Various	Various	Various
Mountain pansy	Viola	Violaceae	*Viola*	*lutea*
Grapes	Grape	Vitaceae	*Vitis*	Various
Tree tumbo	Welwitschia	Welwitschiaceae	*Welwitschia*	*mirabilis*
Guayara	Cycad	Zamiaceae	*Zamia*	*pumila*
Galangal	Ginger	Zingiberaceae	*Alpinia*	*galanga*
Tumeric	Ginger	Zingiberaceae	*Curcuma*	*longa*
Cardamom	Ginger	Zingiberaceae	*Elettaria*	*cardamom*
Ginger	Ginger	Zingiberaceae	*Zingiber*	*officinale*
Lignum vitae	Caltrop	Zygophyllaceae	*Guaiacum*	Various
Creosote	Caltrop	Zygophyllaceae	*Larrea*	*tridentata*
Puncture vine	Caltrop	Zygophyllaceae	*Tribulus*	*terrestris*

References

Chapters 1 and 2. The Hidden World *and* How Enemies Attack Plants

Bartram, William. 1791. *The Travels of William Bartram*. Electronic ed. University of North Carolina at Chapel Hill, 2001. https://docsouth.unc.edu/nc/bartram/bartram.html.

Braun, E. J., and R. J. Howard. 1994. Adhesion of fungal spores and germlings to host-plant surfaces. *Protoplasma* 181(1–4): 202–212.

Chalker-Scott, Linda. 2015. *How Plants Work: The Science behind the Amazing Things Plants Do*. Science for Gardeners. Timber Press.

Chamowitz, Daniel. 2013. *What a Plant Knows: A Field Guide to the Senses*. Scientific American / Farrar, Straus and Giroux.

Cosmos: A Spacetime Odyssey. 2014. Documentary.

Farmer, Edward. 2014. *Leaf Defence*. Oxford University Press.

Gilet, T., and L. Bourouiba. 2014. Rain-induced ejection of pathogens from leaves: revisiting the mechanism of splash-on-film using high-speed visualization. *Integrative and Comparative Biology* 54(6): 974–984.

Gilet, T., and L. Bourouiba. 2015. Fluid fragmentation shapes rain-induced foliar disease transmission. *Journal of the Royal Society Interface* 12. https://doi.org/10.1098/rsif.2014.1092.

Goodall, Jane. 2013. *Seeds of Hope: Wisdom and Wonder from the World of Plants*. Grand Central Publishing.

Howard R. J., M. A. Ferrari, D. H. Roach, and N. P. Money. 1991. Penetration of hard substrates by a fungus employing enormous turgor pressures. *Proceedings of the National Academy of Sciences of the United States of America* 88(24): 11281–11284.

Howard, R. J., and B. Valent. 1996. Breaking and entering: host penetration by the fungal rice blast pathogen *Magnaporthe grisea*. *Annual Review of Microbiology* 50: 491–512.

Hudson, Charles. 2004. *Black Drink: A Native American Tea*. University of Georgia Press.

Janzen, Daniel. 1979. *Herbivores*. New Horizons in the Biology of Plant Defenses. Academic Press.

Karasov, Talia L., Eunyoung Chae, Jacob J. Herman, and Joy Bergelson. 2017. Mechanisms to mitigate the trade-off between growth and defense. *Plant Cell* 29(4): 666–680.

Karban, Richard. 2015. *Plant Sensing and Communication*. Interspecific Interactions. University of Chicago Press.

Mancuso, Stefano. 2018. *The Revolutionary Genius of Plants: A New Understanding of Plant Intelligence and Behavior*. Atria Books.

Mancuso, Stefano, and Alessandra Viola. 2015. *Brilliant Green: The Surprising History and Science of Plant Intelligence*. 2nd ed. Island Press.

Petschenka, G., and A. A. Agrawal. 2016. How herbivores coopt plant defenses: natural selection, specialization, and sequestration. *Current Opinion in Insect Science* 14: 17–24.

Preston, E. 2015. The deadly plant sneeze: high-speed cameras capture how bouncing rain spreads crop disease. *Nautilus*. http://nautil.us/issue/25/water/the-deadly-plant-sneeze.

The Private Life of Plants. 1995. BBC.

Stewart, Amy. 2009. *Wicked Plants: The Weed That Killed Lincoln's Mother and Other Botanical Atrocities*. Algonquin Books.

Stewart, Amy. 2013. *The Drunken Botanist: The Plants That Create the World's Great Drinks*. Algonquin Books.

Walters, Dale R. 2010. *Plant Defense: Warding Off Attack by Pathogens, Herbivores, and Parasitic Plants*. Blackwell Publishing Ltd.

Walters, Dale R. 2017. *Fortress Plant: How to Survive When Everything Wants to Eat You*. 1st ed. Oxford University Press.

War, Abdul Rasheed, Gaurav Kumar Taggar, Barkat Hussain, Monica Sachdeva Taggar, Ramakrishnan M. Nair, and

Hari C. Sharma. 2018. Plant defence against herbivory and insect adaptations. *AoB PLANTS* 10(4): plyo37.
What Plants Talk About. 2013. Nature.
Wohlleben, Peter. 2016. *The Hidden Life of Trees: What They Feel, How They Communicate—Discoveries from a Secret World*. Greystone Books.

Chapter 3. The Surface of the Plant

Bobich, E. G., and P. S. Nobel. 2001. Vegetative reproduction as related to biomechanics, morphology and anatomy of four cholla cactus species in the Sonoran Desert. *Annals of Botany London* 87(4): 485–493.
Deschamps, Cicero, David Gang, Natalia Dudareva, and James E. Simon. 2006. Developmental regulation of phenylpropanoid biosynthesis in leaves and glandular trichomes of basil (*Ocimum basilicum* L.). *International Journal of Plant Sciences* 167(3): 447–454.
Gang, David R., Jihong Wang, Natalia Dudareva, Kyoung Hee Nam, James E. Simon, Efraim Lewisohn, and Eran Pichersky. 2001. An investigation of the storage and biosynthesis of phenylpropenes in sweet basil. *Plant Physiology* 125(2): 539–555.
Gershenzon, J., D. McCaskill, J. I. Rajaonarivony, C. Mihaliak, F. Karp, and R. Croteau. 1992. Isolation of secretory cells from plant glandular trichomes and their use in biosynthetic studies of monoterpenes and other gland products. *Analytical Biochemistry* 200(1): 130–138.
Glover, B. J. 2000. Differentiation in plant epidermal cells. *Journal of Experimental Botany* 51(344): 497–505.
Halpern, M., Dina Raats, and Simcha Lev-Yadun. 2006. Plant biological warfare: thorns inject pathogenic bacteria into herbivores. *Environmental Microbiology* 9(3): 584–592.
Halpern, M., Dina Raats, and Simcha Lev-Yadun. 2014. The potential anti-herbivory role of microorganisms on plant thorns. *Plant Signaling & Behavior* 2(6): 503–504.
Halpern, M., Avivit Waissler, Adi Dror, and Simcha Lev-Yadun. 2011. Biological warfare of the spiny plant introducing pathogenic microorganisms into herbivore's tissues. *Advances in Applied Microbiology* 74: 97–116.
Huang, F., and W. Guo. 2013. Structural and mechanical properties of the spines from *Echinocactus grusonii* cactus. *Journal of Materials Science* 48(16): 5420–5428.
Huchelmann, Alexandre, Marc Boutry, and Charles Hachez. 2017. Plant glandular trichomes: natural cell factories of high biotechnological interest. *Plant Physiology* 175(1): 6–22.
Lev-Yadun, S., and M. Halpern. 2019. Extended phenotype in action: two possible roles for silica needles in plants: not just injuring herbivores but also inserting pathogens into their tissues. *Plant Signaling & Behavior* 14(7): 1–5.
Mauseth, J. D. 2006. Structure-function relationships in highly modified shoots of Cactaceae. *Annals of Botany London* 98(5): 901–926.
Pillemer, E. A., and W. M. Tingey. 1976. Hooked trichomes: a physical plant barrier to a major agricultural pest. *Science* 193(4252): 482–484.
Sapala, Aleksandra, Adam Runions, Anne-Lise Routier-Kierzkowska, Mainak Das Gupta, Lilan Hong, Hugo Hofhuis, Stéphane Verger, Gabriella Mosca, Chun-Biu Li, Angela Hay, et al. 2018. Why plants make puzzle cells, and how their shape emerges. *eLife*. https://doi.org/10.7554/eLife.32794.
Seigler, David S. 1998. *Plant Secondary Metabolism*. Springer.
Shanower, T. G. 2008. Trichomes and insects. In: Capinera, J. L. (ed.), *Encyclopedia of Entomology*. Springer: 3905–3908.
Weinhold, A., and I. T. Baldwin. 2011. Trichome-derived O-acyl sugars are a first meal for caterpillars that tags them for predation. *Proceedings of the National Academy of Sciences of the United States of America* 108(19): 7855–7859.
Xing, Zhenlong, Yongqiang Lu, Wanzhi Cai, Xinzheng Huang, Shengyong Wu, and Zhongren Lei. 2017. Efficiency of trichome-based plant defense in *Phaseolus vulgaris* depends on insect behavior, plant ontogeny, and structure. *Frontiers in Plant Science* 8. https://doi.org/10.3389/fpls.2017.02006.
Yamazaki, Kazuo, Ken-ichi Nakatan, and Keiko Masumoto. 2014. Slug caterpillars of *Parasa lepida* (Cramer, 1799) (Lepidoptera: Limacodidae) become stuck on rose prickles. *Pan-Pacific Entomologist* 90(4): 221–225.
Yeats, T. H., and J. K. C. Rose. 2013. The formation and function of plant cuticles. *Plant Physiology* 163(1): 5–20.

Chapter 4. Minefield

Agrawal, A. A., and K. Konno. 2009. Latex: a model for understanding mechanisms, ecology, and evolution of plant defense against herbivory. *Annual Review of Ecology, Evolution, and Systematics* 40: 311–331.
Arnott, H. J., and M. A. Webb. 1983. Twin crystals of calcium oxalate in the seed coat of the kidney bean. *Protoplasma* 114(1): 23–34.
Baldry, M. G. C. 1983. The bactericidal, fungicidal and sporicidal properties of hydrogen peroxide and peracetic acid. *Journal of Applied Bacteriology* 54(3): 417–423.
Butler, C. B., and J. Christian. 1997. *Treasures of the Longleaf Pines: Naval Stores*. 2nd ed. Tarkel Publishing.
Compton, S. G. 1987. *Aganais speciosa* and *Danaus chrysippus* (Lepidoptera) sabotage the latex defences of their host plants. *Ecological Entomology* 12(1): 115–118.

Coté, G. G. 2009. Diversity and distribution of idioblasts producing calcium oxalate crystals in *Dieffenbachia seguine* (Araceae). *American Journal of Botany* 96(7): 1245–1254.

Dussourd, D. E. 1990. The vein drain; or, how insects outsmart plants. *Natural History* 90(2): 44–49.

Dussourd, D. E., and T. Eisner. 1987. Vein-cutting behavior: insect counterploy to the latex defense of plants. *Science* 237 (4817): 898–901.

Earley, L. S. 2006. *Looking for Longleaf: The Fall and Rise of an American Forest*. University of North Carolina Press.

Franceschi, V. R. 2001. Calcium oxalate in plants. *Trends in Plant Science* 6(7): 331.

Franceschi, V. R., and H. T. Horner. 1980. Calcium oxalate crystals in plants. *Botanical Review* 46 (October): 361–427. https://link.springer.com/article/10.1007/BF02860532#citeas.

Franceschi, V. R., and P. A. Nakata. 2005. Calcium oxalate in plants: formation and function. *Annual Review of Plant Biology* 56: 41–71.

Heath, M. C. 2000. Hypersensitive response-related death. *Plant Molecular Biology* 44(3): 321–334.

Hodgkinson, A. 1977. Oxalic acid metabolism in higher plants. In: Hodgkinson, A. (ed.), *Oxalic Acid in Biology and Medicine*. Academic Press: 131–158.

Horner, H. T., and B. L. Wagner. 1980. The association of druse crystals with the developing stomium of *Capsicum annuum* (Solanaceae) anthers. *American Journal of Botany* 67(9): 1347–1360.

Hücklehoven, R. 2007. Cell wall–associated mechanisms of disease resistance and susceptibility. *Annual Review of Phytopathology* 45: 101–127.

Jáuregui-Zúñiga, David, Juan Pablo Reyes-Grajeda, José David Sepúlveda-Sánchez, John R. Whitaker, and Abel Moreno. 2003. Crystallochemical characterization of calcium oxalate crystals isolated from seed coats of *Phaseolus vulgaris* and leaves of *Vitis vinifera*. *Journal of Plant Physiology* 160(3): 239–245.

Katayama, H., Y. Fujibayashi, S. Nagaoka, and Y. Sugimura. 2007. Cell wall sheath surrounding calcium oxalate crystals in mulberry idioblasts. *Protoplasma* 231(3–4): 245–248.

Keates, S. E., N. M. Tarlyn, F. A. Loewus, and V. R. Franceschi. 2000. L-ascorbic acid and L-galactose are sources of oxalic acid and calcium oxalate in *Pistia stratiotes*. *Phytochemistry* 53(4): 433–440.

Konno, K. 2011. Plant latex and other exudates as plant defense systems: roles of various defense chemicals and proteins contained therein. *Phytochemistry* 72(13): 1510–1530.

Konyar, Sevil Tütüncü, Necla Öztürk, and Feruzan Dane. 2014. Occurrence, types and distribution of calcium oxalate crystals in leaves and stems of some species of poisonous plants. *Botanical Studies* 55, article no. 32. *https://doi.org/10.1186/1999-3110-55-32*.

Korth, Kenneth L., Sarah J. Doege, Sang-Hyuck Park, Fiona L. Goggin, Qin Wang, S. Karen Gomez, Guangjie Liu, Lingling Jia, and Paul A. Nakata. 2006. *Medicago truncatula* mutants demonstrate the role of plant calcium oxalate crystals as an effective defense against chewing insects. *Plant Physiology* 141(1): 188–195.

Lersten, N. R., and H. T. Horner. 2006. Crystal macropattern development in *Prunus serotine* (Rosaceae, Prunoideae) leaves. *Annals of Botany* 97(5): 723–729.

Lewinsohn, T. M. 1991. The geographical distribution of plant latex. *Chemoecology* 2: 64–68.

Li, X. X., and V. R. Franceschi. 1990. Distribution of peroxisomes and glycolate metabolism in relation to calcium oxalate formation in *Lemna minor* L. *European Journal of Cell Biology* 51(1): 9–16.

Libert, B., and V. R. Franceschi. 1987. Oxalate in crop plants. *Journal of Agricultural and Food Chemistry* 35(6): 926–938.

McNair, J. B. 1932. The interrelation between substances in plants: essential oils and resins, cyanogen and oxalate. *American Journal of Botany* 19(3): 255–271.

Miller, L. P. (ed.). 1973. *Phytochemistry*. Vol. 2: *Organic Metabolites*. Van Nostrand Reinhold.

Morel, J. B., and J. L. Dangl. 1997. The hypersensitive response and the induction of cell death in plants. *Cell Death and Differentiation* 4(8): 671–683.

Mur, Luis A. J., Paul Kenton, Amanda J. Lloyd, Helen Ougham, and Elena Prats. 2008. The hypersensitive response: the centenary is upon us but how much do we know? *Journal of Experimental Botany* 59(3): 501–520.

Nakata, P. A. 2003. Advances in our understanding of calcium oxalate crystal formation and function in plants. *Plant Science* 164(6): 901–909.

Nakata, P. A. 2012. Plant calcium oxalate crystal formation, function, and its impact on human health. *Frontiers in Biology* 7: 254–266.

Nuss, R. F., and F. A. Loewus. 1978. Further studies on oxalic acid biosynthesis in oxalate-accumulating plants. *Plant Physiology* 61(4): 590–592.

Outland, R. B. 2004. *Tapping the Pines: The Naval Stores Industry in the American South*. LSU Press.

Passardi, F., C. Cosio, C. Penel, and C. Dunand. 2005. Peroxidases have more functions than a Swiss army knife. *Plant Cell Reports* 24(5): 255–265.

Prychid, Christina J., Rachel Schmidt Jabaily, and Paula J. Rudall. 2008. Cellular ultrastructure and crystal

development in *Amorphophallus* (Araceae). *Annals of Botany* 101(7): 983–995.
Raman, Vijayasankar, Harry T. Horner, and Ikhlas A. Khan. 2014. New and unusual forms of calcium oxalate raphide crystals in the plant kingdom. *Journal of Plant Research* 127(6): 721–730.
Webb, M. A. 1999. Cell-mediated crystallization of calcium oxalate in plants. *Plant Cell* 11(4): 751–761.
Wrench, K. 2014. *Tar Heels: North Carolina's Forgotten Economy; Pitch, Tar, Turpentine & Longleaf Pines*. 1st ed. CreateSpace Independent Publishing Platform.
Yang, J., and F. A. Loewus. 1975. Metabolic conversion of L-ascorbic acid in oxalate-accumulating plants. *Plant Physiology* 56(2): 283–285.
Zalucki, M. P., and S. B. Malcolm. 1999. Plant latex and first-instar monarch larval growth and survival on three North American milkweed species. *Journal of Chemical Ecology* 25: 1827–1842.
Zindler-Frank, E. 1976. Oxalate biosynthesis in relation to photosynthetic pathways and plant productivity: a survey. *Pflanzenphysiologie* 80(1): 1–13.

Chapter 5. Heavy-Duty Construction

Alexander, H. M., C. L. Cummings, L. Kahn, and A. A. Snow. 2001. Seed size variation and predation of seeds produced by wild and crop-wild sunflowers. *American Journal of Botany* 88(4): 623–627.
Annunziata, M. G. 2019. What is lignin made of? new components discovered! *Plant Physiology* 180(3): 1255.
Armstrong, W. P. 2010. Plant fibers. https://www2.palomar.edu/users/warmstrong/traug99.htm.
Barceló, A. Ros, L. V. Gómez Ros, C. Gabaldón, M. López-Serrano, F. Pomar, J. S. Carrión, and M. A. Pedreño. 2004. Basic peroxidases: the gateway for lignin evolution? *Phytochemistry Reviews* 3: 61–78.
Baroux, C., C. Spillane, and U. Grossniklaus. 2002. Evolutionary origins of the endosperm in flowering plants. *Genome Biology* 3(9): 1026.
Baskin, C. C., and J. M. Baskin. 1998. *Seeds: Ecology, Biogeography, and Evolution of Dormancy and Germination*. Academic Press.
Bauer, P., R. Elbaum, and I. M. Weiss. 2011. Calcium and silicon mineralization in land plants: transport, structure and function. *Plant Science* 180(6): 746–756.
Becraft, P. W., and J. Gutierrez-Marcos. 2012. Endosperm development: dynamic processes and cellular innovations underlying sibling altruism. *Wiley Interdisciplinary Reviews: Developmental Biology* 1(4): 579–593.
Bell, L. 1983. *Papyrus, Tapa, Amate and Rice Paper: Papermaking in Africa, the Pacific, Latin America and Southeast Asia*. Liliaceae Press.
Bhuiyan, Nazmul H., Gopalan Selvaraj, Yangdou Wei, and John King. 2009. Role of lignification in plant defense. *Plant Signaling and Behavior* 4(2): 158–159.
Boerjan, W., J. Ralph, and M. Baucher. 2003. Lignin biosynthesis. *Annual Reviews in Plant Biology* 54(1): 519–549.
Boesewinkel, F. D., and F. Bouman. 1995. The seed: structure. In: Johri, B. M. (ed.), *Embryology of Angiosperms*. Springer-Verlag: 567–610.
Brown, J. H., O. J. Reichman, and D. W. Davidson. 1979. Granivory in desert ecosystems. *Annual Review of Ecology and Systematics* 10: 201–227.
Chauhan, D. K., D. K. Tripathi, N. K. Rai, and A. K. Rai. 2011. Detection of biogenic silica in leaf blade, leaf sheath, and stem of Bermuda grass (*Cynodon dactylon*) using LIBS and phytolith analysis. *Food Biophysics* 6: 416–423.
Chee-Sanford, Joanne C., Martin M. Williams II, Adam S. Davis, and Gerald K. Sims. 2006. Do microorganisms influence seed-bank dynamics? *Weed Science* 54(3): 575–587.
Collin, B., J. Meunier, C. Keller, E. Doelsch, and F. Panfili. 2010. Silica distribution in various bamboo species and its effects on plant growth. Paper presented at the American Geophysical Union, fall meeting. *https://ui.adsabs.harvard.edu/abs/2010AGUFM.B51I0464C/abstract*.
Coors, J. G. 1987. Resistance to the European corn borer, *Ostrinia nubilalis* (Hubner), in maize, *Zea mays* L., as affected by soil silica, plant silica, structural carbohydrates, and lignin. In: Gabelman, W. H., and B. C. Loughman (eds.), *Genetic Aspects of Plant Mineral Nutrition*. Developments in Plant and Soil Sciences, vol. 27. Springer: 445–456.
Crawley, M. J. 2000. Seed predators and plant population dynamics. In: Fenner, M. (ed.), *Seeds: The Ecology of Regeneration in Plant Communities*. 2nd ed. CABI Publishing: 167–182.
Dalling, James W., Adam S. Davis, Brian J. Schutte, and A. Elizabeth Arnold. 2011. Seed survival in soil: interacting effects of predation, dormancy and the soil microbial community. *Journal of Ecology* 99(1): 89–95.
Datnoff, L. E., G. H. Snyder, and G. H. Korndörfer (eds.). 2001. *Silicon in Agriculture*, vol. 8.
Davies, N. S., and M. R. Gibling. 2011. Evolution of fixed-channel alluvial plains in response to Carboniferous vegetation. *Nature Geoscience* 21(9): 629–633.

de Gonzalo, Gonzalo, Dana I. Colpa, Mohamed H. M. Habib, and Marco W. Fraaije. 2016. Bacterial enzymes involved in lignin degradation. *Journal of Biotechnology* 236: 110–119.

Eriksson, K. L., R. A. Blanchette, and P. Ander. 1990. *Microbial and Enzymatic Degradation of Wood and Wood Components*. Springer.

Erlanger, S. 2005. After 2,000 years, a seed from ancient Judea sprouts. *New York Times*, June 12. *https://www.nytimes.com/2005/06/12/world/middleeast/after-2000-years-a-seed-from-ancient-judea-sprouts.html*.

Fenner, M., and K. Thompson. 2005. *The Ecology of Seeds*. Cambridge University Press.

Franceschi, Vincent R., Paal Krokene, Erik Christiansen, and Trygve Krekling. 2005. Anatomical and chemical defenses of conifer bark against bark beetles and other pests. *New Phytologist* 167(2): 353–375.

Frasier, S. L. 2015. Mother plants tell their seeds when to sprout. *Scientific American* 312(5): 26.

Friedman, W. E. 1995. Organismal duplication, inclusive fitness theory, and altruism: understanding the evolution of endosperm and the angiosperm reproductive syndrome. *Proceedings of the National Academy of Sciences of the United States of America* 92(9): 3913–3917.

Friedman, W. E. 1998. The evolution of double fertilization and endosperm: an "historical" perspective. *Sexual Plant Reproduction* 11: 6–16.

Fuerst, E. Patrick, Patricia A. Okubara, James V. Anderson, and Craig F. Morris. 2014. Polyphenol oxidase as a biochemical seed defense mechanism. *Frontiers in Plant Science* 5: 689.

Gali-Muhtasib, H. U., C. C. Smith, and J. J. Higgins. 1992. The effect of silica in grasses on the feeding behavior of the prairie vole, *Microtus ochrogaster*. *Ecology* 73(5): 1724–1729.

Glasser, W. 2019. About making lignin great again—some lessons from the past. *Frontiers in Chemistry* 7: 565.

Gobetz, K. E., and S. R. Bozarth. 2001. Implications for late Pleistocene mastodon diet from opal phytoliths in tooth calculus. *Quaternary Research* 55(2): 115–122.

Goussain, M. M., E. Prado, and J. C. Moraes. 2005. Effect of silicon applied to wheat plants on the biology and probing behaviour of the greenbug *Schizaphis graminum* (Rond.) (Hemiptera: Aphididae). *Neotropical Entomology* 34(5): 807–813.

Grzegorz, Janusz, Anna Pawlik, Justyna Sulej, Urszula Swiderska-Burek, Anna Jarosz-Wilkolazka, and Andrzej Paszczynski. 2017. Lignin degradation: microorganisms, enzymes involved, genomes analysis and evolution. *FEMS Microbiology Reviews* 41(6): 941–962.

Hanley, Mike E., Byron B. Lamont, Meredith M. Fairbanks, and Christine M. Rafferty. 2007. Plant structural traits and their role in anti-herbivore defence. *Perspectives in Plant Ecology, Evolution and Systematics* 8(4): 157–178.

Harries, H. C., and C. R. Clement. 2014. Long-distance dispersal of the coconut palm by migration within the coral atoll ecosystem. *Annals of Botany* 113(4): 565–570.

Haygreen, J. G., and J. L. Bowyer. 1982. *Forest Products and Wood Science*. Iowa State University.

Henry, A. G., A. S. Brooks, and D. R. Piperno. 2010. Microfossils in calculus demonstrate consumption of plants and cooked foods in Neanderthal diets (Shanidar III, Iraq; Spy I and II, Belgium). *Proceedings of the National Academy of Sciences of the United States of America* 108(2): 486–491.

Henry, A. G., and D. R. Piperno. 2008. Using plant microfossils from dental calculus to recover human diet: a case study from Tell al-Raqā'i, Syria. *Journal of Archaeological Science* 35(7): 1943–1950.

Honek, Alice, Zdenka Martinkova, and Vojtech Jarosik. 2003. Ground beetles (Carabidae) as seed predators. *European Journal of Entomology* 100(4): 531–544.

Hulme, P. E., and C. W. Benkman. 2002. Granivory. In: Herrera, C. M., and O. Pellmyr (eds.), *Plant Animal Interactions: An Evolutionary Approach*. Blackwell: 132–154.

Hunt, J. W., A. P. Dean, R. E. Webster, G. N. Johnson, and A. R. Ennos. 2008. A novel mechanism by which silica defends grasses against herbivory. *Annals of Botany* 102(4): 653–656.

Hunter, D. 1978. *Papermaking: The History and Technology of an Ancient Craft*. Dover.

Janzen, D. H. 1971. Seed predation by animals. *Annual Review of Ecology and Systematics* 2: 465–492.

Jingjing, Li. 2011. Isolation of lignin from wood. Bachelor's thesis, Saimaa University of Applied Sciences. https://core.ac.uk/download/pdf/38048748.pdf.

Kaufman, Rachel. 2012. 32,000-year-old plant brought back to life—oldest yet. *National Geographic*, February 23. https://www.nationalgeographic.com/science/article/120221-oldest-seeds-regenerated-plants-science?loggedin=true&rnd=1711401891940.

Klemm, Dieter, Brigitte Heublein, Hans-Peter Fink, and Andreas Bohn. 2005. Cellulose: fascinating biopolymer and sustainable raw material. *Angewandte Chemie International Edition* 44(22): 3358–3393.

Kundanati, Lakshminath, Nimesh R. Chahare, Siddhartha Jaddivada, Abhijith G. Karkisaval, Rajeev Sridhar, Nicola M. Pugno, and Namrata Gundiah. 2019. Mechanics of wood boring by beetle mandibles. bioRxiv, 568089. https://doi.org/10.1101/568089.

Kvedaras, Olivia L., Malcolm G. Keeping, François R. Goebel, and Marcus J. Byrne. 2007. Larval performance of the pyralid borer *Eldana saccharina* Walker and stalk damage in sugarcane: influence of plant silicon, cultivar and feeding site. *International Journal of Pest Management* 53(3): 183–194.

Lanning, F. C., and L. N. Eleuterius. 1992. Silica and ash in seeds of cultivated grains and native plants. *Annals of Botany* 69(2): 151–160.

Law, C., and C. Exley. 2011. New insight into silica deposition in horsetail (*Equisetum arvense*). *BMC Plant Biology* 11, article no. 112.

Lewin, J., and B. E. F. Reimann. 1969. Silicon and plant growth. *Annual Review of Plant Physiology* 20: 289–304.

Liu, Q. 2018. Lignins: biosynthesis and biological functions in plants. *International Journal of Molecular Sciences* 19(2): 335.

Long, Rowena L., Marta J. Gorecki, Michael Renton, John K. Scott, Louise Colville, Danica E. Goggin, Lucy E. Commander, David A. Westcott, Hillary Cherry, and William E. Finch-Savage. 2014. The ecophysiology of seed persistence: a mechanistic view of the journey to germination or demise. *Biological Reviews* 90(1): 31–59.

LoPresti, Eric F., Patrick Grof-Tisza, Moria Robinson, Jessie Godfrey, and Richard Karban. 2018. Entrapped sand as a plant defence: effects on herbivore performance and preference. *Ecological Entomology* 43(2): 154–161.

LoPresti, E. F., and R. Karban. 2016. Chewing sandpaper: grit, plant apparency and plant defense in sand-entrapping plants. *Ecology* 97(4): 826–833.

Lu, H., and K. Liu. 2003. Phytoliths of common grasses in the coastal environments of southeastern USA. *Estuarine, Coastal and Shelf Science* 58(3): 587–600.

Lucas, Peter W., Ian M. Turner, Nathaniel J. Dominy, and Nayuta Yamashita. 2000. Mechanical defences to herbivory. *Annals of Botany* 86(5): 913–920.

Lundgren, J., and K. Rosentrater. 2007. The strength of seeds and their destruction by granivorous insects. *Arthropod-Plant Interactions* 1(2): 93–99.

Ma, J. F. 1983. Role of silicon in enhancing the resistance of plants to biotic and abiotic stresses. *Soil Science and Plant Nutrition* 50(1): 11–18.

Ma, J. F. 2003. Functions of silicon in higher plants. In: Müller, W. E. G. (ed.), *Silicon Biomineralization*. Progress in Molecular and Subcellular Biology, vol. 33. Springer: 127–147.

Mares, M. A., and M. L. Rosenzweig. 1978. Granivory in North and South American deserts: rodents, birds, and ants. *Ecology* 59(2): 235–241.

Massey, F. P., A. R. Ennos, and S. E. Hartley. 2006. Silica in grasses as defence against insect herbivores: contrasting effects on folivores and phloem feeder. *Journal of Animal Ecology* 75(2): 595–603.

Massey, F. P., A. E. Ennos, and S. E. Hartley. 2007a. Grasses and the resource availability hypothesis: the importance of silica-based defences. *Journal of Ecology* 95(3): 414–424.

Massey, F. P., A. R. Ennos, and S. E. Hartley. 2007b. Herbivore specific induction of silica-based plant defences. *Oecologia* 152: 677–683.

Massey, F. P., and S. E. Hartley. 2006. Experimental demonstration of the antiherbivore effects of silica in grasses: impacts on foliage digestibility and vole growth rates. *Proceedings of the Royal Society B: Biological Sciences* 273(1599): 2299–2304.

Massey, F. P., and S. E. Hartley. 2009. Physical defences wear you down: progressive and irreversible impacts of silica on insect herbivores. *Journal of Animal Ecology* 78(1): 281–291.

McNaughton, S. J., and J. L. Tarrants. 1983. Grass leaf silicification: natural selection for an inducible defense against herbivores. *Proceedings of the National Academy of Sciences of the United States of America* 80(3): 790–791.

Müller, Jacqueline Müller, Marcus Clauss, Daryl Codron, Ellen Schulz, Jürgen Hummel, Mikael Fortelius, Patrick Kircher, and Jean-Michel Hatt. 2014. Growth and wear of incisor and cheek teeth in domestic rabbits (*Oryctolagus cuniculus*) fed diets of different abrasiveness. *Journal of Experimental Zoology Part A: Ecological Genetics and Physiology* 321(5): 283–298.

Nabhan, G. P., and J. J. Tewksbury. 2001. Seed dispersal: directed deterrence by capsaicin in chillies. *Nature* 412(6845): 403–404.

Nabity, P. D., R. Orpet, S. Miresmailli, M. R. Berenbaum, and E. H. DeLucia. 2012. Silica and nitrogen modulate physical defense against chewing insect herbivores in bioenergy crops *Miscanthus* × *giganteus* and *Panicum virgatum* (Poaceae). *Journal of Economic Entomology* 105(3): 878–883.

Oliveras, Jordi, Crisanto Gómez, Josep M. Bas, and Xavier Espadaler. 2008. Mechanical defence in seeds to avoid predation by a granivorous ant. *Naturwissenschaften* 95(6): 501.

Olsen, Odd-Arne. 2007. *Endosperm: Developmental and Molecular Biology*. Springer Science & Business Media.

O'Reagain, P. J., and M. T. Mentis. 1989. Leaf silification in grasses—a review. *Journal of the Grassland Society of South Africa* 6(1): 37–43.

Oxford University Press. 2016. Rotting away: getting at the evolutionary roots of wood decay. https://phys.org/news/2016-11-evolutionary-roots-wood.html.

Parry, D. W., and F. Smithson. 1964. Types of opaline silica depositions in the leaves of British grasses. *Annals of Botany* 28(1): 169–185.
Pásztory, Z., I. R. Mohácsiné, G. Gorbacheva, and Z. Börcsök. 2016. The utilization of tree bark. *BioResources* 11(3): 7859–7888.
Paul-Camilo Zalamea, James W. Dalling, Carolina Sarmiento, A. Elizabeth Arnold, Carolyn Delevich, Mark A. Berhow, Anyangatia Nbobegang, Sofia Gripenberg, and Adam S. Davis. 2018. Dormancy-defense syndromes and tradeoffs between physical and chemical defenses in seeds of pioneer species. *Ecology* 99(9): 1988–1998.
Pearsall, D. M. 2018. *Case Studies in Paleoethnobotany: Understanding Ancient Lifeways through the Study of Phytoliths, Starch, Macroremains, and Pollen*. Routledge.
Phillips, T. L., R. A. Peppers, and W. A. DiMichele. 1985. Stratigraphic and interregional changes in Pennsylvanian coal-swamp vegetation: environmental inferences. *International Journal of Coal Geology* 5(1–2): 43–109.
Piperno, D. R. 2006. *Phytoliths: A Comprehensive Guide for Archaeologists and Paleoecologists*. Altamira Press.
Popper, Zoë A., Gurvan Michel, Cécile Hervé, David S. Domozych, William G. T. Willats, Maria G. Tuohy, Bernard Kloareg, and Dagmar B. Stengel. 2011. Evolution and diversity of plant cell walls: from algae to flowering plants. *Annual Review of Plant Biology* 62: 567–590.
Prasad, Vandana, Caroline A. E. Strömberg, Habib Alimohammadian, and Ashok Sahni. 2005. Dinosaur coprolites and the early evolution of grasses and grazers. *Science* 310(5751): 1177–1180.
Pulliam, H. R., and M. R. Brand. 1975. The production and utilization of seeds in plains grassland of southeastern Arizona. *Ecology* 56(5): 1158–1166.
Raupp, M. J. 1985. Effects of leaf toughness on mandibular wear of the leaf beetle, *Plagiodera versicolora*. *Ecological Entomology* 10(1): 73–79.
Raven, J. A. 1983. The transport and function of silicon in plants. *Biological Reviews* 58(2): 179–207.
Reynolds, O. L., M. G. Keeping, and J. H. Meyer. 2009. Silicon-augmented resistance of plants to herbivorous insects: a review. *Annals of Applied Biology* 155(2): 171–186.
Robinson, J. M. 1990. Lignin, land plants, and fungi: biological evolution affecting Phanerozoic oxygen balance. *Geology* 18(7): 607–610.
Rodgerson, L. 1998. Mechanical defense in seeds adapted for ant dispersal. *Ecology* 79(5): 1669–1677.
Sahebi, Mahbod, Mohamed M. Hanafi, Abdullah Siti Nor Akmar, Mohd Y. Rafii, Parisa Azizi, F. F. Tengoua, Jamaludin Nurul Mayzaitul Azwa, and M. Shabanimofrad. 2015. Importance of silicon and mechanisms of biosilica formation in plants. *BioMed Research International*, article ID 396010.
Sallabanks, R., and S. P. Courtney. 1992. Frugivory, seed predation and insect-vertebrate interactions. *Annual Review of Entomology* 37: 337–400.
Sallon, Sarah E., Elaine Solowey, Yuval Cohen, Raia Korchinsky, Markus Egli, Ivan Woodhatch, Orit Simchoni, and Mordechai Kislev. 2008. Germination, genetics, and growth of an ancient date seed. *Science* 320(5882): 1464.
Sanson, G. D., S. A. Kerr, and K. A. Gross. 2007. Do silica phytoliths really wear mammalian teeth? *Journal of Archaeological Science* 34(4): 526–531.
Saul, Hayley, Marco Madella, Anders Fischer, Aikaterini Glykou, Sönke Hartz, and Oliver E. Craig. 2013. Phytoliths in pottery reveal the use of spice in European prehistoric cuisine. *PLoS ONE* 8(8): e70583.
Serbet, R., and G. W. Rothwell. 1992. Characterizing the most primitive seed ferns. I. A reconstruction of *Elkinsia polymorpha*. *International Journal of Plant Sciences* 153(4): 602–621.
Shea, S. B. 2018. Behind the scenes: how fungi make nutrients available to the world. U.S. Department of Energy. https://www.energy.gov/science/articles/behind-scenes-how-fungi-make-nutrients-available-world.
Shen-Miller, J., Mary Beth Mudgett, J. William Schopf, Steven Clarke, and Rainer Berger. 1995. Exceptional seed longevity and robust growth: ancient sacred lotus from China. *American Journal of Botany* 82(11): 1367–1380.
Shen-Miller, J., J. William Schopf, Garman Harbottle, Rui-Ji Cao, Shu Ouyang, Kun-Shu Zhou, John R. Southon, and Guo-Hai Liu. 2002. Long-living lotus: germination and soil {gamma}-irradiation of centuries-old fruits, and cultivation, growth, and phenotypic abnormalities of offspring. *American Journal of Botany* 89(2): 236–247.
Sjöström, E. 1993. *Wood Chemistry: Fundamentals and Applications*. Gulf Professional Publishing.
Smith, C. C. 1970. The coevolution of pine squirrels (Tamiasciurus) and conifers. *Ecological Monographs* 40(3): 349–371.
Stakhov, V., Gülsüm Asena Başlı, G. Gyulai, Z. Tóth, A. Güner, Z. Szabó, Lilia Murenyetz, S. Yashina, L. Heszky, and S. Gubin. 2007. Pleistocene-age *Silene stenophylla* seeds excavated in Russia: a scanning electron microscopic analysis. *Botany & Plant Biology*, Chicago, Ill. http://2007.botanyconference.org/engine/search/index.php?func=detail&aid=2131.
Strömberg, C. A. E. 2011. Evolution of grasses and grassland ecosystems. *Annual Review of Earth and Planetary Sciences* 39: 517–544.

Teaford, M. F., P. W. Lucas, P. S. Ungar, and K. E. Glander. 2006. Mechanical defenses in leaves eaten by Costa Rican howling monkeys (*Alouatta palliata*). *American Journal of Physical Anthropology* 129(1): 99–104.

Thummel, Ryan V., William H. Brightly, and Caroline A. E. Stömberg. 2019. Evolution of phytolith deposition in modern bryophytes, and implications for the fossil record and influence on silica cycle in early land plant evolution. *New Phytologist* 221(4): 2273–2285.

Tiansawat, Pimonrat, Adam S. Davis, Mark A. Berhow, Paul-Camilo Zalamea, and James W. Dalling. 2014. investment in seed physical defence is associated with species' light requirement for regeneration and seed persistence: evidence from *Macaranga* species in Borneo. *PLoS ONE* 9(6): e99691.

Tsou, C., and S. A. Mori. 2002. Seed coat anatomy and its relationship to seed dispersal in subfamily Lecythidoideae of the Lecythidaceae (the Brazil nut family). *Botanical Bulletin—Academia Sinica Taipei* 43(1): 37–56.

Van Bockhaven, Jonas, David De Vleesschauwer, and Monica Höfte. 2013. Towards establishing broad-spectrum disease resistance in plants: silicon leads the way. *Journal of Experimental Botany* 64(5): 1281–1293.

Vane, Christopher H., Trevor C. Drage, and Colin E. Snape. 2003. Biodegradation of oak wood during growth of the shiitake mushroom: a molecular approach. *Journal of Agricultural and Food Chemistry* 51(4): 947–956.

Vane, Christopher H., Trevor C. Drage, and Colin E. Snape. 2006. Bark decay by the white-rot fungus *Lentinula edodes*: polysaccharide loss, lignin resistance and the unmasking of suberin. *International Biodeterioration & Biodegradation* 57(1): 14–23.

Vicari, M., and D. R. Bazely. 1993. Do grasses fight back? the case for antiherbivore defenses. *Trends in Ecology & Evolution* 8(4): 137–141.

Wang, Min, Limin Gao, Suyue Dong, Yuming Sun, Qirong Shen, and Shiwei Guo. 2017. Role of silicon on plant-pathogen interactions. *Frontiers in Plant Science* 8: 701.

Wittkowski, Reiner, Joachim Ruther, Heike Drinda, and Foroozan Rafiei-Taghanaki. 1992. Formation of smoke flavor compounds by thermal lignin degradation. In: Teranishi, Roy, Gary R. Takeoka, and Matthias Güntert (eds.), *Flavor Precursors*. ACS Symposium Series, vol. 490: 232–243.

Yamanaka, Shigeru, Kanna Sato, Fuyu Ito, Satoshi Komatsubara, Hiroshi Ohata, and Katsumi Yoshino. 2012. Roles of silica and lignin in horsetail (*Equisetum hyemale*), with special reference to mechanical properties. *Journal of Applied Physics* 111(4): 044703.

Yildiz, Mustafa, Ramazan Beyaz, Mehtap Gursoy, Murat Aycan, Yusuf Koc, and Mustafa Kayan. 2017. Seed dormancy. In: Jiminez-Lopez, J. C. (ed.), *Advances in Seed Biology*. IntechOpen. https://doi.org/10.5772/intechopen.68178.

Zalamea, Paul-Camilo, James W. Dalling, Carolina Sarmiento, A. Elizabeth Arnold, Carolyn Delevich, Mark A. Berhow, Anyangatia Ndobegang, Sofia Gripenberg, and Adam S. Davis. 2018. Dormancy-defense syndromes and tradeoffs between physical and chemical defenses in seeds of pioneer species. *Ecology* 99(9): 1988–1998.

Zhu, Y., and H. Gong. 2014. Beneficial effects of silicon on salt and drought tolerance in plants. *Agronomy for Sustainable Development* 34: 455–472.

Chapter 6. Plumbing Revolution

Bertaccini, A., and B. Duduk. 2009. Phytoplasma and phytoplasma diseases: a review of recent research. *Phytopathologia Mediterranea* 48: 355–378.

Chen, X. Y., and J. Y. Kim. 2009. Callose synthesis in higher plants. *Plant Signaling & Behavior* 4(6): 489–492.

Douglas, A. E. 2003. Nutritional physiology of aphids. *Advances in Insect Physiology* 31: 73–140.

Douglas, A. E. 2006. Phloem-sap feeding by animals: problems and solutions. *Journal of Experimental Botany* 57(4): 747–754.

Gaupels, F., and C. Vlot. 2012. Plant defense and long-distance signaling in the phloem. In: Thompson, G. A., and A. J. E. van Bel (eds.), *Phloem: Molecular Cell Biology, Systemic Communication, Biotic Interactions*. 1st ed. John Wiley & Sons: 227–247.

Goodchild, A. J. P. 1966. Evolution of the alimentary canal in the Hemiptera. *Biological Reviews* 41: 97–140.

Hagel, J. M., A. Onoyovwi, E. C. Yeung, and P. J. Facchini. 2012. Role of phloem metabolites in plant defense. In: Thompson, G. A., and A. J. E. van Bel (eds.), *Phloem: Molecular Cell Biology, Systemic Communication, Biotic Interactions*. 1st ed. John Wiley & Sons: 251–270.

Huang, Y. 2007. Phloem feeding regulates the plant defense pathways responding to both aphid infestation and pathogen infection. In: Xu, Z., J. Li, Y. Xue, and W. Yang (eds.), *Biotechnology and Sustainable Agriculture 2006 and Beyond*. Springer: 215–219.

Klein, M. 2001. Phytoplasma diseases. In: Loebenstein, G., P. H. Berger, A. A. Brunt, and R. H. Lawson (eds.), *Virus and Virus-Like Diseases of Potatoes and Production of Seed-Potatoes*. Springer: 145–157.

Kumari, Shweta, Krishnan Nagendran, Awadhesh Bahadur Rai, Bijendra Singh, Govind Pratap Rao, and Assunta Bertaccini. 2019. Global status of phytoplasma diseases in vegetable crops. *Frontiers in Microbiology* 10: 1349.

Lough, T. J., and W. J. Lucas. 2006. Integrative plant biology: role of phloem long-distance macromolecular trafficking. *Annual Review of Plant Biology* 57(1): 203–232.

Lucas, William J., Andrew Groover, Raffael Lichtenberger, Kaori Furuta, Shri-Ram Yadav, Ykä Helariutta, Xin-Qiang He, Hiroo Fukuda, Julie Kang, Siobhan M. Brady, et al. 2013. The plant vascular system: evolution, development and functions. *Journal of Integrative Plant Biology* 55(4): 294–388.

Lucas, W. J., and C. Liu. 2017. The plant vascular system II: from resource allocation, inter-organ communication and defense, to evolution of the monocot cambium. *Journal of Integrative Plant Biology* 59(6): 354–355.

Marcone, C., and G. P. Rao. 2019. Control of phytoplasma diseases through resistant plants. In: Bertaccini, A., P. Weintraub, G. Rao, and N. Mori (eds.), *Phytoplasmas: Plant Pathogenic Bacteria—II*. Springer: 165–184.

Neergaard, E. de. 1989. Studies of *Didytmella bryoniae* (Auersw.) Rehm: development in the host. *Journal of Phytopathology* 127(2): 107–115.

Pegg, G. F. 1989. Pathogenesis in vascular diseases of plants. In: Tjamos, E. C., and C. H. Beckman (eds.), *Vascular Wilt Diseases of Plants*. NATO ASI Series (Series H: Cell Biology), vol. 28. Springer: 51–94.

Perlman, S. J., M. S. Hunter, and E. Zchori-Fein. 2006. The emerging diversity of *Rickettsia*. *Proceedings of the Royal Society B: Biological Sciences* 273(1598): 2097–2106.

Ploetz, R. C. (ed.). 1990. *Fusarium Wilt of Banana*. APS Press.

Pouzoulet, Jérôme, Alexandria L. Pivovaroff, Louis S. Santiago, and Philippe E. Rolshausen. 2014. Can vessel dimension explain tolerance toward fungal vascular wilt diseases in woody plants? lessons from Dutch elm disease and esca disease in grapevine. *Frontiers in Plant Science* 5: 253.

Shen, A. 2006. Forisome: a smart plant protein inside a phloem system. Paper presented at the conference Cells and Materials: At the Interface between Mathematics, Biology and Engineering, Institute for Pure and Applied Mathematics, UCLA. https://www.ipam.ucla.edu/programs/long-programs/cells-and-materials-at-the-interface-between-mathematics-biology-and-engineering/?tab=overview.

Shigo, A. L. 1984. Compartmentalization: a conceptual framework for understanding how trees grow and defend themselves. *Annual Review of Phytopathology* 22: 189–214.

Stafford, C. A., G. P. Walker, and D. E. Ullman. 2012. Hitching a ride: vector feeding and virus transmission. *Communicative and Integrative Biology* 5(1): 43–49.

Taylor, T. N., E. L. Taylor, and M. Krings. 2009. *Paleobotany: The Biology and Evolution of Fossil Plants*. 2nd ed. Academic Press.

Walling, L. 2008. Avoiding effective defenses: strategies employed by phloem-feeding insects. *Plant Physiology* 146(3): 859–866.

Will, T., A. C. U. Furch, and M. R. Zimmerman. 2013. How phloem-feeding insects face the challenge of phloem-located defenses. *Frontiers in Plant Science* 4(336): 336.

Will, T., W. F. Tjallingii, A. Thönnessen, and A. J. E. van Bel. 2007. Molecular sabotage of plant defense by aphid saliva. *Proceedings of the National Academy of Sciences of the United States of America* 104(25): 10536–10541.

Yadeta, K. A., and B. P. H. J. Thomma. 2013. The xylem as battleground for plant hosts and vascular wilt pathogens. *Frontiers in Plant Science* 4: 97.

Chapter 7. Plant-Speak

Adamec, Lubomír. 2011. The smallest but fastest: ecophysiological characteristics of traps of aquatic carnivorous *Utricularia*. *Plant Signaling and Behavior* 6(5): 640–646.

Alborn, H. T., T. C. J. Turlings, T. H. Jones, G. Stenhagen, J. H. Loughrin, and J. H. Tumlinson. 1997. An elicitor of plant volatiles from beet armyworm oral secretion. *Science* 276(5314): 945–949.

Appel, H. M., and R. B. Cocroft. 2014. Plants respond to leaf vibrations caused by insect herbivore chewing. *Oecologia* 175(4): 1257–1266.

Arimura, G., R. Ozawa, T. Shimoda, T. Nishioka, W. Boland, and J. Takabayashi. 2000. Herbivory-induced volatiles elicit defense genes in lima bean leaves. *Nature* 406(6795): 512–515.

Arimura, G., K. Shiojiri, and R. Karban. 2010. Acquired immunity to herbivory and allelopathy caused by airborne plant emissions. *Phytochemistry* 71(14–15): 1642–1649.

Armstrong, C. M. 2015. Packaging life: the origin of ion-selective channels. *Biophysical Journal* 109(2): 173–177.

Babikova, Zdenka, Lucy Gilbert, Toby J. A. Bruce, Michael Birkett, John C. Caulfield, Christine Woodcock, John A. Pickett, and David Johnson. 2013. Underground signals carried through common mycelial networks warn neighbouring plants of aphid attack. *Ecology Letters* 16(7): 835–843.

Baldwin, Ian T., Rayko Halitschke, Andre Kessler, and Ursula Schittko. 2001. Merging molecular and ecological approaches in plant-insect interactions. *Current Opinion in Plant Biology* 4(1): 351–358.

Baldwin, I. T., and J. C. Schultz. 1983. Rapid changes in tree leaf chemistry induced by damage: evidence for communication between plants. *Science* 221(4607): 277–279.

Baluška, František, Dieter Volkmann, Andrej Hlavacka, Stefano Mancuso, and Peter W. Barlow. 2006. Neurological view of plants and their body plan. In: Baluška, F., S. Mancuso, and D. Volkmann (eds.), *Communication in Plants*. Springer: 19–35.

Barto, E. Kathryn, Monika Hilker, Frank Müller, Brian K. Mohney, Jeffrey D. Weidenhamer, and Matthias C. Rillig. 2011. The fungal fast lane: common mycorrhizal networks extend bioactive zones of allelochemicals in soils. *PLoS ONE* 6(11): e27195.

Barto, E. Kathryn, Jeffrey D. Weidenhamer, Don Cipollini, and Matthias C. Rillig. 2012. Fungal superhighways: do common mycorrhizal networks enhance below ground communication? *Trends in Plant Science* 17(11): 633–637.

Berridge, M. J., P. Lipp, and M. D. Bootman. 2000. The versatility and universality of calcium signaling. *Nature Reviews Molecular Cell Biology* 1: 11–21.

Biedrzycki, Meredith L., Tafari A. Jilany, Susan A. Dudley, and Harsh P. Bais. 2010. Root exudates mediate kin recognition in plants. *Communicative & Integrative Biology* 3(1): 28–35.

Bingham, M. A., and S. W. Simard. 2011. Do mycorrhizal network benefits to survival and growth of interior Douglas-fir seedlings increase with soil moisture stress? *Ecology and Evolution* 1(3): 306–316.

Braam, J. 2004. In touch: plant responses to mechanical stimuli. *New Phytologist* 165(2): 373–389.

Browse, J. 2005. Jasmonate: an oxylipin signal with many roles in plants. *Vitamins and Hormones* 72: 431–456.

Bruin, J., and M. Dicke. 2001. Chemical information transfer between wounded and unwounded plants: backing up the future. *Biochemical Systematics and Ecology* 29(10): 1103–1113.

Callaway, R. M. 2002. The detection of neighbors by plants. *Trends in Ecology & Evolution* 17(3): 104–105.

Canales, J., C. Henriquez-Valencia, and S. Brauchi. 2018. the integration of electrical signals originating in the root of vascular plants. *Frontiers in Plant Science* 10(8): 2173.

Chen, Bin J. W., Heinjo J. During, and Niels P. R. Anten. 2012. Detect thy neighbor: identity recognition at the root level in plants. *Plant Science* 195: 157–167.

Cheong, J., and Y. D. Choi. 2003. Methyl jasmonate as a vital substance in plants. *Trends in Genetics* 19(7): 409–413.

Coats, Joel R., Laura L. Karr, and Charles D. Drewes. 1991. Toxicity and neurotoxic effects of monoterpenoids in insects and earthworms. *Naturally Occurring Pest Bioregulators* 449(20): 305–316.

Conrath, U. 2009. Priming of induced plant defense responses. In: Van Loon, L. C. (ed.), *Plant Innate Immunity*. Advances in Botanical Research, vol. 51. Elsevier: 361–395.

Diarmuid, J. 2008. *Aspirin: The Remarkable Story of a Wonder Drug*. Bloomsbury.

Dudley, Susan A., Guillermo P. Murphy, and Amanda L. File. 2013.Kin recognition and competition in plants. *Functional Ecology* 27(4): 898–906.

Eisner, T. 1964. Catnip: its raison d'être. *Science* 146(3649): 1318–1320.

Elhakeem, Ali, Dimitrije Markovic, Anders Broberg, Niels P. R. Anten, and Velemir Ninkovic. 2018. Aboveground mechanical stimuli affect belowground plant–plant communication. *PLoS ONE* 13(5): e0195646.

Engelberth, Juergen, Hans T. Alborn, Eric A. Schmelz, and James H. Tumlinson. 2004. Airborne signals prime plants against insect herbivore attack. *Proceedings of the National Academy of Sciences of the United States of America* 101(6): 1781–1785.

Falik, Omer, Yonat Mordoch, Lydia Quansah, Aaron Fait, and Ariel Novoplansky. 2011. Rumor has it . . . : relay communication of stress cues in plants. *PLoS ONE* 6(11): e23625.

Farmer, E. E., and C. A. Ryan. 1990. Interplant communication: airborne methyl jasmonate induces synthesis of proteinase inhibitors in plant leaves. *Proceedings of the National Academy of Sciences of the United States of America* 87(19): 7713–7716.

Farré-Armengol, Gerard, Iolanda Filella, Joan Llusia, and Josep Peñuelas. 2016. Bidirectional interaction between phyllospheric microbiotas and plant volatile emissions. *Trends in Plant Science* 21(10): 854–860.

Forterre, Yoël. 2013. Slow, fast and furious: understanding the physics of plant movements. *Journal of Experimental Botany* 64(15): 4745–4760.

Francis, R., and D. J. Read. 1984. Direct transfer of carbon between plants connected by vesicular-arbuscular mycorrhizal mycelium. *Nature* 307: 53–56.

Franklin, K. A. 2009. Light and temperature signal crosstalk in plant development. *Current Opinion in Plant Biology* 12(1): 63–68.

Franzios, Gerasimos, Maria Mirotsou, Emmanouel Hatziapostolou, Jiri Kral, Zacharias G. Scouras, and Penelope Mavragani-Tsipidou. 1997. Insecticidal and genotoxic activities of mint essential oils. *Journal of Agricultural and Food Chemistry* 45(7): 2690–2694.

Fromm, J., and S. Lautner. 2007. Electrical signals and their physiological significance in plants. *Plant, Cell & Environment* 30(3): 249–257.

Gilbert, L., and D. Johnson. 2017. Plant–plant communication through common mycorrhizal networks. In: Becard, Guillaume (ed.), *How Plants Communicate with Their Biotic Environment*. Advances in Botanical Research, vol. 82. Elsevier: 83–97.

Gorzelak, Monika A., Amanda K. Asay, Brian J. Pickles, and Suzanne W. Simard. 2015. Inter-plant communication through mycorrhizal networks mediates complex adaptive behaviour in plant communities. *AoB PLANTS* 7: 1–13.

Halling, D. Brent, Paula Aracena-Parks, and Susan L. Hamilton. 2005. Regulation of voltage-gated Ca^{2+} channels by calmodulin. *Science Signaling* 315: 15.

He, Xinhua, Minggang Xu, Guo Yu Qiu, and Jianbin Zhou. 2009. Use of ^{15}N stable isotope to quantify nitrogen transfer between mycorrhizal plants. *Journal of Plant Ecology* 2(3): 107–118.

Hedrich, R., V. Salvador-Recatalà, and I. Dreyer. 2016. Electrical wiring and long-distance plant communication. *Trends in Plant Science* 21(5): 376–387.

Heil, M., and J. C. S. Bueno. 2007. Within-plant signaling by volatiles leads to induction and priming of an indirect plant defense in nature. *Proceedings of the National Academy of Sciences of the United States of America* 104(13): 5467–5472.

Heil, M., and R. Karban. 2010. Explaining the evolution of plant communication by airborne signals. *Trends in Ecology & Evolution* 25(3): 137–144.

Hines, P. J. 2018. Rapid, long-distance signaling in plants. *Science* 361(6407): 1083–1085.

Karban, R. 2008. Plant behaviour and communication. *Ecology Letters* 11(7): 727–739.

Karban, R. 2010. Neighbors affect resistance to herbivory—a new mechanism. *New Phytologist* 186(3): 564–566.

Karban, R. 2015. *Plant Sensing and Communication*. University of Chicago Press.

Karban, R. 2017. The language of plant communication (and how it compares to animal communication). In: Gagliano, M., J. Ryan, and P. Vieira (eds.), *The Language of Plants*. University of Minnesota Press: 3–26.

Karban, R., M. Huntzinger, and A. C. McCall. 2004. The specificity of eavesdropping on sagebrush by other plants. *Ecology* 85(7): 1846–1852.

Karban, R., J. Maron, G. W. Felton, G. Ervin, and H. Eichenseer. 2003. Herbivore damage to sagebrush induces wild tobacco: evidence for eavesdropping between plants. *Oikos* 100(2): 325–332.

Karban, Richard, Kaori Shiojiri, Mikaela Huntzinger, and Andrew C. McCall. 2006. Damage-induced resistance in sagebrush: volatiles are key to intra- and interplant communication. *Ecology* 87(4): 922–930.

Karban, R., K. Shiojiri, and S. Ishizaki. 2010. An air-transfer experiment confirms the role of volatile cues in communication between plants. *American Naturalist* 176(3): 381–384.

Karban, R., K. Shiojiri, and S. Ishizaki. 2011. Plant communication—why should plants emit volatile cues? *Journal of Plant Interactions* 6(2–3): 81–84.

Karban, Richard, Kaori Shiojiri, Satomi Ishizaki, William C. Wetzel, and Richard Y. Evans. 2013. Kin recognition affects plant communication and defence. *Proceedings of the Royal Society B: Biological Sciences* 280(1756), article no. 20123062.

Karban, R., W. C. Wetzel, K. Shiojiri, E. Pezzola, and J. D. Blande. 2016. Geographic dialects in volatile communication between sagebrush individuals. *Ecology* 97(11): 2917–2924.

Karban, Richard, William C. Wetzel, Kaori Shiojiri, Satomi Ishizaki, Santiago R. Ramirez, and James D. Blande. 2014. Deciphering the language of plant communication: volatile chemotypes of sagebrush. *New Phytologist* 204(2): 380–385.

Karban, Richard, Louie H. Yang, and Kyle F. Edwards. 2014. Volatile communication between plants that affects herbivory: a meta-analysis. *Ecology Letters* 17(1): 44–52.

Katsir, Leron, Hoo Sun Chung, Abraham J. K. Koo, and Gregg A. Howe. 2008. Jasmonate signaling: a conserved mechanism of hormone sensing. *Current Opinion in Plant Biology* 11(4): 428–435.

Kessler, A., and I. T. Baldwin. 2001. Defensive function of herbivore-induced plant volatile emissions in nature. *Science* 291(5511): 2141–2144.

Kirstine, Wayne, Ian Galbally, and Martin Hooper. 2002. Air pollution and the smell of cut grass. In: Reese, Michela (ed.), *Thinking Outside the Square: 16th International Clean Air & Environment Conference*. CASANZ: 433–438.

Mafli, Ali, Jérôme Goudet, Edward E. Farmer. 2012. Plants and tortoises: mutations in the *Arabidopsis jasmonate* pathway increase feeding in a vertebrate herbivore. *Molecular Ecology* 21(10): 2534–2541.

Markovic, Dimitrije, Ilaria Colzi, Cosimo Taiti, Swayamjit Ray, Romain Scalone, Jared Gregory Ali, Stefano Mancuso, and Velemir Ninkovic. 2019. Airborne signals synchronize the defenses of neighboring plants in response to touch. *Journal of Experimental Botany* 70(2): 691–700.

Martin, Diane M., Jonathan Gershenzon, and Jörg Bohlmann. 2016. induction of volatile terpene biosynthesis and diurnal emission by methyl jasmonate in foliage of Norway spruce. *Plant Physiology* 132(3): 1586–1599.

McGowan, Kat. 2013. How plants secretly talk to each other. *Wired*, December 28.

Moran, N. 2007. Osmoregulation of leaf motor cells. *FEBS Letters* 581(12): 2337–2347.

Mousavi, Seyed A. R., Adeline Chauvin, François Pascaud, Stephan Kellenberger, and Edward E. Farmer. 2013.

GLUTAMATE RECEPTOR-LIKE genes mediate leaf-to-leaf wound signaling. *Nature* 500(7463): 422–426.

Neill, Steven, Raimundo Barros, Jo Bright, Radhika Desikan, John Hancock, Judith Harrison, Peter Morris, Dimas Ribeiro, and Ian Wilson. 2008. Nitric oxide, stomatal closure, and abiotic stress. *Journal of Experimental Botany* 59(2): 165–176.

Orians, C. 2005. Herbivores, vascular pathways, and systemic induction: facts and artifacts. *Journal of Chemical Ecology* 31(10): 2231–2242.

Orrock, John L., Brian M. Connolly, Won-Gyu Choi, Peter W. Guiden, Sarah J. Swanson, and Simon Gilroy. 2018. Plants eavesdrop on cues produced by snails and induce costly defenses that affect insect herbivores. *Oecologia* 186: 703–710.

Pearse, I. S. 2012. Interplant volatile signaling in willows: revisiting the original talking trees. *Oecologia* 172: 869–875.

Pearse, Ian S., Lauren M. Porensky, Louie H. Yang, Maureen L. Stanton, Richard Karban, Lisa Bhattacharyya, Rosa Cox, Karin Dove, August Higgins, Corrina Kamoroff, et al. 2012. Complex consequences of herbivory and interplant cues in three annual plants. *PLoS ONE* 7(5): e38105.

Pezzola, E., S. Mancuso, and R. Karban. 2017. Precipitation affects plant communication and defense. *Ecology* 98(6): 1693–1699.

Preston, E. 2013. Learning to speak shrub: using molecular codes, plants cry for help, ward off bugs, and save each other. *Nautilus*, August 26.

Qi, Jinfeng, Saif Ul Malook, Guojing Shen, Lei Gao, Cuiping Zhang, Jing Li, Jingxiong Zhang, Lei Wang, and Jianqiang Wu. 2018. Current understanding of maize and rice defense against insect herbivores. *Plant Diversity* 40(4): 189–195.

Rhoades, D. F. 1983. Responses of alder and willow to attack by tent caterpillars and webworms: evidence for pheromonal sensitivity of willows. In: Hedin, P. A. (ed.), *Plant Resistance to Insects*. ACS Symposium Series, vol. 208. American Chemical Society: 55–68.

Rivas-San, V. M., and J. Plasencia. 2011. Salicylic acid beyond defence: its role in plant growth and development. *Journal of Experimental Botany* 62(10): 3321–3338.

Roshchina, V. V. 2001. *Neurotransmitters in Plant Life*. Science Publishers.

Scala, Alessandra, Silke Allmann, Rossana Mirabella, Michel A. Haring, and Robert C. Schuurink. 2013. Green leaf volatiles: a plant's multifunctional weapon against herbivores and pathogens. *International Journal of Molecular Sciences* 14(9): 17781–17811.

Shepherd, V. A. 2012. At the root of plant neurobiology. In: Volkov, A. G. (ed.), *Plant Electrophysiology*. Springer: 3–43.

Simard, S. W., D. M. Durall, and M. D. Jones. 1997. Carbon allocation and carbon transfer between *Betula papyrifera* and *Pseudotsuga menziesii* seedlings using a ^{13}C pulse-labeling method. *Plant and Soil* 191: 41–55.

Simard, Suzanne W., David A. Perry, Melanie D. Jones, David D. Myrold, Daniel M. Durall, and Randy Molina. 1997. Net transfer of carbon between ectomycorrhizal tree species in the field. *Nature* 388: 579–582.

Song, Yuan Yuan, Mao Ye, Chuanyou Li, Xinhua He, Keyan Zhu-Salzman, Rui Long Wang, Yi Juan Su, Shi Ming Luo, and Ren Sen Zeng. 2014. Hijacking common mycorrhizal networks for herbivore-induced defence signal transfer between tomato plants. *Scientific Reports* 28(4): 3915.

Song, Yuan Yuan, Ren Sen Zeng, Jian Feng Xu, Jun Li, Xiang Shen, and Woldemariam Gebrehiwot Yihdego. 2010. Interplant communication of tomato plants through underground common mycorrhizal networks. *PLoS ONE* 5(10): e13324.

Staudt, Michael, Benjamin Jackson, Hanane El-Aouni, Bruno Buatois, Jean-Philippe Lacroze, Jean-Luc Poëssel, and Marie-Helene Sauge. 2010. Volatile organic compound emissions induced by the aphid *Myzus persicae* differ among resistant and susceptible peach cultivars and a wild relative. *Tree Physiology* 30(10): 1320–1334.

Teste, François P., Suzanne W. Simard, Daniel M. Durall, Robert D. Guy, and Melanie D. Jones. 2009. Access to mycorrhizal networks and roots of trees: importance for seedling survival and resource transfer. *Ecology* 90(10): 2808–2822.

Thaler, J. S., R. Karban, D. E. Ullman, K. Boege, and R. M. Bostock. 2002. Cross-talk between jasmonate and salicylate plant defense pathways: effects on several plant parasites. *Oecologia* 131: 227–235.

Toyota, Masatsugu, Dirk Spencer, Satoe Sawai-Toyota, Wang Jiaqi, Tong Zhang, Abraham J. Koo, Gregg A. Howe, and Simon Gilroy. 2018. Glutamate triggers long-distance, calcium-based plant defense signaling. *Science* 361(6407): 1112–1115.

Uehlein, N., and R. Kaldenhoff. 2008. Aquaporins and plant leaf movements. *Annals of Botany* 101(1): 1–4.

ul Hassan, M. N., Z. Zainal, and I. Ismanizan. 2015. Green leaf volatiles: biosynthesis, biological functions and their applications in biotechnology. *Plant Biotechnology Journal* 13(6): 727–739.

Ullah, Naseem, Muhammad Khurram, Hamid Hussain, Afridi Farhat, Farhat ali Khan, Muhammad Umar Khayam Sahibzada, Umar Khayam, M. Asif Khan,

Muhammad Usman Amin, Umberin Najeeb, et al. 2011. Comparison of phytochemical constituents and antimicrobial activities of *Mentha spicata* from four northern districts of Khyber Pakhtunkhwa. *Journal of Applied Pharmaceutical Science* 1(7): 72–76.

Vivaldo, Gianna, Elisa Masi, Cosimo Taiti, Guido Caldarelli, and Stefano Mancuso. 2017. The network of plants volatile organic compounds. *Scientific Reports* 7(1), article no. 11050.

Volkov, A. G., T. Adesina, and E. Jovanov. 2007. Closing of Venus flytrap by electrical stimulation of motor cells. *Plant Signaling and Behavior* 2(3): 139–145.

Volkov, Alexander G., Holly Carrell, Tejumade Adesina, Vladislav S. Markin, and Emil Jovanov. 2008. Plant electrical memory. *Plant Signaling & Behavior* 3(7): 490–492.

Volkov, Alexander G., Justin C. Foster, Talitha A. Ashby, Ronald K. Walker, Jon A. Johnson, and Vladislav S. Markin. 2010. *Mimosa pudica*: Electrical and mechanical stimulation of plant movements. *Plant, Cell & Environment* 33(2): 163–73.

Wasternack, C. 2007. Jasmonates: an update on biosynthesis, signal transduction and action in plant stress response, growth and development. *Annals of Botany* 100(4): 681–697.

Yang, T., and B. W. Poovaiah. 2003. Calcium/calmodulin-mediated signal network in plants. *Trends in Plant Science* 8(10): 505–512.

Yellen, G. 1998. The moving parts of voltage-gated ion channels. *Quarterly Reviews of Biophysics* 31(3): 239–295.

Zhu, J., and K. Park. 2005. Methyl salicylate, a soybean aphid-induced plant volatile attractive to the predator *Coccinella septempunctata*. *Journal of Chemical Ecology* 31(8): 1733–1746.

Chapter 8. The Enemy of My Enemy

Bentley, B. L. 1977. Extrafloral nectaries and protection by pugnacious bodyguards. *Annual Review of Ecology, Evolution, and Systematics* 8: 407–427.

Bruinsma, Maaike, Maarten A. Posthumus, Roland Mumm, Martin J. Mueller, Joop J. A. van Loon, and Marcel Dicke. 2009. Jasmonic acid-induced volatiles of *Brassica oleracea* attract parasitoids: effects of time and dose, and comparison with induction by herbivores. *Journal of Experimental Botany* 60(9): 2575–2587.

Byk, J., and K. Del-Claro. 2011. Ant-plant interaction in the Neotropical savanna: direct beneficial effects of extrafloral nectar on ant colony fitness. *Population Ecology* 53(2): 327–332.

Chomicki, G., M. Janda, and S. S. Renner. 2017. The assembly of ant-farmed gardens: mutualism specialization following host broadening. *Proceedings of the Royal Society B: Biological Sciences* 284(1850): 1–9.

Dejean, Alain, Pascal Solano, M. Belin-Depoux, P. Cerdan, and Bruno Corbara. 2001. Predatory behavior of patrolling *Allomerus decemarticulatus* workers (Formicidae: Myrmicinae) on their host plant. *Sociobiology* 37(3): 571–578.

Dicke, M., and J. J. A. van Loon. 2003. Multitrophic effects of herbivore-induced plant volatiles in an evolutionary context. *Entomologia Experimentalis et Applicata* 97(3): 237–249.

Dicke, Marcel, Rieta Gols, Daniel Ludeking, and Maarten A. Posthumus. 1999. Jasmonic acid and herbivory differentially induce carnivore-attracting plant volatiles in lima bean plants. *Journal of Chemical Ecology* 25: 1907–1922.

Erb, M., S. Meldau, and G. A. Howe. 2012. Role of phytohormones in insect-specific plant reactions. *Trends in Plant Science* 17(5): 250–259.

Fatouros, N. E. 2012. Plant volatiles induced by herbivore egg deposition affect insects of different trophic levels. *PLoS ONE* 7(8): e43607.

Fernandez, F. 2007. The Myrmicine ant genus Allomerus Mayr (Hymenoptera: Formicidae). *Caldasia* 29(1): 159–175.

Fleming, J. G., and M. D. Summers. 1991. Polydnavirus DNA is integrated in the DNA of its parasitoid wasp host. *Proceedings of the National Academy of Sciences of the United States of America* 88(21): 9770–9774.

Fotso, A. K., R. Hanna, M. Tindo, A. Doumtsop, and P. Nagel. 2015. How plants and honeydew-producing hemipterans affect ant species richness and structure in a tropical forest zone. *Insectes Sociaux* 62: 443–453.

Gorb, E., and S. Gorb. 2003. *Seed Dispersal by Ants in a Deciduous Ecosystem*. Kluwer Academic Publishers.

Grangier, J., A. Dejean, P. Malé, and J. Orivel. 2008. Indirect defense in a highly specific ant-plant mutualism. *Naturwissenschaften* 95(10): 909–916.

Gross, P. 1993. Insect behavioral and morphological defences against parasitoids. *Annual Review of Entomology* 38: 251–273.

Heil, M. 2008. Indirect defence: recent developments and open questions. *Progress in Botany* 178(1): 41–61.

Heil, M., T. Koch, A. Hilpert, B. Fiala, W. Boland, and K. Linsenmair. 2001. Extrafloral nectar production of the ant-associated plant, *Macaranga tanarius*, is an induced, indirect, defensive response elicited by jasmonic acid. *Proceedings of the National Academy of Sciences of the United States of America* 98(3): 1083–1088.

Herniou, Elisabeth A., Elisabeth Huguet, Julien Thézé, Annie Bézier, Georges Periquet, and Jean-Michel Drezen. 2013. When parasitic wasps hijacked viruses: genomic and functional evolution of polydnaviruses. *Philosophical Transactions of the Royal Society B* 368(1626): 20130051.

Janzen, D. H. 1967. Interaction of the bull's-horn acacia (*Acacia cornigera* L.) with an ant inhabitant (*Pseudomyrmex ferruginea* F. Smith) in eastern Mexico. *University of Kansas Science Bulletin* 47(6): 315–558.

Jervis, M. A., and N. A. Kidd. 1986. Host-feeding strategies in hymenopteran parasitoids. *Biological Reviews* 61(4): 395–434.

Kacsoh, B. Z., Z. R. Lynch, N. T. Mortimer, and T. A. Schlenke. 2013. Fruit flies medicate offspring after seeing parasites. *Science* 339(6122): 947–950.

Kessler, A., and I. T. Baldwin. 2001. Defensive function of herbivore-induced plant volatile emissions in nature. *Science* 291(5511): 2141–2144.

Koptur, S. 1992. Extrafloral nectar-mediated interactions between insects and plants. In: Bernays, E. (ed.), *Insect-Plant Interactions*. CRC Press: 81–129.

Koptur, Suzanne, Mónica Palacios-Rios, Cecilia Díaz-Castelazo, William P. Mackay, and Víctor Rico-Gray. 2013. Nectar secretion on fern fronds associated with lower levels of herbivore damage: field experiments with a widespread epiphyte of Mexican cloud forest remnants. *Annals of Botany* 111(6): 1277–1283.

Koptur, S., V. Rico-Gray, and M. Palacios-Rios. 1998. Ant protection of the nectaried fern *Polypodium plebeium* in central Mexico. *American Journal of Botany* 85(5): 736.

Kost, C., and M. Heil. 2006. Herbivore-induced plant volatiles induce an indirect defence in neighbouring plants. *Journal of Ecology* 94(3): 619–628.

Lengyel, Szabolcs, Aaron D. Gove, Andrew M. Latimer, Jonathan D. Majer, and Robert R. Dunn. 2010. Convergent evolution of seed dispersal by ants, and phylogeny and biogeography in flowering plants: a global survey. *Perspectives in Plant Ecology, Evolution and Systematics* 12(1): 43–55.

Leroy, Céline, Alain Jauneau, Angélique Quilichini, Alain Dejean, and Jérôme Orivel. 2008. Comparison between the anatomical and morphological structure of leaf blades and foliar domatia in the ant-plant *Hirtella physophora* (Chrysobalanaceae). *Annals of Botany* 101(4): 501–507.

Lucchi, Andrea, Augusto Loni, Luca Mario Gandini, Pierluigi Scaramozzino, Claudio Ioriatti, Renato Ricciardi, and Peter W. Shearer. 2017. Using herbivore-induced plant volatiles to attract lacewings, hoverflies and parasitoid wasps in vineyards: achievements and constraints. *Bulletin of Insectology* 70(2): 273–282.

Lundgren, J. G. 2009. Nutritional aspects of non-prey foods and the life histories of predaceous Coccinellidae. *Biological Control* 51(2): 294–305.

Marting, P. 2020. Azteca-Cecropia Behavioral Ecology. http://aztecacecropia.com/projects.

McCormick, A. C., S. B. Unsicker, and J. Gershenzon. 2012. The specificity of herbivore-induced plant volatiles in attracting herbivore enemies. *Trends in Plant Science* 17(5): 303–310.

Meiners, T., and M. Hilker. 2000. Induction of plant synomones by oviposition of a phytophagous insect. *Journal of Chemical Ecology* 26(1): 221–232.

Milan, N. F., B. Z. Kacsoh, and T. A. Schlenke. 2012. Alcohol consumption as self-medication against blood-borne parasites in the fruit fly. *Current Biology* 22(6): 488–493.

Mizell, R. F. 2018. *Many Plants Have Extrafloral Nectaries Helpful to Beneficials*. University of Florida IFAS Extension.

Mumm, R., and M. Dicke. 2010. Variation in natural plant products and the attraction of bodyguards involved in indirect plant defense. *Canadian Journal of Zoology* 88(7): 627–668.

Nepi, M. 2017. New perspectives in nectar evolution and ecology: simple alimentary reward or a complex multiorganism interaction? *Acta Agrobotanica* 70(1): 1704.

Ozawa, R., G. Arimura, J. Takabayashi, T. Shimoda, and T. Nishioka. 2000. Involvement of jasmonate- and salicylate-related signaling pathways for the production of specific herbivore-induced volatiles in plants. *Plant & Cell Physiology* 41(4): 391–398.

Paré, P. W., and M. A. Fara. 2008. Natural enemy attraction to plant volatiles. In: Capinera, J. L. (ed.), *Encyclopedia of Entomology*. Springer: 1534–1535.

Pashalidou, F. G. 2015. To be in time: egg deposition enhances plant-mediated detection of young caterpillars by parasitoids. *Oecologia* 177(2): 477–486.

Peeters, C., and D. Wiwatwitaya. 2014. Philidris ants living in Dischidia epiphytes from Thailand. *Asian Myrmecology* 6(1): 49–61.

Pemberton, R. W., and L. Lee. 1996. The influence of extrafloral nectaries on parasitism of an insect herbivore. *American Journal of Botany* 83(9): 1187–1194.

Pemberton, R. W., and N. J. Vandenberg. 1993. Extrafloral nectar feeding by ladybird beetles (Coleoptera: Coccinellidae). *Proceedings of the Entomological Society of Washington* 95: 139–151.

Ponzio, Camille, Rieta Gols, Berhane T. Weldegergis, and Marcel Dicke. 2014. Caterpillar-induced plant volatiles remain a reliable signal for foraging wasps during dual

attack with a plant pathogen or non-host insect herbivore. *Plant, Cell & Environment* 37(8): 1924–1935.
Price, Peter W., Robert F. Denno, Micky D. Eubanks, Deborah L. Finke, and Ian Kaplan. 2011. *Insect Ecology: Behavior, Populations and Communities*. 1st ed. Cambridge University Press.
Revel, M., A. Dejean, R. Céréghino, and O. Roux. 2010. An assassin among predators: the relationship between plant-ants, their host Myrmecophytes and the Reduviidae *Zelus annulosus*. *PLoS ONE* 5(10): e13110.
Rogers, C. E. 1985. Extrafloral nectar: entomological implications. *Bulletin of the Entomological Society of America* 31(3): 15–20.
Ruiz-González, Mario X., Pierre-Jean G. Malé, Céline Leroy, Alain Dejean, Hervé Gryta, Patricia Jargeat, Angélique Quilichini, and Jérôme Orivel. 2010. Specific, non-nutritional association between an ascomycete fungus and Allomerus plant-ants. *Biology Letters* 7(3): 475–479.
Singer, M. S., K. C. Mace, and E. A. Bernays. 2009. Self-medication as adaptive plasticity: increased ingestion of plant toxins by parasitized caterpillars. *PLoS ONE* 4(3): e4796.
Thaler, J. S. 1999. Jasmonate-inducible plant defences cause increased parasitism of herbivores. *Nature* 399(6737): 686–688.
Thaler, J. S. 2002. Effect of jasmonate-induced plant responses on the natural enemies of herbivores. *Journal of Animal Ecology* 71(1): 141–150.
Turlings, T. C., P. J. McCall, H. T. Alborn, and J. H. Tumlinson. 1993. An elicitor in caterpillar oral secretions that induces corn seedlings to emit chemical signals attractive to parasitic wasps. *Journal of Chemical Ecology* 19(3): 411–425.
van Poecke, R. M. P., and M. Dicke. 2004. Indirect defence of plants against herbivores: using *Arabidopsis thaliana* as a model plant. *Plant Biology* 6(4): 387–401.
Wei, Jianing, Joop J. A. van Loon, Rieta Gols, Tila R. Menzel, Na Li, Le Kang, and Marcel Dicke. 2014. Reciprocal crosstalk between jasmonate and salicylate defence-signalling pathways modulates plant volatile emission and herbivore host-selection behavior. *Journal of Experimental Botany* 65(12): 3289–3298.
Whitfield, J. 2001. Making crops cry for help. *Nature* 410(6830): 736–737.
Zhang, Peng-Jun, Colette Broekgaarden, Si-Jun Zheng, Tjeerd A. L. Snoeren, Joop J. A. van Loon, Rieta Gols, and Marcel Dicke. 2013. Jasmonate and ethylene signaling mediate whitefly-induced interference with indirect plant defense in *Arabidopsis thaliana*. *New Phytologist* 197(4): 1291–1299.
Zhu, Feng, Colette Broekgaarden, Berhane T. Weldegergis, Jeffrey A. Harvey, Ben Vosman, Marcel Dicke, and Erik H. Poelman. 2015. Parasitism overrides herbivore identity allowing hyperparasitoids to locate their parasitoid host using herbivore-induced plant volatiles. *Molecular Ecology* 24(11): 2886–2899.

Chapter 9. Hiding in Plain Sight

Archetti, M. 2000. The origin of autumn colours by coevolution. *Journal of Theoretical Biology* 205(4): 625–630.
Augner, M. 1994. Should a plant always signal its defence against herbivores? *Oikos* 70(3): 322–332.
Barlow, Bryan. 2012. Mistletoes—cryptic mimicry. Australian National Botanic Gardens, Centre for Australian National Biodiversity Research. https://www.anbg.gov.au/mistletoe/mimicry.html.
Barlow, B., and D. Wiens. 1977. Host-parasite resemblance in Australian mistletoes: the case for cryptic mimicry. *Evolution* 31(1): 69–84. https://doi.org/10.2307/2407546.
Barrett, S. 1983. Crop mimicry in weeds. *Economic Botany* 37(3): 255–282.
Barrett, S. C. H. 1987. Mimicry in plants. *Scientific American* 257(3): 76–85.
Blick, R. A. J., K. C. Burns, and A. T. Moles. 2012. Predicting network topology of mistletoe-host interactions: do mistletoes really mimic their hosts? *Oikos* 121(5): 761–771.
Briscoe, A. D., and L. Chittka. 2001. The evolution of color vision in insects. *Annual Review of Entomology* 46(1): 471–510.
Brown, V. K., and J. H. Lawton. 1991. Herbivory and the evolution of leaf size and shape. *Philosophical Transactions of the Royal Society B* 333(1267): 89–96.
Bruce, Toby J. A., Lester J. Wadhams, and Christine M. Woodcock. 2005. Insect host location: a volatile situation. *Trends in Plant Science* 10(6): 269–274.
Burchell, W. J. 1822. *Travels in the Interior of Southern Africa*. Vol. 1. Longman, Hurst, Rees, Orne and Brown.
Burns, K. C. 2010. Is crypsis a common defensive strategy in plants? speculation on signal deception in the New Zealand flora. *Plant Signaling and Behavior* 5(1): 9–13.
Campitelli, B. E., I. Stehlik, and J. R. Stinchcombe. 2008. Leaf variegation is associated with reduced herbivore damage in *Hydrophyllum virginianum*. *Botany* 86(3): 306–313.
Canyon, D., and C. J. Hill. 1997. Mistletoe host-resemblance: a study of herbivory, nitrogen and moisture in two Australian mistletoes and their host trees. *Australian Journal of Ecology* 22(4): 395–403.
Cole, D. T., and N. A. Cole. 2005. *Lithops: Flowering Stones*. Cactus & Co. Libri.
Ehleringer, J. R., I. Ullmann, O. L. Lange, G. D. Farquhar, I. R. Cowan, E. D. Schulze, and H. Ziegler. 1986. Mistletoes:

a hypothesis concerning morphological and chemical avoidance of herbivory. *Oecologia* 70(2): 234–237. https://doi.org/10.1007/BF00379245.

Endler, J. A. 1981. An overview of the relationships between mimicry and crypsis. *Biological Journal of the Linnean Society* 16(1): 25–31.

Erskine, William, Joseph Smartt, and Fred J. Muehlbauer. 1994. Mimicry of lentil and the domestication of common vetch and grass pea. *Economic Botany* 48(3): 326–332.

Gilbert, Lawrence E. 1973. Ecological consequences of a coevolved mutualism between butterflies and plants. In: Gilbert, Lawrence E., and Peter H. Raven (eds.), *Coevolution of Animals and Plants*. Symposium V, First International Congress of Systematic and Evolutionary Biology. University of Texas Press: 210–240.

Gilbert, L. E. 1982. The coevolution of a butterfly and a vine. *Scientific American* 247(2): 110–121.

Givnish, T. 1990. Leaf mottling: relation to growth form and leaf phenology and possible role as camouflage. *Functional Ecology* 4(4): 463–474.

Juergens, N. 1996. Psammophorous plants and other adaptations to desert ecosystems with high incidence of sandstorms. *Journal of Botanical Taxonomy and Geobotany* 107(5–6): 345–359.

Kellner, A., C. M. Ritz, P. Schlittenhardt, and F. H. Hellwig. 2011. Genetic differentiation in the genus *Lithops* L. (Ruschioideae, Aizoaceae) reveals a high level of convergent evolution and reflects geographic distribution. *Plant Biology* 13(2): 368–380.

Lev-Yadun, S. 2006. Defensive coloration in plants: a review of current ideas about anti-herbivore coloration strategies. In: Teixeira da Silva, Jaime A. (ed.), *Floriculture, Ornamental and Plant Biotechnology: Advances and Topical Issues*, vol. 4. Global Science Books: 292–299.

Lev-Yadun, S. 2009. Aposematic (warning) coloration in plants. In: Baluka, F. (ed.), *Plant-Environment Interactions: Signaling and Communication in Plants*. Springer: 167–202.

Lev-Yadun, S. 2014a. Defensive masquerade by plants. *Biological Journal of the Linnean Society* 113(4): 1162–1166.

Lev-Yadun, S. 2014b. The proposed anti-herbivory roles of white leaf variegation. *Progress in Botany* 76: 241–269.

Lev-Yadun, S. 2016. *Defensive (Anti-Herbivory) Coloration in Land Plants: Anti-Herbivory Plant Coloration and Morphology*. Springer.

Lev-Yadun, S., and K. S. Gould. 2008. Role of anthocyanins in plant defence. In: Winefield, C., K. Davies, and K. Gould (eds.), *Anthocyanins: Biosynthesis, Functions, and Applications*. Springer: 22–28.

Lev-Yadun, S., and M. Gutman. 2013. Carrion odor and cattle grazing: evidence for plant defense by carrion odor. *Communicative and Integrative Biology* 6(6): e26111.

Lev-Yadun, S., and M. Inbar. 2002. Defensive ant, aphid and caterpillar mimicry in plants. *Biological Journal of the Linnean Society* 77(3): 393–398.

Lev-Yadun, S., G. Ne'eman, and U. Shanas. 2009. A sheep in wolf's clothing: do carrion and dung odours of flowers not only attract pollinators but also deter herbivores? *BioEssays* 31(1): 84–88.

LoPresti, E. F., P. Grof-Tisza, M. Robinson, J. Godfrey, and R. Karban. 2017. Entrapped sand as a plant defence: effects on herbivore performance and preference. *Ecological Entomology* 43(2): 154–161.

LoPresti, E. F., and R. Karban. 2016. Chewing sandpaper: grit, plant apparency and plant defense in sand-entrapping plants. *Ecology* 97(4): 826–833.

Manetas, Y. 2006. Why some leaves are anthocyanic and why most anthocyanic leaves are red? *Flora* 201(3): 163–177.

Mcelroy, Scott J. 2014. Vavilovian mimicry: Nikolai Vavilov and his little-known impact on weed science. *Weed Science* 62(2): 207–216.

Neinhuis, Christoph, Ute Müller-Doblies, and Dietrich Müller-Doblies. 1996. *Psammophora* and other sand-coated plants from southern Africa. *Feddes Repertorium* 107(5–6): 549–555.

Niemelä, P., and J. Tuomi. 1987. Does the leaf morphology of some plants mimic caterpillar damage? *Oikos* 50(2): 256–257.

Niu, Y., G. Chen, D.-L. Peng, B. Song, Y. Yang, Z.-M. Li, and H. Sun. 2014. Grey leaves in an alpine plant: a cryptic colouration to avoid attack? *New Phytologist* 203(3): 953–963.

Niu, Yang, Zhe Chen, Martin Stevens, and Hang Sun. 2017. Divergence in cryptic leaf colour provides local camouflage in an alpine plant. *Proceedings of the Royal Society B: Biological Sciences* 284(1864), article no. 20171654.

Niu, Y., H. Sun, and M. Stevens. 2018. Plant camouflage: ecology, evolution, and implications. *Trends in Ecology & Evolution* 3(8): 608–618.

Pannell, John R., and Edward E. Farmer. 2016. Mimicry in plants. *Current Biology* 26(17): 784–785.

Ruxton, G. D., T. N. Sherratt, and M. P. Speed. 2004. *Avoiding Attack: The Evolutionary Ecology of Crypsis, Warning Signals & Mimicry*. Oxford University Press.

Shapiro, A. M. 1981a. Egg-mimics of *Streptanthus* (Cruciferae) deter oviposition by *Pieris sisymbrii* (Lepidoptera: Pieridae). *Oecologia* 48(1): 142–143.

Shapiro, A. M. 1981b. The pierid red-egg syndrome. *American Naturalist* 117(3): 276–294.

Skelhorn, John, Hannah M. Rowland, and Graeme D. Ruxton. 2010. The evolution and ecology of masquerade. *Biological Journal of the Linnean Society* 99(1): 1–8.

Soltau, U., S. Dötterl, and S. Liede-Schumann. 2009. Leaf variegation in *Caladium steudneriifolium* (Araceae): a case of mimicry? *Evolutionary Ecology* 23(4): 503–512.

Stone, B. C. 1979. Protective coloration of young leaves in certain Malaysian palms. *Biotropica* 11(2): 126.

Strauss, S. Y., and N. I. Cacho. 2013. Nowhere to run, nowhere to hide: the importance of enemies and apparency in adaptation to harsh soil environments. *American Naturalist* 182(1): E1–E14.

Turner, J. S., and M. D. Picker. 1993. Thermal ecology of an embedded dwarf succulent from southern Africa (*Lithops* spp.: Mesembryanthemaceae). *Journal of Arid Environments* 24(4): 361–385.

Wiens, D. 1978. Mimicry in plants. In: Hecht, M. K., W. C. Steere, and B. Wallace (eds.), *Evolutionary Biology*, vol. 11. Springer: 365–403.

Williams, K. S., and L. E. Gilbert. 1981. Insects as selective agents on plant vegetative morphology: egg mimicry reduces egg laying by butterflies. *Science* 212(4493): 467–469.

Williamson, G. B. 1982. Plant mimicry: evolutionary constraints. *Biological Journal of the Linnean Society* 18(1): 49–58. https://doi.org/10.1111/j.1095-8312.1982.tb02033.x.

Zabka, H., and G. Tembrock. 1986. Mimicry and crypsis: a behavioural approach to classification. *Behavioural Processes* 13(1–2): 159–176.

Chapter 10. True Colors

Archetti, M. 2000. The origin of autumn colours by coevolution. *Journal of Theoretical Biology* 205(4): 625–630.

Archetti, M. 2007. Autumn colours and the nutrient retranslocation hypothesis: a theoretical assessment. *Journal of Theoretical Biology* 244(4): 714–721.

Archetti, Marco, Thomas F. Döring, Snorre B. Hagen, Nicole M. Hughes, Simon R. Leather, David W. Lee, Simcha Lev-Yadun, Yiannis Manetas, Helen J. Ougham, Paul G. Schaberg, et al. 2009. Unravelling the evolution of autumn colours: an interdisciplinary approach. *Trends in Ecology & Evolution* 24(3): 166–173.

Ayres, Matthew P., Thomas P. Clausen, Stephen F. MacLean Jr., Ahyna M. Redman, and Paul B. Reichardt. 1997. Diversity of structure and antiherbivore activity in condensed tannins. *Ecology* 78(6): 1696–1712.

Barbehenn, R. V., and C. P. Constabel. 2011. Tannins in plant-herbivore interactions. *Phytochemistry* 72(13): 1551–1565.

Bernays, E. A., G. C. Driver, and M. Bilgener. 1989. Herbivores and plant tannins. *Advances in Ecological Research* 19: 263–302.

Blount, J. D., M. P. Speed, G. D. Ruxton, and P. A. Stephens. 2009. Warning displays may function as honest signals of toxicity. *Proceedings of the Royal Society B: Biological Sciences* 276(1658): 871–877.

Burger, J., and G. E. Edwards. 1996. Photosynthetic efficiency, and photodamage by UV and visible radiation, in red versus green leaf coleus varieties. *Plant Cell Physiology* 37(3): 395–399.

Canham, H. O. 2011. Hemlock and hide: the tanbark industry in old New York. *Northern Woodlands* 69(Summer): 36–41.

Chalker-Scott, L. 2008. Environmental significance of anthocyanins in plant stress responses. *Photochemistry and Photobiology* 70(1): 1–9.

Chin, S., C. A. Behm, and U. Mathesius. 2018. Functions of flavonoids in plant-nematode interactions. *Plants* 7(4): 85.

Clausen, T. P., P. B. Reichardt, J. P. Bryant, and F. Provenza. 1992. Condensed tannins in plant defense: a perspective on classical theories. In: Hemingway, R. W., and P. E. Laks (eds.), *Plant Polyphenols: Basic Life Sciences*, vol. 59. Springer: 639–651.

Clement, J. S., T. J. Mabry, H. Wyler, and A. S. Dreiding. 1994. Chemical review and evolutionary significance of the betalains. In: Behnke, H. D., and T. J. Mabry (eds.), *Caryophyllales: Evolution and Systematics*. Springer: 247–261.

Close, D. C., and C. L. Beadle. 2003. The ecophysiology of foliar anthocyanin. *Botanical Review* 69(2): 149–161.

Constabel, C. P., K. Yoshida, and V. Walker. 2014. Diverse ecological roles of plant tannins: plant defense and beyond. In: Romani, A., V. Lattanzio, and S. Quideau (eds.), *Recent Advances in Polyphenol Research*. John Wiley & Sons: 464.

Dai, J., and R. J. Mumper. 2010. Plant phenolics: extraction, analysis and their antioxidant and anticancer properties. *Molecules* 15(10): 7313–7352.

Demmig-Adams, B., and W. W. Adams II. 1992. Photoprotection and other responses of plants to light stress. *Annual Review of Plant Physiology and Plant Molecular Biology* 43: 599–626.

Falcone Ferreyra, M. L., S. P. Rius, and P. Casati. 2018. Flavonoids: biosynthesis, biological functions, and biotechnological applications. *Frontiers in Plant Science* 28(3): 222.

Feeny, P. 1976. Plant apparency and chemical defense. In: Wallace, J. W., and R. L. Mansell (eds.), *Biochemical Interaction between Plants and Insects*. Recent Advances in Phytochemistry, vol. 10. Springer: 1–40.

Findlay, Stuart, Margaret Carreiro, Vera Krischik, and Clive G. Jones. 1996. Effects of damage to living plants on leaf litter quality. *Ecological Applications* 6(1): 269–275.

Gallagher, B. 2015. The hidden warning of fall colors: did autumn red and yellows evolve to repel insects? *Nautilus*, August 13. https://nautil.us/the-hidden-warning-of-fall-colors-235586.

Galvão, V. C., and C. Fankhauser. 2015. Sensing the light environment in plants: photoreceptors and early signaling steps. *Current Opinion in Neurobiology* 34: 46–53.

Givnish, T. 1990. Leaf mottling: relation to growth form and leaf phenology and possible role as camouflage. *Functional Ecology* 4(4): 463–474.

Gould, Kevin, Kevin M. Davies, and Chris Winefield. 2009. *Anthocyanins: Biosynthesis, Functions, and Applications*. Springer.

Gould, Kevin S., Christian Jay-Allemand, Barry A. Logan, Yves Baissac, and Luc P. R. Bidel. 2018. When are foliar anthocyanins useful to plants? re-evaluation of the photoprotection hypothesis using *Arabidopsis thaliana* mutants that differ in anthocyanin accumulation. *Environmental and Experimental Botany* 154: 11–22.

Gronquist, Matthew, Alexander Bezzerides, Athula Attygalle, Jerrold Meinwald, Maria Eisner, and Thomas Eisner. 2001. Attractive and defensive functions of the ultraviolet pigments of a flower (*Hypericum calycinum*). *Proceedings of the National Academy of Sciences of the United States of America* 98(24): 13745–13750.

Hagen, S. B., S. Debeausse, N. G. Yoccoz, and I. Folstad. 2004. Autumn coloration as a signal of tree condition. *Proceedings of the Royal Society B: Biological Sciences* 271(suppl. 4): S184–S185.

Hagen, S. B., I. Folstad, and S. W. Jakobsen. 2003. Autumn colouration and herbivore resistance in mountain birch (*Betula pubescens*). *Ecology Letters* 6(9): 807–811.

Hagerman, A. E., and C. T. Robbins. 1987. Implications of soluble tannin-protein complexes for tannin analysis and plant defense mechanisms. *Journal of Chemical Ecology* 13(5): 1243–1259.

Hamilton, W. D., and S. P. Brown. 2001. Autumn tree colours as a handicap signal. *Proceedings of the Royal Society B: Biological Sciences* 268(1475): 1489–1493.

Hatier, J. H., and K. S. Gould. 2008. Foliar anthocyanins as modulators of stress signals. *Journal of Theoretical Biology* 253(3): 625–627.

Hättenschwiler, S., and P. M. Vitousek. 2000. The role of polyphenols in terrestrial ecosystem nutrient cycling. *Trends in Ecology & Evolution* 15(6): 238–243.

Hausen, B. M., and K. H. Schulz. 1977. Occupational contact dermatitis due to croton (*Codiaeum variegatum* (L.) A. Juss var. *pictum* (Lodd.) Muell. Arg.). Sensitization by plants of the Euphorbiaceae. *Contact Dermatitis* 3(6): 289–292.

Havaux, Michel. 2013. Carotenoid oxidation products as stress signals in plants. *Plant Journal* 79(4): 597–606.

Hernes, P. J., and J. I. Hedges. 2000. Determination of condensed tannin monomers in environmental samples by capillary gas chromatography of acid depolymerization extracts. *Analytical Chemistry* 72(20): 5115–5124.

Holopainen, J. K., and P. Peltonen 2002. Bright autumn colours of deciduous trees attract aphids: nutrient retranslocation hypothesis. *Oikos* 99(1): 184–188.

Kraus, Tamara, R. Dahlgren, and R. Zasoski. 2003. Tannins in nutrient dynamics of forest ecosystems—a review. *Plant and Soil* 256(1): 41–66.

Kuiters, A. T. 1990. Role of phenolic substances from decomposing forest litter in plant-soil interactions. *Acta Botanica Neerlandica* 39(4): 329–348.

Latowski, D., P. Kuczyńska, and K. Strzałka. 2011. Xanthophyll cycle—a mechanism protecting plants against oxidative stress. *Redox Report* 16(2): 78–90.

Lev-Yadun, S. 2006. Defensive functions of white coloration in coastal and dune plants. *Israel Journal of Plant Sciences* 54: 317–325.

Lev-Yadun, S. 2009. Aposematic (warning) coloration in plants. In: Baluska, F. (ed.), *Plant-Environment Interactions: From Sensory Plant Biology to Active Behavior*, vol. 2. Springer: 167–202.

Lev-Yadun, S. 2014. Potential defence from herbivory by "dazzle effects" and "trickery coloration" of leaf variegation. *Biological Journal of the Linnean Society* 111(3): 692–697.

Lev-Yadun, S. 2015. The proposed anti-herbivory roles of white leaf variegation. *Progress in Botany* 76: 241–269.

Lev-Yadun, Simcha, Amots Dafni, Moshe A. Flaishman, Moshe Inbar, Ido Izhaki, Gadi Katzir, and Gidi Ne'eman. 2004. Plant coloration undermines herbivorous insect camouflage. *BioEssays* 26(10): 1126–1130.

Lev-Yadun S., and K. S. Gould. 2008. Role of anthocyanins in plant defence. In: Winefield, C., K. Davies, and K. Gould (eds.), *Anthocyanins: Biosynthesis, Functions, and Applications*. Springer: 22–28.

Lu, Yanfen, Qi Chen, Yufen Bu, Rui Luo, Suxiao Hao, Jie Zhang, Ji Tian, and Yuncong Yao. 2017. Flavonoid accumulation plays an important role in the rust resistance of *Malus* plant leaves. *Frontiers in Plant Science* 8, article no. 1286.

Madritch, M. D., and R. L. Lindroth. 2015. Condensed tannins increase nitrogen recovery by trees following insect defoliation. *New Phytologist* 208(2): 410–420.

Manetas, Y. 2006. Why some leaves are anthocyanic and why most anthocyanic leaves are red? *Flora* 201(3): 163–177.

Maskato, Yamit, Stav Talal, Tamar Keasar, and Eran Gefen. 2014. Red foliage color reliably indicates low host quality and increased metabolic load for development of an herbivorous insect. *Arthropod-Plant Interactions* 8(4): 285–292.

Mathesius, U. 2018. Flavonoid functions in plants and their interactions with other organisms. *Plants* 7(2): 30.

Mierziak, J., K. Kostyn, and A. Kulma. 2014. Flavonoids as important molecules of plant interactions with the environment. *Molecules* 19(10): 16240–16265.

Muller, P., X.-P. Li, and K. K. Niyogi. 2001. Non-photochemical quenching: a response to excess light energy. *Plant Physiology* 125(4): 1558–1566.

Northup, R. R., Z. Yu, R. A. Dahlgren, and K. A. Vogt. 1995. Polyphenol control of nitrogen release from pine litter. *Nature* 377(6546): 227–229.

Ogunwenmo, K. O., O. A. Idowu, C. Innocent, E. B. Esan, and O. A. Oyelana. 2007. Cultivars of *Codiaeum variegatum* (L.) Blume (Euphorbiaceae) show variability in phytochemical and cytological characteristic. *African Journal of Biotechnology* 6(20): 2400–2405.

Patisaul, H. B., and W. Jefferson. 2010. The pros and cons of phytoestrogens. *Frontiers in Neuroendocrinology* 31(4): 400–419.

Pourcel, Lucille, Jean-Marc Routaboul, Véronique Cheynier, Loïc Lepiniec, and Isabelle Debeaujon. 2007. Flavonoid oxidation in plants: from biochemical properties to physiological functions. *Trends in Plant Science* 12(1): 29–36.

Rhoades, D. F., and R. G. Cates. 1976. Toward a general theory of plant antiherbivore chemistry. In: Wallace, J. W., and R. L. Mansell (eds.), *Biochemical Interaction between Plants and Insects.* Recent Advances in Phytochemistry, vol. 10. Springer: 168–213.

Robbins, C. T., S. Mole, A. E. Hagerman, and T. A. Hanley. 1987. Role of tannins in defending plants against ruminants: reduction in dry matter digestion? *Ecology* 68(6): 1606–1615.

Salminen, Juha-Pekka, and Maarit Karonen. 2011. Chemical ecology of tannins and other phenolics: we need a change in approach. *Functional Ecology* 25(2): 325–338.

Samanta, A., G. Das, and S. K. Das. 2011. Roles of flavonoids in plants. *International Journal of Pharmaceutical Science and Technology* 6(1): 12–35.

Santos-Buelga, C., N. Mateus, and V. de Freitas. 2014. Anthocyanins: Plant pigments and beyond. *Journal of Agricultural and Food Chemistry* 62(29): 6879–6884.

Schaefer, H. M., and G. Rolshausen. 2006. Plants on red alert: do insects pay attention? *Bioessays* 28(1): 65–71.

Steyn, W. J., S. J. E. Wand, D. M. Holcroft, and G. Jacobs. 2002. Anthocyanins in vegetative tissues: a proposed unified function in photoprotection. *New Phytologist* 155(3): 349–361.

Tanaka, Y., N. Sasaki, and A. Ohmiya. 2008. Biosynthesis of plant pigments: anthocyanins, betalains and carotenoids. *Plant Journal* 54(4): 733–749.

Treutter, D. 2005. Significance of flavonoids in plant resistance and enhancement of their biosynthesis. *Plant Biology* 7(6): 581–591.

Treutter, D. 2006. Significance of flavonoids in plant resistance: a review. *Environmental Chemistry Letters* 4: 147–157.

U.S. Forest Service. 2020. Tannins. *https://www.fs.usda.gov/wildflowers/ethnobotany/tannins.shtml.*

Vicente, O., and M. Boscaiu. 2018. Flavonoids: antioxidant compounds for plant defence and for a healthy human diet. *Notulae Botanicae Horti Agrobotanici Cluj-Napoca* 46(1): 14–21.

Yonekura-Sakakibara, K. Higashi, and R. Nakabayashi. 2019. The origin and evolution of plant flavonoid metabolism. *Frontiers in Plant Science* 10, article no. 943.

Zucker, W. V. 1983. Tannins: does structure determine function? an ecological perspective. *American Naturalist* 121(3): 335–365.

Chapter 11. Feel the Burn

Abdou, A., A. Abou-Zeid, M. R. El-Sherbeeny, and Z. H. Abou-El-Gheat. 1972. Antimicrobial activities of *Allium sativum*, *Allium cepa*, *Raphanus sativus*, *Capsicum frutescens*, *Eruca sativa*, *Allium kurrat* on bacteria. *Qualitas Plantarum et Materiae Vegetabiles* 22(1): 29–35.

Agerbirk, N., M. De Vos, J. H. Kim, and G. Jander. 2009. Indole glucosinolate breakdown and its biological effects. *Phytochemical Review* 8(1): 101–120.

Ahn, Seung-Joon, F. R. Badenes, M. Reichelt, A. Svatoš, B. Schneider, J. Gershenzon, and D. G. Heckel. 2011. Metabolic detoxification of capsaicin by UDP-glycosyltransferase in three *Helicoverpa* species. *Insect Biochemistry and Physiology* 78(2): 104–118.

Ahn, Seung-Joon, F. R. Badenes-Pérez, and D. G. Heckel. 2011. A host-plant specialist, *Helicoverpa assulta*, is more tolerant to capsaicin from *Capsicum annuum* than other noctuid species. *Journal of Insect Physiology* 57(9): 1212–1219.

Albert, V. A., and Tien-Hao Chang. 2014. Evolution of a hot genome. *Proceedings of the National Academy of Sciences of the United States of America* 111(14): 5069–5070.

Alcorn, J. B. 1984. *Huastec Mayan Ethnobotany*. University of Texas Press.

Al-Delaimy, K. S., and S. H. Ali. 1970. Antibacterial action of vegetable extracts on the growth of pathogenic bacteria. *Journal of the Science of Food and Agriculture* 21(2): 110–112.

Andréasson, E., L. B. Jørgensen, A. S. Höglund, L. Rask, and J. Meijer. 2001. Different myrosinase and idioblast distribution in *Arabidopsis* and *Brassica napus*. *Plant Physiology* 127(4): 1750–1763.

Bacon, Karleigh, Renee Boyer, Cynthia Denbow, Sean O'Keefe, Andrew Neilson, and Robert Williams. 2017. Evaluation of different solvents to extract antibacterial compounds from jalapeño peppers. *Food Science and Nutrition* 5(3): 497–503.

Bakri, I. M., and C. W. Douglas. 2005. Inhibitory effect of garlic extract on oral bacteria. *Archives of Oral Biology* 50(7): 645–51.

Bartlet, E., G. Kiddle, I. Williams, and R. Wallsgrove. 1999. Wound-induced increases in the glucosinolate content of oilseed rape and their effect on subsequent herbivory by a crucifer specialist. *Entomologia Experimentalis et Applicata* 91(1): 163–167.

Basith, Shaherin, M. Cui, S. Hong, and S. Choi. 2016. Harnessing the therapeutic potential of capsaicin and its analogues in pain and other diseases. *Molecules* 21(8): 966.

Bayan, L., P. H. Koulivand, and A. Gorji. 2014. Garlic: a review of potential therapeutic effects. *Avicenna Journal of Phytomedicine* 4(1): 1–14.

Bednarek, P., M. Piślewska-Bednarek, A. Svatos, B. Schneider, J. Doubsky, M. Mansurova, M. Humphry, C. Consonni, R. Panstruga, A. Sanchez-Vallet, A. Molina, and P. Schulze-Lefert. 2009. A glucosinolate metabolism pathway in living plant cells mediates broad-spectrum antifungal defense. *Science* 323(5910): 101–106.

Beekwilder, Jules, Wessel van Leeuwen, Nicole M. van Dam, Monica Bertossi, Valentina Grandi, Luca Mizzi, Mikhail Soloviev, Laszlo Szabados, Jos W. Molthoff, Bert Schipper, et al. 2008. The impact of the absence of aliphatic glucosinolates on insect herbivory in *Arabidopsis*. *PLoS ONE* 3(4): e2068.

Billing, Jennifer, and P. W. Sherman. 1998. Antimicrobial functions of spices: why some like it hot. *Quarterly Review of Biology* 73(1): 3–49.

Block, Eric. 2010. *Garlic and Other Alliums: The Lore and the Science*. Royal Society of Chemistry.

Borlinghaus, J., F. Albrecht, M. C. H. Gruhlke, I. D. Nwachukwu, and A. J. Slusarenko. 2014. Allicin: chemistry and biological properties. *Molecules* 19(8): 12591–12618.

Bosland, P. W. 1994. Chiles: history, cultivation, and uses. *Developments in Food Science* 34.

Bosland, P. W., and E. J. Votava. 2012. *Peppers: Vegetable and Spice Capsicums*. CABI.

Braby, M. F., and J. W. Trueman. 2006. Evolution of larval host plant associations and adaptive radiation in pierid butterflies. *Journal of Evolutionary Biology* 19(5): 1677–1690.

Brown, P. D., J. G. Tokuhisa, M. Reichelt, and J. Gershenzon. 2003. Variation of glucosinolate accumulation among different organs and developmental stages of *Arabidopsis thaliana*. *Phytochemistry* 62(3): 471–481.

Ceylan, E., and D. Y. C. Fung. 2007. Antimicrobial activity of spices. *Journal of Rapid Methods and Automation in Microbiology* 12(1): 1–55.

Chan, Theodore C., Gary M. Vilke, Jack Clausen, Richard F. Clark, Paul Schmidt, Thomas Snowden, and Tom Neuman. 2002. The effect of oleoresin capsicum "pepper" spray inhalation on respiratory function. *Journal of Forensic Sciences* 47(2): 299–304.

Chapa-Oliver, A. M., and L. Mejía-Teniente. 2016. Capsaicin: from plants to a cancer-suppressing agent. *Molecules* 21(8): 931.

Chew, F. S. 1988. Biological effects of glucosinolates. In: Cutler, H. G. (ed.), *Biologically Active Natural Products*. American Chemical Society: 155–181.

Cichewicz, R. H., and P. A. Thorpe. 1996. The antimicrobial properties of chile peppers (*Capsicum* species) and their uses in Mayan medicine. *Journal of Ethnopharmacology* 52(2): 61–70.

Crocker, D. R., and S. M. Perry. 1990. Plant chemistry and bird repellents. *IBIS: International Journal of Bird Science* 132(2): 300–308.

Czarra, Fred. 2009. *Spices: A Global History*. Reaktion Books.

Dalby, Andrew. 2000. *Dangerous Tastes: The Story of Spices*. University of California Press.

Dennis, A. J., E. W. Schupp, R. J. Green, and D. A. Westcott (eds.). 2007. *Seed Dispersal: Theory and Its Application in a Changing World*. CAB International.

DeWitt, Dave, and P. W. Bosland. 1993. *The Pepper Garden*. Ten Speed Press.

DeWitt, Dave, and P. W. Bosland. 2014. *The Complete Chile Pepper Book: A Gardener's Guide to Choosing, Growing, Preserving, and Cooking*. Timber Press.

DeWitt, Dave, and J. Lamson. 2015. *The Field Guide to Peppers*. Timber Press.

Ettenberg, Jodi. 2019. A brief history of chili peppers. *Legal Nomads*. https://www.legalnomads.com/history-chili-peppers/.

Everaerts, Wouter, Maarten Gees, Yeranddy A. Alpizar, Ricard Farre, Cindy Leten, Aurelia Apetrei, Ilse Dewachter, Fred van Leuven, Rudi Vennekens, Dirk De Ridder, et al. 2011. The capsaicin receptor TRPV1 is a crucial mediator of the noxious effects of mustard oil. *Current Biology* 21(4): 316–321.

Falk, K. L., J. G. Tokuhisa, and J. Gershenzon. 2007. The effect of sulfur nutrition on plant glucosinolate content: physiology and molecular mechanisms. *Plant Biology* 9(5): 573–581.

Farbman, Karen S., E. Barnett, G. Bolduc, and J. O. Klein. 1993. Antibacterial activity of garlic and onions: a historical perspective. *Pediatric Infectious Disease Journal* 12(7): 613.

Fattori, Victor, Miriam S. N. Hohmann, Ana C. Rossaneis, Felipe A. Pinho-Ribeiro, and Waldiceu A. Verri Jr. 2016. Capsaicin: current understanding of its mechanisms and therapy of pain and other pre-clinical and clinical uses. *Molecules* 21(7): 844.

Gimsing, A. L., J. C. Sorensen, L. Tovgaard, A. M. Jorgensen, and H. C. Hansen. 2006. Degradation kinetics of glucosinolates in soil. *Environmental Toxicology and Chemistry* 25(8): 2038–2044.

Goncagul, Gulsen, and Erol Ayaz. 2010. Antimicrobial effect of garlic (*Allium sativum*). *Recent Patents on Anti-infective Drug Discovery* 5(1): 91–93.

Gummin, David D., James B. Mowry, Michael C. Beuhler, Daniel A. Spyker, Laura J. Rivers, Ryan Feldman, Kaitlyn Brown, P. T. Pham Nathaniel, Alvin C. Bronstein, and Julie A. Weber. 2017. Annual report of the American Association of Poison Control Centers' National Poison Data System (NPDS): 35th annual report. *Clinical Toxicology* 56(12): 1213–1415.

Gupta, R., and N. K. Sharmaj. 2008. A study of the nematicidal activity of allicin—an active principle in garlic, *Allium sativum* L., against root-knot nematode, *Meloidogyne incognita* (Kofoid and White, 1919) Chitwood, 1949. *International Journal of Pest Management* 39(4): 390–392.

Haak, David C., Leslie A. McGinnis, Douglas J. Levey, and Joshua J. Tewksbury. 2011. Why are not all chilies hot? a trade-off limits pungency. *Proceedings of the Royal Society B: Biological Sciences* 10(1098): 1–6.

Halkier, B. A., and J. Gershenzon. 2006. Biology and biochemistry of glucosinolates. *Annual Review of Plant Biology* 57: 303–333.

Heiser, C. B., and P. G. Smith. 1953. The cultivated capsicum peppers. *Economic Botany* 7(3): 214–227.

Higdon, J. V., B. Delage, D. E. Williams, and R. H Dashwood. 2007. Cruciferous vegetables and human cancer risk: epidemiologic evidence and mechanistic basis. *Pharmacological Research* 55(3): 224–236.

Higdon, J., V. J. Drake, B. Delage, and K. Ried. 2016. *Garlic and Organosulfur Compounds*. Linus Pauling Institute, Oregon State University.

Hill, Theresa A., Hamid Ashrafi, Sebastian Reyes-Chin-Wo, JiQiang Yao, Kevin Stoffel, Maria-Jose Truco, Alexander Kozik, Richard W. Michelmore, and Allen Van Deynze. 2013. Characterization of *Capsicum annuum* genetic diversity and population structure based on parallel polymorphism discovery with a 30K unigene pepper GeneChip. *PLoS ONE* 8(2): e56200.

Hopkins, R. J., Nicole M. van Dam, and Joop J. A. van Loon. 2009. Role of glucosinolates in insect-plant relationships and multitrophic interactions. *Annual Review of Entomology* 54(1): 57–83.

Jensen, P. G., P. D. Curtis, J. A. Dunn, R. E. Austic, and M. E. Richmond. 2003. Field evaluation of capsaicin as a rodent aversion agent for poultry feed. *Pest Management Science* 59(9): 1007–1015.

Ji, Xiaoxue, Jingjing Li, Zhen Meng, Sa Dong, Shouan Zhang, and Kang Qiao. 2019. Inhibitory effect of allicin against *Meloidogyne incognita* and *Botrytis cinerea* in tomato. *Scientia Horticulturae* 253: 203–208.

Kazana, E., T. W. Pope, L. Tibbles, M. Bridges, J. A. Pickett, A. M. Bones, G. Powell, and J. T. Rossiter. 2007. The cabbage aphid: a walking mustard oil bomb. *Proceedings of the Royal Society B: Biological Sciences* 274(1623): 2271–2277.

Keay, John. 2007. *The Spice Route: A History*. University of California Press.

Kim, Seungill, Minkyu Park, Seon-In Yeom, Yong-Min Kim, Je Min Lee, Hyun-Ah Lee, Eunyoung Seo, Jaeyoung Choi, Kyeongchae Cheong, Ki-Tae Kim, et al. 2014. Genome sequence of the hot pepper provides insights into the evolution of pungency in *Capsicum* species. *Nature Genetics* 46(3): 270–278.

Kim-Katz, Susan Y., Ilene B. Anderson, Thomas E. Kearney, Conan MacDougall, Karen S. Hudmon, and Paul D. Blanc. 2010. Topical antacid therapy for capsaicin-induced dermal pain: a poison center telephone-directed study.

American Journal of Emergency Medicine 28(5): 596–602.

Kliebenstein, D. J., J. Kroymann, and T. Mitchell-Olds. 2005. The glucosinolate-myrosinase system in an ecological and evolutionary context. *Current Opinions in Plant Biology* 8(3): 264–271.

Kruse, C., R. Jost, M. Lipschis, B. Kopp, M. Hartmann, and R. Hell. 2007. Sulfur-enhanced defence: effects of sulfur metabolism, nitrogen supply, and pathogen lifestyle. *Plant Biology* 9(5): 608–619.

Lankau, R. 2008. A chemical trait creates a genetic trade-off between intra- and interspecific competitive ability. *Ecology* 89(5): 1181–1187.

Lipka, V., J. Dittgen, P. Bednarek, R. Bhat, M. Wiermer, M. Stein, J. Landtag, W. Brandt, S. Rosahl, D. Scheel, F. Llorente, A. Molina, J. Parker, S. Somerville, and P. Schulze-Lefert. 2005. Pre- and postinvasion defenses both contribute to nonhost resistance in *Arabidopsis*. *Science* 310(5751): 1180–1183.

Louda, S., and S. Mole. 1991. Glucosinolates: chemistry and ecology. In: Rosenthal, G. A., and M. R. Berenbaum (eds.), *Herbivores: Their Interaction with Secondary Plant Metabolites*. Academic Press: 300–311.

Lüthy, B., and P. Matile. 1984. The mustard oil bomb: rectified analysis of the subcellular organisation of the myrosinase system. *Biochemie und Physiologie der Pflanzen* 179(1–2): 5–12.

Lüthy, J., and M. H. Benn. 1977. Thiocyanate formation from glucosinolates: a study of the autolysis of allylglucosinolate in *Thlaspi arvense* L. seed flour extracts. *Canadian Journal of Biochemistry and Physiology* 55(10): 1028–1031.

Marini, Emanuela, Gloria Magi, Marina Mingoia, Armanda Pugnaloni, and Bruna Facinelli. 2015. Antimicrobial and Anti-Virulence Activity of Capsaicin Against Erythromycin-Resistant, Cell-Invasive Group A Streptococci. *Frontiers in Microbiology* 13(6), article no. 1281.

Matile, P. 1980. The mustard oil bomb compartmentation of the myrosinase system. *Plant Physiology and Biochemistry* 175(8–9): 722–731.

McClements, D. J., and Eric Decker (eds.). 2009. *Designing Functional Foods: Measuring and Controlling Food Structure Breakdown and Nutrient Absorption*. 1st ed. Woodhead Publishing.

Molina-Torres, J., A. García-Chávez, and E. Ramírez-Chávez. 1999. Antimicrobial properties of alkamides present in flavouring plants traditionally used in Mesoamerica: affinin and capsaicin. *Journal of Ethnopharmacology* 64(3): 241–248.

Müller, René, Martin de Vos, Joel Y. Sun, Ida E. Sønderby, Barbara A. Halkier, Ute Wittstock, and Georg Jander. 2010. Differential effects of indole and aliphatic glucosinolates on lepidopteran herbivores. *Journal of Chemical Ecology* 36(8): 905–913.

Nasrawi, C. W., and R. M. Pangborn. 1990. Temporal effectiveness of mouth-rinsing on capsaicin mouth-burn. *Physiological Behavior* 47(4): 617–623.

Olszewska, J., and E. Tęgowska. 2011. Opposite effect of capsaicin and capsazepine on behavioral thermoregulation in insects. *Journal of Comparative Physiology A* 197(1): 1021–1026.

Omolo, Morrine A., Zen-Zi Wong, Amanda K. Mergen, Jennifer C. Hastings, Nina C. Le, Holly A. Reiland, Kyle Andrew Case, and David J. Baumler. 2014. Antimicrobial properties of chili peppers. *Journal of Infectious Diseases and Therapy* 2(4): 1–8.

O'Neill, J., Christina Brock, Anne Estrup Olesen, Trine Andresen, Matias Nilsson, and Anthony H. Dickenson. 2012. Unravelling the mystery of capsaicin: a tool to understand and treat pain. *Pharmacological Review* 64(4): 939–971.

Parker, J. K. 2015. Introduction to aroma compounds in foods. In: Parker, J. K., J. S. Elmore, and L. Methven (eds.), *Flavour Development, Analysis and Perception in Food and Beverages*. Woodhead Publishing Series in Food Science, Technology and Nutrition 273. Elsevier: 3–30.

Perry, Linda, Ruth Dickau, Sonia Zarrillo, Irene Holst, Deborah M. Pearsall, Dolores R. Piperno, Mary Jane Berman, Richard G. Cooke, Kurt Rademaker, Anthony J. Ranere, et al. 2007. Starch fossils and the domestication and dispersal of chili peppers (*Capsicum* spp. L.) in the Americas. *Science* 315(5814): 986–988.

Petrovska, Biljana Bauer, and Svetlana Cekovska. 2010. Extracts from the history and medical properties of garlic. *Pharmacognosy Review* 4(7): 106–110.

Pickersgill, B. 1997. Genetic resources and breeding of *Capsicum* spp. *Euphytica* 96(1): 129–133.

Presilla, Maricel. 2017. *Peppers of the Americas: The Remarkable Capsicums That Forever Changed Flavor*. Lorena Jones Books.

Purugganan, M. D., and Dorian Q. Fuller. 2009. The nature of selection during plant domestication. *Nature* 457(7231): 843–848.

Qin, Cheng, Changshui Yu, Yaou Shen, Xiaodong Fang, Lang Chen, Jiumeng Min, Jiaowen Cheng, Shancen Zhao, Meng Xu, Yong Luo, et al. 2014. Whole-genome sequencing of cultivated and wild peppers provides insights into *Capsicum* domestication and specialization. *Proceedings of the National Academy of Sciences of the United States of America* 111(14): 5135–5140.

Qiu, Jiazhang, Xiaodi Niu, Jianfeng Wang, Yan Xing, Bingfeng Leng, Jing Dong, Hongen Li, Mingjing Luo, Yu Zhang, Xiaohan Dai, et al. 2012. Capsaicin protects mice from community-associated methicillin-resistant *Staphylococcus aureus* pneumonia. *PLoS ONE* 7(3): e33032.

Rask, L., E. Andréasson, B. Ekbom, S. Eriksson, B. Pontoppidan, and J. Meijer. 2000. Myrosinase: gene family evolution and herbivore defense in Brassicaceae. *Plant Molecular Biology* 42(1): 93–114.

Ratzka, E., H. Vogel, D. J. Kliebenstein, T. Mitchell-Olds, and J. Kroymann. 2002. Disarming the mustard oil bomb. *Proceedings of the National Academy of Sciences of the United States of America* 99(17): 11223–11228.

Reilly, C., D. Crouch, and G. Yost. 2001. Quantitative analysis of capsaicinoids in fresh peppers, oleoresin capsicum and pepper spray products. *Journal of Forensic Sciences* 46(3): 502–509.

Reiter, Jana, Natalja Levina, Mark van der Linden, Martin Gruhlke, Christian Martin, and Alan J. Slusarenko. 2017. Diallylthiosulfinate (Allicin), a volatile antimicrobial from garlic (*Allium sativum*), kills human lung pathogenic bacteria, including MDR strains, as a vapor. *Molecules* 22(10): E1711.

Rivlin, Richard S. 2001. Historical perspective on the use of garlic. *Journal of Nutrition* 131(3): 951S–954S.

Rodríguez-Maturino, Alfonso, Daniel Gonzalez-Mendoza, Aura Valenzuela-Solorio, Rosalba Troncoso-Rojas, Onecimo Grimaldo, Mónica Aviles-Marin, and Lourdes Cervantes-Díaz. 2012. Antioxidant activity and bioactive compounds of chiltepín (*Capsicum annuum* var. *glabriusculum*) and habanero (*Capsicum chinense*): a comparative study. *Journal of Medicinal Plant Research* 6(9): 1758–1763.

Sallam, Kh. I., M. Ishioroshi, and K. Samejimab. 2004. Antioxidant and antimicrobial effects of garlic in chicken sausage. *Lebensmittel-Wissenschaft + Technologie* 37(8): 849–855.

Schenk, P. M., K. Kazan, I. Wilson, J. P. Anderson, T. Richmond, S. C. Somerville, and J. M. Manners. 2000. Coordinated plant defense responses in *Arabidopsis* revealed by microarray analysis. *Proceedings of the National Academy of Sciences of the United States of America* 97(21): 11655–11660.

Sealy, R. L., M. R. Evans, and C. S. Rothrock. 2007. The effect of a garlic extract and root substrate on soilborne fungal pathogens. *HortTechnology* 17(2): 169–173.

Sherman, Paul W., and Jennifer Billing. 1999. Darwinian gastronomy: why we use spices; spices taste good because they are good for us. *BioScience* 49(6): 453–463.

Shroff, R., F. Vergar, A. Muck, A. Svatoš, and J. Gershenzon. 2008. Nonuniform distribution of glucosinolates in *Arabidopsis thaliana* leaves has important consequences for plant defense. *Proceedings of the National Academy of Sciences of the United States of America* 105(16): 6196–6201.

Siemens, D. H., and T. Mitchell-Olds. 1996. Glucosinolates and herbivory by specialists (Coleoptera: Chrysomelidae, Lepidoptera: Plutellidae): consequences of concentration and induced resistance. *Environmental Entomology* 25(6): 1344–1353.

Slusarenko, A., and A. V. Patel. 2008. Control of plant diseases by natural products: allicin from garlic as a case study. In: Collinge, David B., Lisa Munk, and B. M. Cooke (eds.), *Sustainable Disease Management in a European Context*. Springer: 313–322.

Smith, J., and I. Greaves. 2002. The use of chemical incapacitant sprays: a review. *Journal of Trauma and Acute Care Surgery* 52(3): 595–600.

Sønderby, I. E., F. Geu-Flores, and B. Halkier. 2010. Biosynthesis of glucosinolates—gene discovery and beyond. *Trends in Plant Science* 15(5): 283–290.

Story, M. Gina, and Lillian Cruz-Orengo. 2007. Feel the burn: the linked sensations of temperature and pain come from a family of membrane proteins that can tell neurons to fire when heated or hot-peppered. *American Scientist* 95(4): 326–333.

Stutz, R. S., L. Verschuur, O. Leimar, and U. A. Bergvall. 2019. A mechanistic understanding of repellent function against mammalian herbivores. *Ecological Processes* 8(1): 25–32.

Taylor, T. M. (ed.). 2015. *Handbook of Natural Antimicrobials for Food Safety and Quality*. Elsevier and Woodhead Publishing.

Tęgowska, E., B. Grajpel, and B. Piechowicz. 2004. Does red pepper contain an insecticidal compound for Colorado beetle? *Conference: Breeding for Plant Resistance to Pests and Diseases* 28(10): 121–127.

Tewksbury, J. J., C. Manchego, D. C. Haak, and D. J. Levey. 2006. Where did the chili get its spice? biogeography of capsaicinoid production in ancestral wild chili species. *Journal of Chemical Ecology* 32(3): 547–564.

Tewksbury, J. J., and G. P. Nabhan. 2001. Seed dispersal: directed deterrence by capsaicin in chilies. *Nature* 412(6845): 403–404.

Tewksbury, J. J., Karen M. Reagan, Noelle J. Machnicki, Tomás A. Carlo, David C. Haak, Alejandra Lorena Calderón Peñaloza, and Douglas J. Levey. 2008. Evolutionary ecology of pungency in wild chilies. *Proceedings of the*

National Academy of Sciences of the United States of America 105(33): 11808–11811.

Tominaga, M., and D. Julius. 2000. Capsaicin receptor in the pain pathway. *Japanese Journal of Pharmacology* 83(1): 20–24.

Troncoso-Rojas, R., and M. E. Tiznado-Hernández. 2014. *Alternaria alternata* (black rot, black spot). In: Bautista-Baños, Silvia (ed.), *Postharvest Decay: Control Strategies*. Elsevier: 147–188.

Turner, Jack. 2008. *Spice: The History of a Temptation*. Knopf Doubleday Publishing Group.

Vesaluoma, M., L. Müller, J. Gallar, A. Lambiase, J. Moilanen, T. Hack, C. Belmonte, and T. Tervo. 2000. Effects of oleoresin capsicum pepper spray on human corneal morphology and sensitivity. *Investigative Ophthalmology & Visual Science* 41(8): 2138–2147.

War, Abdul Rashid, Michael Gabriel Paulraj, Tariq Ahmad, Abdul Ahad Buhroo, Barkat Hussain, Savarimuthu Ignacimuthu, and Hari Chand Sharma. 2018. Plant defence against herbivory and insect adaptations. *AoB PLANTS* 10(4). https://academic.oup.com/aobpla/article/10/4/ply037/5036447.

Watkins, R. W., D. P. Cowan, and E. L. Gill. 1996. Plant secondary chemicals as non-lethal vertebrate repellents. *Proceedings of the Vertebrate Pest Conference* 17(17): 186–192.

Wittstock, U., N. Agerbirk, E. J. Stauber, C. E. Olsen, M. Hippler, T. Mitchell-Olds, J. Gershenzon, and H. Vogel. 2004. Successful herbivore attack due to metabolic diversion of a plant chemical defense. *Proceedings of the National Academy of Sciences of the United States of America* 101(14): 4859–4864.

Wittstock, U., and M. Burow. 2010. Glucosinolate breakdown in Arabidopsis: mechanism, regulation and biological significance. *Arabidopsis Book* 8: e0134.

Xue, J., M. Jørgensen, U. Pihlgren, and L. Rask. 1995. The myrosinase gene family in *Arabidopsis thaliana*: gene organization, expression and evolution. *Plant Molecular Biology* 27(5): 911–922.

Yenigun, O. M., and M. Thanassi. 2019. Capsaicin: an uncommon exposure and unusual treatment. *Clinical Practice and Cases in Emergency Medicine* 3(3): 219–221.

Yeung, M. F., and W. Y. Tang. 2015. Clinicopathological effects of pepper (oleoresin capsicum) spray. *Hong Kong Medical Journal* 21(6): 542–552.

Zanette, R. A. 2011. *In vitro* susceptibility of *Pythium insidiosum* to garlic extract. *African Journal of Microbiology Research* 5(29).

Zollman, T. M., R. M. Bragg, and D. A. Harrison. 2000. Clinical effects of oleoresin capsicum (pepper spray) on the human cornea and conjunctiva. *Ophthalmology* 107(12): 2186–2189.

Zwanenburg, B. 2004. Thioaldehyde and thioketone s-oxides and s-imides (sulfines and derivatives). In: Padwa, A. (ed.), *Heteroatom Analogues of Aldehydes and Ketones*. Science of Synthesis 27. Thieme: 135–176.

Chapter 12. Energizing and Paralyzing

Almeida, A. A., A. Farah, D. A. Silva, E. A. Nunan, and M. B. Glória. 2006. Antibacterial activity of coffee extracts and selected coffee chemical compounds against enterobacteria. *Journal of Agricultural and Food Chemistry* 54 (23): 8738–8743.

Borota, Daniel, Elizabeth Murray, Gizem Keceli, Allen Chang, Joseph M. Watabe, Maria Ly, John P. Toscano, and Michael A. Yassa. 2014. Post-study caffeine administration enhances memory consolidation in humans. *Nature Neuroscience* 17(2): 201–203.

Carpenter, Murray. 2015. Here's the buzz on America's forgotten native "tea" plant. NPR. https://www.npr.org/sections/thesalt/2015/08/04/429071993/heres-the-buzz-on-americas-forgotten-native-tea-plant.

Crane, Michael. 2015. Yaupon, the only caffeine source native to the U.S., has potential to explode. *Nutritional Outlook*. http://www.nutritionaloutlook.com/trends-business/yaupon-only-caffeine-source-native-us-has-potential-explode.

Dash, Swati Sucharita, and Sathyanarayana N. Gummadi. 2008. Inhibitory effect of caffeine on growth of various bacterial strains. *Research Journal of Microbiology* 3(6): 457–465.

Gaula, Jonathan, and Kelly Donegan. 2015. Caffeine and its effect on bacteria growth. *Rutgers University Journal of Biological Sciences* 1: 4–8.

Healthline Media, Inc. 2019. The effects of caffeine on your body. https://www.healthline.com/health/caffeine-effects-on-body#7.

Ibrahim, Salihu, M. Y. Shukor, M. A. Syed, N. A. A. Rahman, K. A. Khalil, A. Khalid, and S. A. Ahmad. 2014. Bacterial degradation of caffeine: a review. *Asian Journal of Plant Biology* 2(1): 24–33.

Kaufman, Barry. 2019. The untold story of the humble yaupon holly. *Palmetto Bluff*. http://discover.palmettobluff.com/the-untold-story-of-the-humble-yaupon-holly/.

Nathanson, J. A. 1984. Caffeine and related methylxanthines: possible naturally occurring pesticides. *Science* 226(4671): 184–187.

Paul, Kari. 2017. International Coffee Day: Americans drink more coffee than soda, tea and juice combined. *MarketWatch*. https://www.marketwatch.com/story/

international-coffee-day-americans-drink-more-coffee-than-soda-tea-and-juice-combined-2017-09-29.

Pruthviraj, Pawar, B. Suchita, K. Shital, and K. Shilpa. 2011. Evaluation of antibacterial action of caffeine. *International Journal of Research in Ayurveda and Pharmacy* 2(4): 1354–1357.

Sledz, W., E. Los, A. Paczek, J. Rischka, A. Motyka, S. Zoledowska, J. Piosik, and F. Lojkowska. 2015. Antibacterial activity of caffeine against plant pathogenic bacteria. *Acta Biochimica Polonica* 62(3): 605–612.

Srikandi, Fardiaz. 1995. Antimicrobial activity of coffee (*Coffea robusta*) extract. *ASEAN Food Journal* 10(3): 103–106.

World of Molecules. 2019. Caffeine. https://www.worldofmolecules.com/drugs/caffeine.htm.

Wright, Geraldine, D. D. Baker, Mary J. Palmer, and Daniel Stabler. 2013. Caffeine in floral nectar enhances a pollinator's memory of reward. *Science* 339(6124): 1202–1204.

Zimmer, Karl. 2014. How caffeine evolved to help plants survive and help people wake up. *New York Times*, September 4. https://www.nytimes.com/2014/09/04/science/how-caffeine-evolved-to-help-plants-survive-and-help-people-wake-up.html.

Chapter 13. Clean Enough to Kill

Abreu Guirado, Orlando A. 2005. Medicinal potential of the genus *Sapindus* L. (Sapindaceae) and of the species *Sapindus saponaria* L. *Cuban Journal of Medicinal Plants* 10(3–4).

Benfer, Adam. 2019. Foods indigenous to the Western Hemisphere. American Indian Health and Diet Project. http://www.aihd.ku.edu/foods/buffaloberry.html.

Boruah, B., and M. Gogoi. 2013. Plant based natural surfactants. *Asian Journal of Home Science* 8(2): 759–762.

Canadian Biodiversity Information Facility. 2019. *Phytolacca americana*. Poisonous Plants Information System. http://www.cbif.gc.ca/eng/species-bank/canadian-poisonous-plants-information-system/all-plants-scientific-name/phytolacca-americana/?id=1370403266967.

Challinora, V. L., P. G. Parsons, S. Chap, E. F. White, J. T. Blanchfield, R. P. Lehmann, and J. J. De Voss. 2012. Steroidal saponins from the roots of *Smilax* sp.: structure and bioactivity. *Steroids* 77(5): 504–511.

Cheeke, Peter R. 1989. *Toxicants of Plant Origin*. Vol. 2: *Glycosides*. CRC Press.

ChemicalSafetyFacts.org. 2019. Surfactants. https://www.chemicalsafetyfacts.org/surfactants/.

Chen, Yu-Fen, C. Yang, M. Chang, Y. Ciou, and Y. Huang. 2010. Foam properties and detergent abilities of the saponins from *Camellia oleifera*. *International Journal of Molecular Sciences* 11(11): 4417–4425.

Cornell University College of Agriculture and Life Sciences. 2018. Saponins. http://poisonousplants.ansci.cornell.edu/toxicagents/saponin.html.

Faizal, A., and D. Geelen. 2013. Saponins and their role in biological processes in plants. *Phytochemical Review* 12(4): 877–893.

Gauna, Forest Jay. 2019. Soap plant (*Chlorogalum pomeridianum*). U.S. Forest Service Plant of the Week. https://www.fs.fed.us/wildflowers/plant-of-the-week/chlorogalum_pomeridianum.shtml.

Koodalingam, A., P. Mullainadhan, and M. Arumugam. 2009. Antimosquito activity of aqueous kernel extract of soapnut *Sapindus emarginatus*: impact on various developmental stages of three vector mosquito species and nontarget aquatic insects. *Parasitology Research* 105(5): 1425–1434.

Kregiel, Dorota, Joanna Berlowska, Izabela Witonska, Hubert Antolak, Charalampos Proestos, Mirko Babic, Ljiljana Babic, and Bolin Zhang. 2017. Saponin-based, biological-active surfactants from plants. In: Najjar, Reza (ed.), *Application and Characterization of Surfactants*. IntechOpen. https://www.intechopen.com/books/application-and-characterization-of-surfactants/saponin-based-biological-active-surfactants-from-plants.

Kritzon, Chuck. 2003. Fishing with poisons. *Bulletin of Primitive Technology* 25: 35–38.

Krzyzanowska, Justyna, Mariusz Kowalczyk, and Wieslaw Oleszek. 2014. Analysis of plant saponins. In: *Encyclopedia of Analytical Chemistry: Applications, Theory and Instrumentation*. John Wiley & Sons, Ltd. https://onlinelibrary.wiley.com/doi/book/10.1002/9780470027318.

Liu, Rui, and Baojun Xu. 2015. Inhibitory effects of phenolics and saponins from commonly consumed food legumes in China against digestive enzymes pancreatic lipase and α-glycosidase. *International Journal of Food Properties* 18(10): 2246–2255, https://doi.org/10.1080/10942912.2014.971178.

Mandal, Dattatreya. 2016. The short history of soap—from ancient Mesopotamia to Proctor & Gamble. Realm of History. https://www.realmofhistory.com/2016/08/10/origin-soap-ancient-mesopotamia-2800-bc/.

Mascarenhas, Maria Emilia, Cibani Ramesh Mandrekar, Pratisksha Bharat Marathe, and Luena Joey Morais. 2017. Phytochemical screening of selected species from Convolvulaceae. *International Journal of Current Pharmaceutical Research* 9(6): 94–97.

Mosquin, Daniel. 2015. Botany photo of the day: *Saponaria officinalis*. UCB Botanical Garden. https://botanyphoto

.botanicalgarden.ubc.ca/2015/09/saponaria-officinalis/.

MPG North. 2019. Canada buffaloberry. https://mpgnorth.com/field-guide/elaeagnaceae/canada-buffaloberry.

The Naturopathic Herbalist. 2019. Saponins. https://thenaturopathicherbalist.com/plant-constituents/saponins/.

Oelke, E. A., D. H. Putnam, T. M. Teynor, and E. S. Oplinger. 1992. Quinoa. In: *Alternative Field Crops Manual*. https://hort.purdue.edu/newcrop/afcm/quinoa.html.

Osbourn, A. E., J. P. Wubben, R. E. Melton, J. P. Carter, and M. J. Daniels 1998. Saponins and plant defense. In: Romeo, J. T., K. R. Downum, and R. Verpoorte (eds.), *Phytochemical Signals and Plant-Microbe Interactions*. Recent Advances in Phytochemistry (Proceedings of the Phytochemical Society of North America), vol. 32. Springer: 1–15.

Osbourn, Anne. 1996. Saponins and plant defence: a soap story. *Trends in Plant Science* 1(1): 4–9.

Plants for a Future. 2019. Soap plants. https://pfaf.org/user/cmspage.aspx?pageid=49.

Podolak, Irma, A. Galanty, and D. Sobolewska. 2010. Saponins as cytotoxic agents: a review. *Phytochemistry Reviews* 9(3): 425–474.

Resnik, Silvia. 2004. Quillaia extracts Type 1 and Type 2 chemical and technical assessment. Food and Agriculture Organization of the United Nations. http://www.fao.org/fileadmin/templates/agns/pdf/jecfa/cta/61/QUILLAIA.pdf.

Schmidt, Barbara M., and Diana M. Klaser Cheng (eds.). 2018. *Ethnobotany: A Phytochemical Perspective*. Wiley-Blackwell.

Schmitt, C., B. Grassl, G. Lespes, J. Desbrières, V. Pellerin, S. Reynaud, J. Gigault, and V. A. Hackley. 2014. Saponins: a renewable and biodegradable surfactant from its microwave-assisted extraction to the synthesis of monodisperse lattices. *Biomacromolecules* 15(3): 856–862.

Shibata, S. 1977. Saponins with biological and pharmacological activity. In: Wagner, H., and P. Wolff (eds.), *New Natural Products and Plant Drugs with Pharmacological, Biological or Therapeutical Activity*. Proceedings in Life Sciences. Springer: 177–196.

Sobolewska, Danuta, Klaudia Michalska, Irma Podolak, and Karolina Grabowska. 2014. Steroidal saponins from the genus *Allium*. *Phytochemistry Reviews* 15(1): 1–35.

St. Onge, Jeremy. 2002. Fish poison use in the Americas. Survival.com. http://www.survival.com/library/articles/fish-poisons/.

Tirmenstein, D. A. 1990. *Sapindus saponaria* var. *Drummondii*. In: Fire Effects Information System. U.S. Department of Agriculture, Forest Service, Rocky Mountain Research Station, Fire Sciences Laboratory. https://www.fs.fed.us/database/feis/plants/tree/sapsapd/all.html.

United States Department of Agriculture, Forest Service. 2019. Soaps. https://www.fs.fed.us/wildflowers/ethnobotany/soaps.shtml.

Wina, Elizabeth, Stefan Muetzel, and Klaus Becker. 2005. The impact of saponins or saponin-containing plant materials on ruminant production—a review. *Journal of Agricultural and Food Chemistry* 53(21): 8093–8105.

Zhang, J. 2014. Amphiphilic molecules. In: Drioli, E., and L. Giorno (eds.), *Encyclopedia of Membranes*. Springer: 1–4.

Chapter 14. Herbicide Rain and Superweed Reign

Baucom, R. S. 2016. The remarkable repeated evolution of herbicide resistance. *American Journal of Botany* 103(2): 181–183.

Benbrook, C. M. 2016. Trends in glyphosate herbicide use in the United States and globally. *Environmental Science Europe* 28(1): 3.

Bernards, M. L., R. J. Crespo, G. R. Kruger, R. Gaussoin, and P. J. Tranel. 2012. A waterhemp (*Amaranthus tuberculatus*) population resistant to 2,4-D. *Weed Science* 60(3): 379–384.

Boerboom, C., and M. Owen. 2006. *Facts about Glyphosate Resistant Weeds*. Purdue Extension.

Cobb, A. H., and J. P. H. Reade. 2011. *Herbicide and Plant Physiology*. John Wiley & Sons.

Délye, C., C. Deulvot, and B. Chauvel. 2013. DNA analysis of herbarium specimens of the grass weed *Alopecurus myosuroides* reveals herbicide resistance pre-dated herbicides. *PLOS ONE* 8(10): e75117.

Délye, C., M. Jasieniuk, and V. Le Corre. 2013. Deciphering the evolution of herbicide resistance in weeds. *Trends in Genetics* 29(11): 649–658.

Gaines, Todd A., Wenli Zhang, Dafu Wang, Bekir Bukun, Stephen T. Chisholm, Dale L. Shaner, Scott J. Nissen, William L. Patzoldt, Patrick J. Tranel, A. Stanley Culpepper, et al. 2010. Gene amplification confers glyphosate resistance in *Amaranthus palmeri*. *Proceedings of the National Academy of Sciences of the United States of America* 107(3): 1029–1034.

Gilbert, N. 2013. Case studies: a hard look at GM crops. *Nature* 497(7447): 24–26.

Gillam, Carey. 2014. Farmers fight explosion of "superweeds." *Scientific American*. https://www.scientificamerican.com/article/farmers-fight-explosion-of-superweeds/.

Givens, Wade A., David R. Shaw, William G. Johnson, Stephen C. Weller, Bryan G. Young, Robert G. Wilson, Micheal D. K. Owen, and David Jordan. 2009. A grower survey of herbicide use patterns in glyphosate

resistant cropping systems. *Weed Technology* 23(1): 156–161.
Grover, R., J. D. Wolt, A. J. Cessna, and H. B. Schiefer. 1997. Environmental fate of trifluralin. *Reviews of Environmental Contamination and Toxicology* 153: 1–64.
Hébert, Marie-Pier, Vincent Fugère, and Andrew Gonzalez. 2018. The overlooked impact of rising glyphosate use on phosphorus loading in agricultural watersheds. *Frontiers in Ecology and the Environment* 17(1): 48–56.
Hicks, Helen David Comont, Shaun R. Coutts, Laura Crook, Richard Hull, Ken Norris, Paul Neve, Dylan Z. Childs, and Robert P. Freckleton. 2018. The factors driving evolved herbicide resistance at a national scale. *Nature Ecology & Evolution* 2(3): 529–536.
Klementova, S., and L. Keltnerova. 2015. Triazine herbicides in the environment. In: Price, Andrew (ed.), *Herbicides: Physiology of Action and Safety*. IntechOpen Ltd. https://www.intechopen.com/chapters/48620.
Knobloch, Frieda. 2000. *The Culture of Wilderness: Agriculture as Colonization in the American West*. University of North Carolina Press.
Mcelroy, J. S. 2014. Vavilovian mimicry: Nikolai Vavilov and his little-known impact on weed science. *Weed Science* 62(2): 207–216.
Mithila, J., J. C. Hall, W. G. Johnson, K. B. Kelly, and D. E. Riechers. 2011. Evolution of resistance to auxinic herbicides: historical perspectives, mechanisms of resistance, and implications for broadleaf weed management in agronomic crops. *Weed Science* 59(4): 445–457.
Murphy, B. P., and P. J. Tranel. 2019. Target-site mutations conferring herbicide resistance. *Plants* 8(10): 382.
Nadin, P. (ed.). 2007. *The Use of Plant Protection Products in the European Union: Data 1992–2003*. Eurostat.
Pokorny, Robert. 1941. New compounds: some chlorophenoxyacetic acids. *Journal of the American Chemical Society* 63(6): 1768.
Quastel, J. H. 1950. 2,4-Dichlorophenoxyacetic Acid (2,4-D) as a Selective Herbicide. In: *Agricultural Control Chemicals*. American Chemical Society: 244–249.
Schütte, Gesine, Michael Eckerstorfer, Valentina Rastelli, Wolfram Reichenbecher, Sara Restrepo-Vassalli, Marja Ruohonen-Lehto, Anne-Gabrielle Wuest Saucy, and Martha Mertens. 2017. Herbicide resistance and biodiversity: agronomic and environmental aspects of genetically modified herbicide-resistant plants. *Environmental Science Europe* 29(1): 5.
Schwendiman, A., J. H. Torrie, and G. M. Briggs. 1943. Effects of Sinox, a selective weed spray, on legume seedlings, weeds, and crop yields. *Journal of the American Society of Agronomy* 35(10): 901–908.
Service, R. 2013. What happens when weed killers stop killing? *Science* 341(6152): 1329.
Timmons, F. L. 1970. A history of weed control in the United States and Canada. *Weed Science* 18(2): 294–307.
Union of Concerned Scientists. 2013. The rise of superweeds—and what to do about it. www.ucsusa.org/superweeds.
Vencill, William K., Robert L. Nichols, Theodore M. Webster, John K. Soteres, Carol Mallory-Smith, Nilda R. Burgos, William G. Johnson, and Marilyn R. McClelland. 2012. Herbicide resistance: toward an understanding of resistance development and the impact of herbicide-resistant crops. *Weed Science* 60(SP1): 2–30.
Vereecken, H. 2005. Mobility and leaching of glyphosate: a review. *Pest Management Science* 61(12): 1139–1151.
Vrbničanin, Sava, D. Pavlović, and D. Božić. 2017. Weed resistance to herbicides. In: Pacanoski, Z. (ed.), *Herbicide Resistance in Weeds and Crops*. IntechOpen. https://www.intechopen.com/chapters/55149.
Whitehead, C. W., and C. M. Switzer. 1963. The differential response of strains of wild carrot to 2,4-D and related herbicides. *Canadian Journal of Plant Science* 43(3): 255–262.

Chapter 15. The Burning Question

Bellingham, P. J., and A. D. Sparrow. 2000. Resprouting as a life history strategy in woody plant communities. *Oikos* 89(2): 409–416.
Bond, W. J., and J. E. Keeley. 2005. Fire as a global "herbivore": the ecology and evolution of flammable ecosystems. *Trends in Ecology & Evolution* 20(7): 387–394.
Bond, W. J., and J. J. Midgley. 1995. Kill thy neighbour: an individualistic argument for the evolution of flammability. *Oikos* 73(1): 79–85.
Bond, W. J., and J. J. Midgley. 2001. Ecology of sprouting in woody plants: the persistence niche. *Trends in Ecology & Evolution* 16(1): 45–51.
Canadell, J., and P. H. Zedler. 1994. Underground structures of woody plants in Mediterranean ecosystems of Australia, California, and Chile. In: Kalin Arroya, M. T., P. H. Zedler, and M. D. Fox (eds.), *Ecology and Biogeography of Mediterranean Ecosystems in Chile, California, and Australia*. Springer: 170–210.
Carr, D. J., R. Jahnke, and S. G. M. Carr. 1984. Initiation, development, and anatomy of lignotubers in some species of *Eucalyptus*. *Australian Journal of Botany* 32(4): 415–437.
Daniell, J. W., W. E. Chappell, and H. B. Couch. 1969. Effect of sublethal and lethal temperatures on plant cells. *Plant Physiology* 44(12): 1684–1689.

DeWoody, Jennifer, Carol A. Rowe, Valerie D. Hipkins, and Karen E. Mock. 2008. "Pando" lives: molecular genetic evidence of a giant aspen clone in central Utah. *Western North American Naturalist* 68(4): 493–497.

Drewa, P. B., W. J. Platt, and E. B. Moser. 2002. Fire effects on resprouting of shrubs in headwaters of southeastern longleaf pine savannas. *Ecology* 83(3): 755–767.

Fengel, D., and G. Wegener. 2003. *Wood: Chemistry, Ultrastructure, Reactions*. Kessel.

Flematti, Gavin R., Emilio L. Ghisalberti, Kingsley W. Dixon, and Robert D. Trengove. 2004. A compound from smoke that promotes seed germination. *Science* 305(5686): 977.

Guo, Yongxia, Zuyu Zheng, James J. La Clair, Joanne Chory, and Joseph P. Noel. 2013. Smoke-derived karrikin perception by the α/β-hydrolase KAI2 from *Arabidopsis*. *Proceedings of the National Academy of Sciences of the United States of America* 110(20): 8284–8289.

Hengst, G. E., and J. O. Dawson. 1993. Bark thermal properties of selected central hardwood species. In: Gillespie, Andrew R., George R. Parker, Phillip E. Pope, and George Rink (eds.), *Proceedings of the 9th Central Hardwood Forest Conference*. Gen. Tech. Rep. NC-161. U.S. Department of Agriculture, Forest Service, North Central Forest Experiment Station: 55–75.

James, S. 1984. Lignotubers and burls—their structure, function, and ecological significance in Mediterranean ecosystems. *Botanical Review* 50(3): 225–266.

Jenik, J. 1994. Clonal growth in woody plants: a review. *Folia Geobotanica et Phytotaxonomica* 29: 291–306.

Kain, Günther, Marius-Catalin Barbu, Stefan Hinterreriter, Klaus Richter, and A. Petutschnigg. 2012. Thermal insulation materials out of tree barks. *Holztechnologie* 53(4): 31–37.

Kain, Günther, Marius-Catalin Barbu, Stefan Hinterreriter, Klaus Richter, and A. Petutschnigg. 2013. Substantial bark use as insulation material. *Forest Products Journal* 62(6): 480–487.

Lamont, Byron B., D. C. Le Maitre, R. M. Cowling, and N. J. Enright. 1991. Canopy seed storage in woody plants. *Botanical Review* 57(4): 277–317.

Meier, A. R., M. R. Saunders, and C. H. Michler. 2012. Epicormic buds in trees: a review of bud establishment, development and dormancy release. *Tree Physiology* 32(5): 565–584.

Mesléard, F., and J. Lepart. 1989. Continuous basal sprouting from a lignotuber: *Arbutus unedo* L. and *Erica arborea* L., as woody Mediterranean examples. *Oecologia* 80(1): 127–131.

Minnich, R. A. 1983. Fire mosaics in southern California and northern Baja California. *Science* 219(4590): 1287–1294.

Molinas, M. L., and D. Verdaguer. 1993. Lignotuber ontogeny in the cork-oak (*Quercus suber*; Fagaceae), II. Germination and young seedling. *American Journal of Botany* 80(2): 182–191.

Pausas, J. G., and J. E. Keeley. 2009. A burning story: the role of fire in the history of life. *BioScience* 59(7): 593–601.

Pausas, J. G., and J. E. Keeley. 2017. Epicormic resprouting in fire-prone ecosystems. *Trends in Plant Science* 22(12): 1008–1015.

Pyne, S. J. 2002. How plants use fire (and are used by it). PBS NOVA Online. https://www.pbs.org/wgbh/nova/fire/plants.html.

Savage, M., and J. N. Mast. 2005. How resilient are southwestern ponderosa pine forests after crown fires? *Canadian Journal of Forest Research* 35(4): 967–977.

Suleiman, B. M., J. Larfeldt, B. Leckner, and M. Gustavsson. 1999. Thermal conductivity and diffusivity of wood. *Wood Science and Technology* 33(6): 465–473.

Tredici, P. D. 2001. Sprouting in temperate trees: a morphological and ecological review. *Botanical Review* 67(2): 121–140.

Tributsch, H., and S. Fiechter. 2008. The material strategy of fire-resistant tree barks. High Performance Structures and Materials IV. *WIT Transactions on the Built Environment* 97: 43–52.

Ward, D. 2004. Grasstrees show fire history. *Australian Geographic* 75 (July–September): 28. *https://shop.australiangeographic.com.au/products/australian-geographic-issue-075-2004-july-september*.

Chapter 16. Every Drop Counts

Adams, Henry D., Melanie J. B. Zeppel, William R. L. Anderegg, Henrik Hartmann, Simon M. Landhäusser, David T. Tissue, Travis E. Huxman, Patrick J. Hudson, Trenton E. Franz, Craig D. Allen, et al. 2017. A multi-species synthesis of physiological mechanisms in drought-induced tree mortality. *Nature Ecology & Evolution* 1(9): 1285–1291.

Alvarez, Juliana Magalhães, Joecildo Francisco Rocha, and Silvia Rodrigues Machado. 2008. Bulliform cells in *Loudetiopsis chrysothrix* (Nees) Conert and *Tristachya leiostachya* Nees (Poaceae): structure in relation to function. *Brazilian Archives of Biology and Technology* 51(1): 113–119.

Ashraf, M. 2010. Inducing drought tolerance in plants: recent advances. *Biotechnology Advances* 2(1): 169–183.

Basu, Supratim, Venkategowda Ramegowda, Anuj Kumar, and Andy Pereira. 2016. Plant adaptation to drought stress. *F1000 Research* 5: F1000 Faculty Rev-1554.

Brunner, Ivano, Claude Herzog, Melissa A. Dawes, Matthias Arend, and Christoph Sperisen. 2015. How tree roots respond to drought. *Frontiers in Plant Science* 29(6): 547.

Chaves, M. M., J. P. Maroco, and J. S. Pereira. 2003. Understanding plant responses to drought—from genes to the whole plant. *Functional Plant Biology* 30(3): 239–264.

Christin, Pascal-Antoine, Monica Arakaki, Colin P. Osborne, Andrea Bräutigam, Rowan F. Sage, Julian M. Hibberd, Steven Kelly, Sarah Covshoff, Gane Ka-Shu Wong, Lillian Hancock, et al. 2014. Shared origins of a key enzyme during the evolution of C_4 and CAM metabolism. *Journal of Experimental Botany* 65(13): 3609–3621.

Christin, Pascal-Antoine, Colin P. Osborne, David S. Chatelet, J. Travis Columbus, Guillaume Besnard, Trevor R. Hodkinson, Laura M. Garrison, Maria S. Vorontsova, and Erika J. Edwards. 2013. Anatomical enablers and the evolution of C_4 photosynthesis in grasses. *Proceedings of the National Academy of Sciences of the United States of America* 110(4): 1381–1386.

Dietrich, D. 2018. Hydrotropism: how roots search for water. *Journal of Experimental Botany* 69(11): 2759–2771.

Dortort, F. 2011. *The Timber Press Guide to Succulent Plants of the World: A Comprehensive Reference to More Than 2000 Species*. Timber Press.

Evans, Margaret, Xavier Aubriot, David Hearn, Maxime Lanciaux, Sebastien Lavergne, Corinne Cruaud, Porter P. Lowry II, and Thomas Haevermans. 2014. Insights on the evolution of plant succulence from a remarkable radiation in Madagascar (Euphorbia). *Systematic Biology* 63(5): 697–711.

Fahn, A., and D. F. Cutler. 1992. *Xerophytes*. Gebrüder Borntraeger.

Fang, Y., and L. Xiong. 2015. General mechanisms of drought response and their application in drought resistance improvement in plants. *Cellular and Molecular Life Sciences* 72(4): 673–689.

Hammer, S. 2010. *Lithops: Treasures of the Veld*. British Cactus & Succulent Society.

Iljin, W. S. 1957. Drought resistance in plants and physiological processes. *Annual Review of Plant Physiology* 8: 257–274.

Ju, Jie, Hao Bai, Yongmei Zheng, Tianyi Zhao, Ruochen Fang, and Lei Jiang. 2012. A multi-structural and multi-functional integrated fog collection system in cactus. *Nature Communications* 3: 1247.

Kelsey, Rick G., D. Gallego, F. J. Sánchez-García, and J. A. Pajares. 2014. Ethanol accumulation during severe drought may signal tree vulnerability to detection and attack by bark beetles. *Canadian Journal of Forest Research* 44(6): 554–561.

Khait, Itzhak, Ohad Lewin-Epstein, Raz Sharon, Kfir Saban, Revital Goldstein, Yehuda Anikster, Yarden Zeron, Chen Agassy, Shaked Nizan, Gayl Sharabi, et al. 2019. Plants emit informative airborne sounds under stress. *bioRxiv*.

Khait, I., U. Obolski, Y. Yovel, and L. Hadany. 2019. Sound perception in plants. *Seminars in Cell & Developmental Biology* 92: 134–138.

Khait, Itzhak, Raz Sharon, Ran Perelman, Arjan Boonman, Yossi Yovel, Lilach Hadany. 2019. Plants emit remotely detectable ultrasounds that can reveal plant stress. *bioRxiv*.

Khan, Aziz, Xudong Pan, Ullah Najeeb, Daniel Kean Yuen Tan, Shah Fahad, Rizwan Zahoor, and Honghai Luo. 2018. Coping with drought: stress and adaptive mechanisms, and management through cultural and molecular alternatives in cotton as vital constituents for plant stress resilience and fitness. *Biological Research* 51(1): 47.

McDowell, Nate, William T. Pockman, Craig D. Allen, David D. Breshears, Neil Cobb, Thomas Kolb, Jennifer Plaut, John Sperry, Adam West, David G. Williams, et al. 2008. Mechanisms of plant survival and mortality during drought: why do some plants survive while others succumb to drought? *New Phytologist* 178(4): 719–739.

Nakashima, K., and K. Suenaga. 2017. Toward the genetic improvement of drought tolerance in crops. *Japan Agricultural Research Quarterly* 51(1): 1–10.

Ogburn, R. M., and E. J. Edwards. Chapter 4—the ecological water-use strategies of succulent plants. *Advances in Botanical Research* 55: 179–225.

Pavek, Diane S. 1994. *Parkinsonia florida*. In: Fire Effects Information System. U.S. Department of Agriculture, Forest Service, Rocky Mountain Research Station, Fire Sciences Laboratory. https://www.fs.usda.gov/database/feis/plants/tree/parflo/all.html.

Sage, R. F. 2001. Environmental and evolutionary preconditions for the origin and diversification of the C_4 photosynthetic syndrome. *Plant Biology* 3(3): 202–213.

Sage, R. F. 2004. The evolution of C_4 photosynthesis. *New Phytologist* 161(2): 341–370.

Sage, R. F., P. A. Christin, and E. J. Edwards. 2011. The C_4 plant lineages of planet Earth. *Journal of Experimental Botany* 62(9): 3155–3169.

Sage, R. F., and R. K. Monson (eds.). 1999. *C_4 Plant Biology*. Academic Press.

Sage, R. F., T. L. Sage, and F. Kocacinar. 2012. Photorespiration and the evolution of C_4 photosynthesis. *Annual Review of Plant Biology* 63: 19–47.

Turner, J. S., and M. D. Pircker. 1993. Thermal ecology of an embedded dwarf succulent from southern Africa

(*Lithops* spp.: Mesembryanthemaceae). *Journal of Arid Environments* 24(4): 361–385.

Varshney, R. K., R. Tuberosa, and F. Tardieu. 2018. Progress in understanding drought tolerance: from alleles to cropping systems. *Journal of Experimental Botany* 69(13): 3175–3179.

Xu, Zhenzhu, Guangsheng Zhou, and Hideyuki Shimizu. 2010. Plant responses to drought and rewatering. *Plant Signaling and Behavior* 5(6): 649–654.

Chapter 17. Ice, Snow, and Plantifreeze

Anderson, Dave, Chris Martin, and Emily Quirk. 2021. Bend but don't break—how trees survive northern winters. Society for the Protection of New Hampshire Forests. https://forestsociety.org/something-wild/bend-dont-break-how-trees-survive-northern-winters.

Chapter 18. Wind-gineering

Ahmad, Zuraida, and Paridah M. Tahir. 2010. Oil palm trunk fiber as a bio-waste resource for concrete reinforcement. *International Journal of Mechanical and Materials Engineering* 5(2): 199–207.

Albion, Robert Greenhalgh. 1926. *Forests and Sea Power: The Timber Problem of the Royal Navy 1652–1862*. Harvard University Press.

Alves, Luciana F., Fernando R. Martins, and Flavio A. M. Santos. 2004. Allometry of a neotropical palm, *Euterpe edulis*. *Acta Botanica Brasilica* 18(2). http://dx.doi.org/10.1590/S0102-33062004000200016.

Baker, R. L., and B. E. Dahl. 1981. Determining vigor of natural and planted stands of sea oats on the Texas Gulf Coast. *Southwestern Naturalist* 26(2): 117–123.

Bonnesoeur, Vivien, Thiéry Constant, Bruno Moulia, and Meriem Fournier. 2016. Forest trees filter chronic wind-signals to acclimate to high winds. *New Phytologist* 210(3): 850–860. https://doi.org/10.1111/nph.13836.

Bourmaud, Alain, Johnny Beaugrand, Darshil U. Shah, Vincent Placet, and Christophe Baley. 2018. Towards the design of high-performance plant fibre composites. *Progress in Materials Science* 97: 347–408. https://www.sciencedirect.com/science/article/pii/S0079642518300653.

Candeias, Matt. 2017. How do palms survive hurricanes? *In Defense of Plants*. http://www.indefenseofplants.com/blog/2017/9/10/how-do-palms-survive-hurricanes.

Carlton, William R. 1939. New England masts and the king's navy. *New England Quarterly* 12(1): 4–18.

Coutand, Catherine. 2010. Mechanosensing and thigmomorphogenesis, a physiological and biomechanical point of view. *Plant Science* 179(3): 168–182.

Coutts, M. P. 1983 Root architecture and tree stability. In: Atkinson D., K. K. S. Bhat, M. P. Coutts, P. A. Mason, and D. J. Read (eds.), *Tree Root Systems and Their Mycorrhizas*. Developments in Plant and Soil Sciences, vol. 7. Springer: 171–188.

Crook, M. J., and A. R. Ennos. 1996. The anchorage mechanics of deep rooted larch, *Larix europea* × *L. japonica*. *Journal of Experimental Botany* 47(10): 1509–1517. https://doi.org/10.1093/jxb/47.10.1509.

Dahl, B. E., and D. W. Woodard. 1977. Construction of Texas coastal foredunes with sea oats (*Uniola paniculata*) and bitter panicum (*Panicum amarum*). *International Journal of Biometeorology* 21(3): 267–275.

Danjon, Frédéric, Thierry Fourcaud, and Didier Bert. 2005. Root architecture and wind-firmness of mature *Pinus pinaster*. *New Phytologist* 168(2): 387–400. https://nph.onlinelibrary.wiley.com/doi/full/10.1111/j.1469-8137.2005.01497.x.

Davey, James. 2011. Securing the sinews of sea power: British intervention in the Baltic 1780–1815. *International History Review* 33(2): 161–184. https://www.researchgate.net/publication/233215156_Securing_the_Sinews_of_Sea_Power_British_Intervention_in_the_Baltic_1780-1815.

de Langre, Emmanuel. 2008. Effects of wind on plants. *Annual Review of Fluid Mechanics* 40(1): 141–168. https://www.annualreviews.org/doi/full/10.1146/annurev.fluid.40.111406.102135.

Dureya, Mary, and Eliana Kampf. 2007. Wind and tree: lessons learned from hurricanes. University of Florida IFAS Extension Publication no. FORS 118. http://hoaexpert.net/millpond/wp-content/uploads/2015/10/Trees-and-Hurricanes1.pdf.

Ennos, A. R. 1997. Wind as an ecological factor. *Trends in Ecology & Evolution* 12(3): 108–111. https://www.sciencedirect.com/science/article/abs/pii/S0169534796100665.

Ennos, A. R. 1999. The aerodynamics and hydrodynamics of plants. *Journal of Experimental Biology* 202: 3281–3284. https://jeb.biologists.org/content/jexbio/202/23/3281.full.pdf.

Fournier, M., T. Alméras, B. Clair, and J. Gril. 2014. Biomechanical action and biological functions. In: Gardiner, B., J. Barnett, P. Saranpää, and J. Gril (eds.), *The Biology of Reaction Wood*. Springer Series in Wood Science. Springer: 139–171.

Friel, Ian. 1995. *The Good Ship: Ships, Shipbuilding and Technology in England, 1200–1520*. Johns Hopkins University Press.

Gardiner, Barry, Peter Berry, and Bruno Moulia. 2016. Review: wind impacts on plant growth, mechanics and damage. *Plant Science* 245: 94–118. https://doi.org/10.1016/j.plantsci.2016.01.006.

Grace, J. 1988. Plant response to wind. *Agriculture, Ecosystems & Environment* 22–23: 71–88. https://www.sciencedirect.com/science/article/pii/0167880988900084

Griffith, Patrick, Ericka Witcher, Larry Noblick, and Chad Husby. 2013. Palm stem shape correlates with hurricane tolerance, in a manner consistent with natural selection. *Palms* 57(3): 115–122.

Harris, John M. 1989. *Spiral Grain and Wave Phenomena in Wood Formation*. Springer.

Jaffe, M. J. 1973. Thigmomorphogenesis: the response of plant growth and development to mechanical stimulation. *Planta* 114(2): 143–157.

Kenchington, Trevor. 1993. The structures of English wooden ships: William Sutherland's ship, circa 1710. *Northern Mariner* 3(1): 1–43. https://pdfs.semanticscholar.org/c60f/6d0e79ae82dc342da5e05eebf40066a449f9.pdf.

Kent, H. S. K. 1955. The Anglo-Norwegian timber trade in the eighteenth century. *Economic History Review*, n.s., 8(1): 62–74.

Lawton, Robert O. 1982. Wind stress and elfin stature in a montane rain forest tree: an adaptive explanation. *American Journal of Botany* 69(8): 1224–1230. https://bsapubs.onlinelibrary.wiley.com/doi/abs/10.1002/j.1537-2197.1982.tb13367.x.

McLaren, Kurt, Luke Denneko, Edmund Tanner, Peter Bellingham, and John Healey. 2019. Reconstructing the effects of hurricanes over 155 years on the structure and diversity of trees in two tropical montane rainforests in Jamaica. *Agricultural and Forest Meteorology* 276–277(15): 107621. https://doi.org/10.1016/j.agrformet.2019.107621.

Monshausen, Gabriele B., and Elizabeth S. Haswell. 2013. A force of nature: molecular mechanisms of mechanoperception in plants. *Journal of Experimental Botany* (64)15: 4663–4680. https://doi.org/10.1093/jxb/ert204.

Moore, John, Barry Gardiner, and Damien Sellier. 2018. Tree mechanics and wind loading. In: Geitmann, A., and J. Gril (eds.), *Plant Biomechanics: From Structure to Function at Multiple Scales*. Springer: 79–106.

Moulia, Bruno, Catherine Coutand, and Jean-Louis Julien. 2015. Mechanosensitive control of plant growth: bearing the load, sensing, transducing, and responding. *Frontiers in Plant Science* 6. https://doi.org/10.3389/fpls.2015.00052.

National Park Service. 2017. USS Constitution. Boston National Historical Park. https://www.nps.gov/bost/learn/historyculture/ussconst.htm.

Nodjimbadem, Katie. 2015. The 390-year-old tree that survived the bombing of Hiroshima. *Smithsonian*, August 4. https://www.smithsonianmag.com/history/390-year-old-tree-survived-bombing-hiroshima-180956157/#c5mmYwL2WeRyDugD.99.

Plant Life. 2019. Thigmomorphogenesis. http://lifeofplant.blogspot.com/2011/01/thigmomorphogenesis.html.

Rankin, Joe. 2014. Krummholz: the high life of crooked wood. Northern Woodlands. https://northernwoodlands.org/outside_story/article/krummholz-wood.

Shadow, R. A. 2007. Plant fact sheet for sea oats (*Uniola paniculata*). USDA Natural Resources Conservation Service, East Texas Plant Material Center. https://plants.usda.gov/factsheet/pdf/fs_unpa.pdf.

Simons, Paul. 2017. How palm trees stand tall in the face of a hurricane. *The Guardian*, September 17. https://www.theguardian.com/environment/2017/sep/17/palm-trees-stand-tall-hurricane-irma-caribbean-plantwatch.

Snell, C. W. 1983. *A History of the Naval Live Oaks Reservation Program, 1794–1880: A Forgotten Chapter in the History of American Conservation*. Special History Study. Gulf Islands National Seashore, National Park Service.

Snyder, R. A., and C. L. Boss. 2002. Recovery and stability in barrier island plant communities. *Journal of Coastal Research* 18(3): 530–536.

Telewski, Frank W. 2006. A unified hypothesis of mechanoperception in plants. *American Journal of Botany* 93(10): 1466–1476. https://bsapubs.onlinelibrary.wiley.com/doi/full/10.3732/ajb.93.10.1466.

Telewski, Frank W. 2016. Flexure wood: mechanical stress induced secondary xylem formation. In: Kim, Y. S., R. Funada, and A. P. Singh (eds.), *Secondary Xylem Biology: Origins, Functions, and Applications*. Academic Press: 73–91.

Telewski, Frank W., and Mordecai J. Jaffe. 1986. Thigmomorphogenesis: field and laboratory studies of *Abies fraseri* in response to wind or mechanical perturbation. *Physiologia Plantarum* 66(2): 211–218. https://doi.org/10.1111/j.1399-3054.1986.tb02411.x.

U.S. Navy. 2019. I am Old Ironsides. https://allhands.navy.mil/Features/Constitution/.

Chapter 19. Sink or Swim

Armstrong, W., R. Brandle, and M. B. Jackson. 1994. Mechanisms of flood tolerance in plants. *Botanica Neerlandica* 43(4): 307–358.

Bailey-Serres, Julia, Takeshi Fukao, Daniel J. Gibbs, Michael J. Holdsworth, Seung Cho Lee, Francesco Licausi, Pierdomenico Perata, Laurentius A. C. J. Voesenek, and Joost T. van Dongen. 2012. Making sense of low oxygen sensing. *Trends in Plant Science* 17(3): 129–138.

Blom, C. W. P. M., G. M. Bögemann, P. Laan, A. J. M. van der Sman, H. M. van de Steeg, and L. A. C. J. Voesenek. 1990.

Adaptations to flooding in plants from river areas. *Aquatic Botany* 38(1): 29–47.

Blom, C. W. P. M., L. A. C. J. Voesenek, M. Banga, W. M. H. G. Engelaar, J. H. G. M. Rijnders, H. M. van de Steeg, and E. J. W. Visser. 1994. Physiological ecology of riverside species: adaptive responses of plants to submergence. *Annals of Botany* 74(3): 253–263.

Carter, M. F., and J. B. Grace. 1990. Relationships between flooding tolerance, life history, and short-term competitive performance in three species of *Polygonum*. *American Journal of Botany* 77(3): 381–387.

Colmer, T. D., and L. A. C. J. Voesenek. 2009. Flooding tolerance: suites of plant traits in variable environments. *Functional Plant Biology* 36(8): 665–681.

Comis, D. 1997. Aerenchyma: lifelines for living underwater. *Agricultural Research* 45(8): 4–8.

Coops, H., and G. v. D. Velde. 1995. Seed dispersal, germination and seedling growth of six helophyte species in relation to water-level zonation. *Freshwater Biology* 34(1): 13–20.

Crawford, R. M. M. 1992. Oxygen availability as an ecological limit to plant distribution. *Advances in Ecological Research* 23: 93–185.

Drew, M. C. 1990. Sensing soil oxygen. *Plant, Cell & Environment* 13(7): 681–693.

Engelaar, W. M. H. G., M. W. van Bruggen, W. P. M. van den Hoek, M. A. H. Huyser, and W. P. M. Blom. 1993. Root porosities and radial oxygen losses of *Rumex* and *Plantago* species as influenced by soil pore diameter and soil aeration. *New Phytologist* 125(3): 565–574.

Ernst, W. H. O. 1990. Ecophysiology of plants in waterlogged and flooded environments. *Aquatic Botany* 38(1): 73–90.

Gunawardena, A. H. L. A. N., D. M. E. Pearce, M. B. Jackson, C. R. Hawes, and D. E. Evans. 2001. Characterization of programmed cell death during aerenchyma formation induced by ethylene or hypoxia in roots of maize (*Zea mays* L.). *Planta* 212(2): 205–214.

Jackson, M. B. 1983. Plant and crop responses to waterlogging of the soil. *Aspects of Applied Biology* 4: 99–116.

Jackson, M. B. 1990. Hormones and developmental change in plants subjected to submergence or soil waterlogging. *Aquatic Botany* 38(1): 49–72.

Jackson, M. B. 2019. The impact of flooding stress on plants and crops. https://plantstress.com/water/.

Jackson, M. B., and W. Armstrong. 1999. Formation of aerenchyma and the process of plant ventilation in relation to soil flooding and submergence. *Plant Biology* 1(3): 274–287.

Jackson, M. B., D. D. Davies, and H. Lambers. 1991. *Plant Life under Oxygen Deprivation*. SPB Academic Publishing.

Jackson, M. B., M. C. Drew, and S. C. Giffard. 1981. Effects of applying ethylene to the root system of *Zea mays* L. on growth and nutrient concentration in relation to flooding. *Physiologia Plantarum* 52(1): 23–28.

Jackson, M. B., K. Ishizawa, and O. Ito. 2009. Evolution and mechanisms of plant tolerance to flooding stress. *Annals of Botany* 103(2): 137–142.Jackson, M. B., and P. C. Ram. 2003. Physiological and molecular basis of susceptibility and tolerance of rice plants to complete submergence. *Annals of Botany* 91(2): 227–241.

Jackson, M. B., and B. Ricard. 2003. Physiology, biochemistry and molecular biology of plant root systems subjected to flooding of the soil. In: de Kroon, H., and E. J. W. Visser (eds.), *Root Ecology*. Springer: 193–213.

Jackson, M. B., L. R. Saker, C. M. Crisp, M. A. Else, and F. Janowiak. 2003. Ionic and pH signalling from roots to shoots of flooded tomato plants in relation to stomatal closure. *Plant and Soil* 253: 103–113.

Justin, S. H. F. W., and W. Armstrong. 1987. The anatomical characteristics of roots and plant response to soil flooding. *New Phytologist* 106(3): 465–495.

Kasote, D. M., S. S. Katyare, M. V. Hegde, and H. Bae. 2015. Significance of antioxidant potential of plants and its relevance to therapeutic applications. *International Journal of Biological Science* 11(8): 982–991.

Koch, M. S., and A. Mendelssohn. 1989. Sulphide as a soil phytotoxin: differential responses in two marsh species. *Journal of Ecology* 77(2): 565–578.

Kuzovkina, J., M. Knee, and M. F. Quigley. 2004. Soil compaction and flooding effects on the growth of twelve *Salix* L. species. *Journal of Environmental Horticulture* 22(3): 155–160.

Laanbroek, H. J. 1990. Bacterial cycling of minerals that affect plant growth in waterlogged soils: a review. *Aquatic Botany* 38(1): 109–125.

Loreti, E., H. van Veen, and P. Perata. 2016. Plant responses to flooding stress. *Current Opinion in Plant Biology* 33: 64–71.

Kozlowski, T. T. (ed.). 1985. *Flooding and Plant Growth*. Physiological Ecology: A Series of Monographs, Texts, and Treatises. *Biologia Plantarum* 27(338), https://doi.org/10.1007/BF02879874.

Mori, Scott. 2013. The Amazon water lily: adapted to the river's rise and fall. Plant Talk: Inside the New York Botanical Garden.

Perata, P., and A. Alpi. 1993. Plant responses to anaerobiosis. *Plant Science* 93(1–2): 1–17.

Saglio, P., M. C. Drew, and A. Pradet. 1988. Metabolic adaptation to anoxia induced by low (2-4 kPa partial pressure) oxygen pretreatment (hypoxia) in root tips of *Zea mays*. *Plant Physiology* 86(1): 61–66.

Sasidharan, R., and L. A. C. J. Voesenek. 2015. Ethylene-mediated acclimations to flooding stress. *Plant Physiology* 169(1): 3–12.

Seymour, R. S., and P. Schultze-Motel. 1997. Heat-producing flowers. *Endeavor* 21(3): 125–129.

Shahzad, Zaigham, Matthieu Canut, Colette Tournaire-Roux, Alexandre Martinière, Yann Boursiac, Olivier Loudet, and Christophe Maurel. 2016. A potassium-dependent oxygen sensing pathway regulates plant root hydraulics. *Cell* 167(1): 87–98.

Sman, A. J. M. V. D., L. A. C. J. Voesenek, C. W. P. M. Blom, F. J. M. Harren, and J. Reuss. 1991. The role of ethylene in shoot elongation with respect to survival and seed output of flooded *Rumex maritimus* L. plants. *Functional Ecology* 5(2): 304–313.

Smits, A. J. M., P. Laan, R. H. Their, and G. V. D. Velde. 1990. Root aerenchyma, oxygen leakage patterns and alcoholic fermentation ability of the roots of some nymphaeid and isoetid macrophytes in relation to the sediment type of their habitat. *Aquatic Botany* 38(1): 3–17.

University of Massachusetts Amherst. Natural History of the American Cranberry, Vaccinium macrocarpon Ait. https://www.umass.edu/cranberry/downloads/nathist.pdf.

Vartapetian, B. B., and M. B. Jackson. 1997. Plant adaptations to anaerobic stress. *Annals of Botany* 79(1): 3–20.

Wang, G. B., and L. Cao. 2012. Formation and function of aerenchyma in baldcypress (*Taxodium distichum* (L.) Rich.) and Chinese tallow tree (*Sapium sebiferum* (L.) Roxb.) under flooding. *South African Journal of Botany* 81: 71–78.

Yamauchi, T., T. D. Colmer, O. Pedersen, and M. Nakazono. 2018. Regulation of root traits for internal aeration and tolerance to soil waterlogging-flooding stress. *Plant Physiology* 176(2): 1118–1130.

Yeung, E., J. Bailey-Serres, and R. Sasidharan. 2019. After the deluge: plant revival post-flooding. *Trends in Plant Science* 24(5): 443–454.

Yoon, C. K. 1996. Heat of lotus attracts insects and scientists. *New York Times*, October 6.

Zarembinski, T. I., and A. Theologis. 1993. Anaerobiosis and plant growth hormones induce two genes encoding 1-aminocyclopropane-1-carboxylate synthase in rice (*Oryza sativa* L.). *Molecular Biology of the Cell* 4(4): 353–443.

Chapter 20. Life in the Barrens

Agarie, Sakae, Toshifumi Shimoda, Yumi Shimizu, Kathleen Baumann, Haruki Sunagawa, Ayumu Kondo, Osamu Ueno, Teruhisa Nakahara, Akihiro Nose, and John C. Cushman. 2007. Salt tolerance, salt accumulation, and ionic homeostasis in an epidermal bladder-cell-less mutant of the common ice plant *Mesembryanthemum crystallinum*. *Journal of Experimental Botany* 58(8): 1957–1967.

Alexander, E. B., R. G. Coleman, T. Keeler-Wolf, and S. P. Harrison. 2007. *Serpentine Geoecology of Western North America: Geology, Soils, and Vegetation*. Oxford University Press.

Anderson, C. W. N., R. R. Brooks, A. Chiarucci, C. J. LaCoste, M. Leblanc, B. H. Robinson, R. Simcock, and R. B. Stewart. 1999. Phytomining for nickel, thallium and gold. *Journal of Geochemical Exploration* 67(1–3): 407–415.

Anderson, R. C. and E. S. Menges. 1997. Effects of fire on sandhill herbs: nutrients, mycorrhizae, and biomass allocation. *American Journal of Botany* 84(7): 938–948.

Arabas, K. B. 2000. Spatial and temporal relationships among fire frequency, vegetation, and soil depth in an eastern North American serpentine barren. *Journal of the Torrey Botanical Society* 127(1): 51–65.

Aronson, J. 1989. *HALOPH: a data base of salt tolerant plants of the world*. Office of Arid Land Studies, University of Arizona.

Ayyappan, Durai, Ganesan Sathiyaraj, and Konganapuram Chellappan Ravindran. 2015. Phytoextraction of heavy metals by *Sesuvium portulacastrum* L. a salt marsh halophyte from tannery effluent. *International Journal of Phytoremediation* 18(5): 453–459.

Baker, A., J. Proctor, and R. Reeves (eds.). 1992. *Vegetation of Ultramafic (Serpentine) Soils: Proceedings of the First International Conference on Serpentine Ecology*. Intercept.

Balsamo, R. A., M. E. Adams, and W. W. Thomson. 1995. Electrophysiology of the salt glands of *Avicennia germinans*. *International Journal of Plant Science* 156(5): 658–667.

Bennett, T. H., T. J. Flowers, and L. Bromham. 2013. Repeated evolution of salt-tolerance in grasses. *Biology Letters* 9(2): 20130029.

Bergman, B., C. Johansson, and E. Söderbäck. 1992. The *Nostoc-Gunnera* symbiosis. *New Phytologist* 122(3): 379.

Bernstein, L. 1975. Effects of salinity and sodicity on plant growth. *Annual Review of Phytopathology* 13: 295–312.

Blumwald, E. 2000. Sodium transport and salt tolerance in plants. *Current Opinion in Cell Biology* 12(4): 431–434.

Bolan, N. S., R. Naidu, S. Mahimairaja, and S. Baskaran. 1994. Influence of low-molecular-weight organic acids on the solubilization of phosphates. *Biology and Fertility of Soils* 18: 311.

Booker, John, Bill Keogh, Daniel Chu, Jenny Conner, and Irene Hooper. 1998. Mangroves. Seacamp Association, Inc. https://www.nhmi.org/mangroves/.

Bothe, H., and A. Słomka. 2017. Divergent biology of facultative heavy metal plants. *Journal of Plant Physiology* 219: 45–61.

Bouchenak-Khelladi, Yanis, George Anthony Verboom, Vincent Savolainen, and Trevor R. Hodkinson. 2010. Biogeography of the grasses (Poaceae): a phylogenetic approach to reveal evolutionary history in geographical space and geological time. *Botanical Journal of the Linnean Society* 162(4): 543–557.

Boyle, Brad, Nicole Hopkins, Zhenyuan Lu, Juan Antonio Raygoza Garay, Dmitry Mozzherin, Tony Rees, Naim Matasci, Martha L. Narro, William H. Piel, Sheldon J. Mckay, et al. 2013. The taxonomic name resolution service: an online tool for automated standardization of plant names. *BMC Bioinformatics* 14, article no. 16.

Bringezu, K., O. Lichtenberger, I. Leopold, and D. Neumann. 1999. Heavy metal tolerance of *Silene vulgaris*. *Journal of Plant Physiology* 154(4): 536–546.

Bromham, L., and T. H. Bennett. 2014. Salt tolerance evolves more frequently in C_4 grass lineages. *Journal of Evolutionary Biology* 27(3): 653–659.

Bromham, Lindell, C. Haris Saslis-Lagoudakis, Thomas H. Bennett, and Timothy J. Flowers. 2013. Soil alkalinity and salt tolerance: adapting to multiple stresses. *Biology Letters* 9(5): 20130642.

Brooks, R. R. 1987. *Serpentine and Its Vegetation: A Multidisciplinary Approach*. Dioscorides Press.

Bui, E. N. 2013. Soil salinity: a neglected factor in plant ecology and biogeography. *Journal of Arid Environments* 92: 14–25.

Burley, S. T., K. A. Harper, and J. T. Lundholm. 2010. Vegetation composition, structure and soil properties across coastal forest-barren ecotones. *Plant Ecology* 211(2): 279–296.

Cambrollé, J., S. Redondo-Gómez, E. Mateos-Naranjo, and M. E. Figueroa. 2008. Comparison of the role of two *Spartina* species in terms of phytostabilization and bioaccumulation of metals in the estuarine sediment. *Marine Pollution Bulletin* 56(12): 2037–2042.

Cambrollé, J., J. M. Mancilla-Leytón, S. Muñoz-Vallés, E. Figueroa-Luque, T. Luque, and M. E. Figueroa. 2013. Evaluation of zinc tolerance and accumulation potential of the coastal shrub *Limoniastrum monopetalum* (L.) Boiss. *Environmental and Experimental Botany* 85: 50–57.

Cassaniti, C., D. Romano, M. E. C. M. Hop, and T. J. Flowers. 2013. Growing floricultural crops with brackish water. *Environmental and Experimental Botany* 92: 165–175.

Chalker-Scott, Linda. 2009. Mycorrhizae: so, what the heck are they, anyway? *Master Gardener* (Winter): 3–6.

Chan, Raymund, Bruce G. Baldwin, and Robert Ornduff. 2001. Goldfields revisited: a molecular phylogenetic perspective on the evolution of *Lasthenia* (Compositae: Heliantheae *sensu* lato). *International Journal of Plant Science* 162(6): 1347–1360.

Chapin, F. S., III, K. Autumn, and F. Pugnaire. 1993. Evolution of suites of traits in response to environmental stress. *American Naturalist* 142: S78–S92.

Cheeseman, J. M. 2015. The evolution of halophytes, glycophytes and crops, and its implications for food security under saline conditions. *New Phytologist* 206(2): 557–570.

Crawford, Daniel J., Robert Ornduff, and Michael C. Vasey. 1985. Allozyme variation within and between *Lasthenia minor* and its derivative species, *L. maritima* (Asteraceae). *American Journal of Botany* 72(8): 1177–1184.

Dajic, Z. 2006. Salt stress. In: Madhava Rao, K. V., A. S. Raghavendra, and K. Janardhan Reddy (eds.), *Physiology and Molecular Biology of Stress Tolerance in Plants*. Springer: 41–82.

D'Amato, Peter. 2013. *The Savage Garden: Cultivating Carnivorous Plants*. Ten Speed Press.

Darwin, Charles. 1875. *Insectivorous Plants*. D. Appleton and Company.

Deinlein, Ulrich, Aaron B. Stephan, Tomoaki Horie, Wei Luo, Guohua Xu, and Julian I. Schroeder. 2014. Plant salt tolerance mechanisms. *Trends in Plant Science* 19(6): 371–379.

Denison, R. F. 2000. Legume sanctions and the evolution of symbiotic cooperation by rhizobia. *American Naturalist* 156(6): 567–576.

Depew, M. W., and P. H. Tillman. 2006. Commercial application of halophytic turfs for golf and landscape developments utilizing hyper-saline irrigation. In: Khan, M. A., and D. J. Weber (eds.), *Ecophysiology of High Salinity Tolerant Plants*. Springer: 255–278.

Duarte, B., D. Santos, J. C. Marques, and I. Caçador. 2013. Ecophysiological adaptations of two halophytes to salt stress: photosynthesis, PS II photochemistry and antioxidant feedback—implications for resilience in climate change. *Plant Physiology and Biochemistry* 67: 178–188.

Duarte, Bernardo, Noomene Sleimi, and Isabel Caçador. 2014. Biophysical and biochemical constraints imposed by salt stress: learning from halophytes. *Frontiers in Plant Science* 5: 746.

Eisa, Sayed, Sayed Hussin, Nicole Geissler, and Hans-Werner Koyro. 2012. Effect of NaCl salinity on water relations, photosynthesis and chemical composition of quinoa (*Chenopodium quinoa* Willd.) as a potential cash crop halophyte. *Australian Journal of Crop Science* 6(2): 357–368.

Elmerich, Claudine, and William E. Newton (eds.). 2007. *Associative and Endophytic Nitrogen-Fixing Bacteria and Cyanobacterial Associations*. Nitrogen Fixation: Origins, Applications, and Research Progress, vol. 5. Springer.

Flowers, T. J. 1972. Salt tolerance in *Suaeda maritima* (L.) Dum: the effect of sodium chloride on growth, respiration, and soluble enzymes in a comparative study with *Pisum sativum* L. *Journal of Experimental Botany* 23(75): 310–321.

Flowers, T. J. 2004. Improving crop salt tolerance. *Journal of Experimental Botany* 55(396): 307–319.

Flowers, T. J., and T. D. Colmer. 2008. Salinity tolerance in halophytes. *New Phytologist* 179(4): 945–963.

Flowers, T. J., H. K. Galal, and L. Bromham. 2010. Evolution of halophytes: multiple origins of salt tolerance in land plants. *Functional Plant Biology* 37(7): 604–612.

Flowers, T. J., P. F. Troke, and A. R. Yeo. 1977. The mechanism of salt tolerance in halophytes. *Annual Review of Plant Physiology* 28: 89–121.

Flowers, T. J., and A. R. Yeo. 1995. Breeding for salinity resistance in crop plants: where next? *Australian Journal of Plant Physiology* 22(6): 875–884.

Forman, T., and T. Richard, 1979. *Pine Barrens: ecosystem and landscape*. Academic Press.

Francis, C., Y. Wong, and J. C. Meek. 2002. Establishment of a functional symbiosis between the cyanobacterium *Nostoc punctiforme* and the bryophyte *Anthoceros punctatus* requires genes involved in nitrogen control and initiation of heterocyst differentiation. *Microbiology* 148(1): 315–323.

Fukushima, Kenji, Xiaodong Fang, David Alvarez-Ponce, Huimin Cai, Lorenzo Carretero-Paulet, Cui Chen, Tien-Hao Chang, Kimberly M. Farr, Tomomichi Fujita, Yuji Hiwatashi, et al. 2017. Genome of the pitcher plant *Cephalotus* reveals genetic changes associated with carnivory. *Nature Ecology & Evolution* 1(3): 59.

Gage, Daniel J. 2017. Infection and invasion of roots by symbiotic, nitrogen-fixing rhizobia during nodulation of temperate legumes. *Microbiology and Molecular Biology Reviews* 68(2): 280–300.

Geurts, René. 2012. Mycorrhizal symbiosis: ancient signalling mechanisms co-opted. *Current Biology* 22(23): R997–R999.

Gill, S. S., and N. Tuteja. 2010. Reactive oxygen species and antioxidant machinery in abiotic stress tolerance in crop plants. *Plant Physiology and Biochemistry* 48(12): 909–930.

Glenn, E. P., J. J. Brown, and E. Blumwald. 1999. Salt tolerance and crop potential of halophytes. *Critical Reviews in Plant Sciences* 18(2): 227–255.

Gong, Qingqiu, Pinghua Li, Shisong Ma, S. Indu Rupassara, and Hans J. Bohnert. 2005. Salinity stress adaptation competence in the extremophile *Thellungiella halophila* in comparison with its relative *Arabidopsis thaliana*. *Plant Journal* 44(5): 826–839.

Gonzaga, M. I. S., J. A. G. Santos, and L. Q. Ma. 2006. Arsenic phytoextraction and hyperaccumulation by fern species. *Scientia Agricola* 63(1).

Gray, L. Joseph, Kristie Shubin, Hays Cummins, Donna McCollum, Tyler Bruns, and Eleanor Comiskey. 2010. Sacrificial leaf hypothesis of mangroves. *ISME/GLOMIS Electronic Journal* 8(4): 7–8.

Grieve, Catherine M., Stephen R. Grattan, and Eugene V. Maas. 2012. Plant salt tolerance. In: Wallender, W. W., and K. K. Tanji (eds.), *Agricultural Assessment and Management*. 2nd ed. ASCE Manual and Reports on Engineering Practice no. 71. ASCE: 405–459.

Gupta, B., and B. Huang. 2014. Mechanism of salinity tolerance in plants: physiological, biochemical, and molecular characterization. *International Journal of Genomics* 2014: article ID 701596.

Hall, I. R., G. Brown, and A. Zambonelli. 2007. *Taming the Truffle: The History, Lore, and Science of the Ultimate Mushroom*. Timber Press.

Hanin, Moez, Chantal Ebel, Mariama Ngom, Laurent Laplaze, and Khaled Masmoudi. 2016. New insights on plant salt tolerance mechanisms and their potential use for breeding. *Frontiers in Plant Science* 7: 1787.

Hasanuzzaman, Mirza, Kamrun Nahar, Mahabub Alam, Prasanta C. Bhowmik, Amzad Hossain, Motior M. Rahman, Majeti Narasimha Vara Prasad, Munir Ozturk, and Masayuki Fujita. 2014. Potential use of halophytes to remediate saline soils. *BioMed Research International* 2014, article ID 589341.

Heath, Katy D. 2008. Stabilizing mechanisms in a legume-rhizobium mutualism. *Evolution* 63(3): 652–662.

Husby, Chad E., José Delatorre, Vittorio Oreste, Steven F. Oberbauer, Danielle T. Palow, Lázaro Novara, and Alfredo Grau. 2011. Salinity tolerance ecophysiology of *Equisetum giganteum* in South America: a study of 11 sites providing a natural gradient of salinity stress. *AoB Plants* 2011: plr022.

Jacobsen, T., and R. M. Adams. 1958. Salt and silt in ancient Mesopotamian agriculture. *Science*, n.s., 128(3334): 1251–1258.

Jaffré, T., Y. Pillon, S. Thomine, and S. Merlot. 2013. The metal hyperaccumulators from New Caledonia can broaden our understanding of nickel accumulation in plants. *Frontiers in Plant Science* 4: 279.

Kiers, E. T. 2006. Measured sanctions: legume hosts detect quantitative variation in rhizobium cooperation and punish accordingly. *Evolutionary Ecology Research* 8(6): 1077–1086.

Kiers, E. Toby, Robert A. Rousseau, Stuart A. West, and R. Ford Denison. 2003. Host sanctions and the legume-rhizobium mutualism. *Nature* 425(6953): 78–81.

Knauth, L. P. 1998. Salinity history of the Earth's early ocean. *Nature* 395(6702): 554–555.

Kruckeberg, A. 1984. *California Serpentines: Flora, Vegetation, Geology, Soils, and Management Problems*. University of California Press.

Kruckeberg, A. R. 1992. Plant life of western North American ultramafics. In: Roberts, B. A., and J. Proctor (eds.), *The Ecology of Areas with Serpentinized Rocks: A World View*. Kluwer: 31–74.

Kruckeberg, A. R. 2002. *Geology and Plant Life: the Effects of Landforms and Rock Types on Plants*. University of Washington Press.

Lange, Bastien, Antony van der Ent, Alan John Martin Baker, Guillaume Echevarria, Grégory Mahy, François Malaisse, Pierre Meerts, Olivier Pourret, Nathalie Verbruggen, and Michel-Pierre Faucon. 2017. Copper and cobalt accumulation in plants: a critical assessment of the current state of knowledge. *New Phytologist* 213(2): 537–551.

Liang, W. 2018. Plant salt-tolerance mechanism: a review. *Biochemical and Biophysical Research Communications* 495(1): 286–291.

LoPresti, E. F. 2014. Chenopod salt bladders deter insect herbivores. *Oecologia* 174(3): 921–930.

Lowry, David B., Megan C. Hall, David E. Salt, and John H. Willis. 2009. Genetic and physiological basis of adaptive salt tolerance divergence between coastal and inland *Mimulus guttatus*. *New Phytologist* 183(3): 776–788.

Mahajan, S., and N. Tuteja. 2005. Cold, salinity and drought stresses: an overview. *Archives of Biochemistry and Biophysics* 444(2): 139–158.

Manousaki, E., and N. Kalogerakis. 2010. Halophytes present new opportunities in phytoremediation of heavy metals and saline soils. *Industrial & Engineering Chemistry Research* 50(2): 656–660.

Mansberg, L., and T. R. Wentworth. 1984. Vegetation and soils of a serpentine barren in western North Carolina. *Bulletin of the Torrey Botanical Club* 111(3): 273–286.

Martin, Parniske. 2008. Arbuscular mycorrhiza: the mother of plant root endosymbioses. *Nature Reviews Microbiology* 6(10): 763–775.

Mateos-Naranjo, E., S. Redondo-Gómez, L. Andrades-Moreno, and A. J. Davy. 2010. Growth and photosynthetic responses of the cordgrass *Spartina maritima* to CO_2 enrichment and salinity. *Chemosphere* 81(6): 725–731.

Menzel, U., and H. Lieth. 2003. Halophyte database version 2.0. In: Lieth, H., and M. Mochtchenko (eds.), *Cash Crop Halophytes: Recent Studies*. Kluwer Academic Publishers: 221–250.

Milindasuta, B. 1975. Developmental anatomy of coralloid roots in cycads. *American Journal of Botany* 62(5): 468–472.

Morris, Logan, Kassandra Yun, Allison Rutter, and Barbara A. Zeeb. 2019. Characterization of excreted salt from the recretohalophytes *Distichlis spicata* and *Spartina pectinate*. *Journal of Environmental Quality Abstract—Plant and Environment Interaction* 48(6): 1775–1780.

Munns, R. 1993. Physiological processes limiting plant growth in saline soil: some dogmas and hypothesis. *Plant, Cell & Environment* 16(1): 15–24.

Munns, R. 2002. Comparative physiology of salt and water stress. *Plant, Cell & Environment* 25(2): 239–250.

Munns, R., and A. Termaat. 1986. Whole-plant responses to salinity. *Australian Journal of Plant Physiology* 13(1): 143–160.

Munns, R., and M. Tester. 2008. Mechanisms of salinity tolerance. *Annual Review of Plant Biology* 59: 651–681.

Nguyen, N. T., H. Saneoka, R. Suwa, and K. Fujita. 2009. The interaction among provenances of *Melaleuca leucadendra* (weeping paperbark), salt, and aluminum. *Forest Science* 55(5): 443–454.

Oldroyd, Giles. 2008. Coordinating nodule morphogenesis with rhizobial infection in legumes. *Annual Review of Plant Biology* 59: 519–546.

Osborne, C. P., and R. P. Freckleton. 2009. Ecological selection pressures for C_4 photosynthesis in the grasses. *Proceedings of the Royal Society B: Biological Sciences* 276(1663): 1753–1760.

Panta, Suresh, Tim Flowers, Peter Lane, Richard Doyle, Gabriel Haros, and Sergey Shabala. 2014. Halophyte agriculture: success stories. *Environmental & Experimental Botany* 107: 71–83.

Parida, A. K., and A. B. Das. 2005. Salt tolerance and salinity effects on plants: a review. *Ecotoxicology and Environmental Safety* 60(3): 324–349.

Parniske, Martin. 2000. Intracellular accommodation of microbes by plants: a common developmental program for symbiosis and disease? *Current Opinions in Plant Biology* 3(4): 320–328.

Peer, Wendy Ann, Ivan R. Baxter, Elizabeth L. Richards, John L. Freeman, and Angus S. Murphy. 2005. Phytoremediation and hyperaccumulator plants. In: Tamas, M. J., and E. Martinoia (eds.), *Molecular Biology of Metal Homeostasis and Detoxification.* Topics in Current Genetics 14. Springer.

Poppinga, S., S. R. H. Hartmeyer, T. Masselter, I. Hartmeyer, and T. Speck. 2013. Trap diversity and evolution in the family Droseraceae. *Plant Signaling and Behavior* 8(7): e24685.

Poschenrieder, C., R. Tolrá, and J. Barceló. 2006. Can metals defend plants against biotic stress? *Trends in Plant Science* 11(6): 288–295.

Prasad, M. N. V., and H. M. de Oliveira Freitas. 2003. Metal hyperaccumulation in plants: biodiversity prospecting for phytoremediation technology. *Electronic Journal of Biotechnology* 6(3): 285–321.

Rascio, Nicoletta, and Flavia Navari-Izzo. 2011. Heavy metal hyperaccumulating plants: how and why do they do it? and what makes them so interesting? *Plant Science* 180(2): 169–181.

Raven, J. A., and D. Edwards. 2001. Roots: evolutionary origins and biogeochemical significance. *Journal of Experimental Botany* 52: 381–401.

Reich, Peter B., David S. Ellsworth, Michael B. Walters, J ames M. Vose, Charles Gresham, John C. Volin, and William D. Bowman. 1999. Generality of leaf trait relationships: a test across six biomes. *Ecology* 80(6): 1955–1969.

Reimond, R. J., and W. H. Queen. 1974. *Ecology of Halophytes*. Academic Press.

Rengasamy, P. 2006. World salinization with emphasis on Australia. *Journal of Experimental Botany* 57(5): 1017–1023.

Rozema, J., and T. Flowers. 2008. Crops for a salinized world. *Science*, n.s., 322(5907): 1478–1480.

Rozema, J., H. Gude, and G. Pollack. 1981. An ecophysiological study of the salt secretion of four halophytes. *New Phytologist* 89(2): 207–217.

Rozema, J., and H. Schat. 2013. Salt tolerance of halophytes, research questions reviewed in the perspective of saline agriculture. *Environmental and Experimental Botany* 92: 83–95.

Ruan, C.-J., J. A. T. da Silva, S. Mopper, P. Qin, and S. Lutts. 2010. Halophyte improvement for a salinized world. *Critical Reviews in Plant Sciences* 29(6): 329–359.

Sachs, Joel L. 2004. The evolution of cooperation. *Quarterly Review of Biology* 79(2): 135–160.

Sage, R. F. 2001. Environmental and evolutionary preconditions for the origin and diversification of the C_4 photosynthetic syndrome. *Plant Biology* 3(3): 202–213.

Sage, R. F. 2004. The evolution of C_4 photosynthesis. *New Phytologist* 161(2): 341–370.

Sage, R. F., P.-A. Christin, and E. J. Edwards. 2011. The C(4) plant lineages of planet Earth. *Journal of Experimental Botany* 62(9): 3155–3169.

Sage, R. F., and R. K. Monson (eds.). 1999. *C_4 Plant Biology*. Academic Press.

Sage, R. F., T. L. Sage, and F. Kocacinar. 2012. Photorespiration and the evolution of C_4 photosynthesis. *Annual Review of Plant Biology* 63: 19–47.

Sahi, C., A. Singh, E. Blumwald, and A. Grover. 2006. Beyond osmolytes and transporters: novel plant salt-stress tolerance-related genes from transcriptional profiling data. *Physiologia Plantarum* 127(1): 1–9.

Santos, Joaquim, Mohammed Al-Azzawi, James Aronson, and Timothy J. Flowers. 2016. eHALOPH a database of salt-tolerant plants: helping put halophytes to work. *Plant Cell Physiology*. 57(1): e10.

Saslis-Lagoudakis, C. Haris, Camile Moray, and Lindell Bromham. 2015. Evolution of salt tolerance in angiosperms: a phylogenetic approach. https://www.researchgate.net/profile/C_Haris_Saslis-Lagoudakis/publication/269992839_Evolution_of_salt_tolerance_in_angiosperms_A_phylogenetic_approach/links/55edfc9108aef559dc438b02/Evolution-of-salt-tolerance-in-angiosperms-A-phylogenetic-approach.pdf.

Saslis-Lagoudakis, C. Haris, Vincent Savolainen, Elizabeth M. Williamson, Félix Forest, Steven J. Wagstaff, Sushim R. Baral, Mark F. Watson, Colin A. Pendry, and Julie A. Hawkins. 2012. Phylogenies reveal predictive power of traditional medicine in bioprospecting. *Proceedings of the National Academy of Sciences of the United States of America* 109(39): 15835–15840.

Schat, H., S. S. Sharma, and R. Vooijs. 1997. Heavy metal–induced accumulation of free proline in a metal-tolerant and a nontolerant ecotype of *Silene vulgaris*. *Physiologia Plantarum* 101(3): 477–482.

Schedlbauer, J. L., and V. L Pistoia. 2013. Water relations of an encroaching vine and two dominant C_4 grasses in the serpentine barrens of southeastern Pennsylvania. *Journal of the Torrey Botanical Society* 140(4): 493–505.

Shabala, S., J. Bose, and R. Hedrich. 2014. Salt bladders: do they matter? *Trends in Plant Science* 19(11): 687–691.

Shanmugam, V., J. Lo, and K. Yeh. 2013. Control of Zn Uptake in *Arabidopsis halleri*: a balance between Zn and Fe. *Frontiers in Plant Science* 4: 281.

Sharma, S. S., and K.-J. Dietz. 2006. The significance of amino acids and amino acid–derived molecules in plant responses and adaptation to heavy metal stress. *Journal of Experimental Botany* 57(4): 711–726.

Smith, S. A., J. M. Beaulieu, A. Stamatakis, and M. J. Donoghue. 2011. Understanding angiosperm diversification using small and large phylogenetic trees. *American Journal of Botany* 98(3): 404–414.

Stewart, G. R., and J. A. Lee. 1974. The role of proline accumulation in halophytes. *Planta* 120(3): 279–289.

Strömberg, C. A. E. 2011. Evolution of grasses and grassland ecosystems. *Annual Review of Earth and Planetary Sciences* 39: 517–544.

Teakle, N. L., and S. D. Tyerman. 2010. Mechanisms of Cl(-) transport contributing to salt tolerance. *Plant, Cell & Environment* 33(4): 566–589.

Tedrow, J. C. F. 1952. Soil conditions in the Pine Barrens of New Jersey. *Bartonia* 26: 28–35.

Vagnoli, L., M. C. Margheri, G. Allotta, and R. Materassi. 1992. Morphological and physiological properties of symbiotic cyanobacteria. *New Phytologist* 120(2): 243–249.

Van Auken, O. W., and J. K. Bush. 1998. Spatial relationships of *Helianthus paradoxus* (Compositae) and associated salt marsh plants. *Southwestern Naturalist* 43(3): 313–320.

Van Breemen, Nico, Roger Finlay, Ulla Lundström, Antoine G. Jongmans, Reiner Giesler, and Mats Olsson. 2000. Mycorrhizal weathering: a true case of mineral plant nutrition? *Biogeochemistry* 49(1): 53–67.

Ventura, Y., and M. Sagi. 2013. Halophyte crop cultivation: the case for *Salicornia* and *Sarcocornia*. *Environmental and Experimental Botany* 92: 144–153.

Volkov, V., and M. J. Beilby. 2017. Editorial: salinity tolerance in plants: mechanisms and regulation of ion transport. *Frontiers in Plant Science* 8: 7–10.

Waisel, Y., A. Eshel, and M. Agami. 1986. Salt balance of leaves of the mangrove *Avicennia marina*. *Physiologia Plantarum* 67(1): 67–72.

Wang, Chang-Quan, Ji-Qiang Zhao, Min Chen, and Bao-Shan Wang. 2006. Identification of betacyanin and effects of environmental factos on its accumulation in halophyte *Suaeda salsa*. *Journal of Plant Physiology and Molecular Biology* 32(2): 195–201.

Wang, S.-M., J.-L. Zhang, and T. J. Flowers. 2007. Low affinity Na^+ uptake in the halophyte *Suaeda maritima*. *Plant Physiology* 145(2): 559–571.

Welch, M. E., and L. H. Rieseberg. 2002. Habitat divergence between a homoploid hybrid sunflower species, *Helianthus paradoxus* (Asteraceae), and its progenitors. *American Journal of Botany* 89(3): 472–478.

Wright, I. J., P. B. Reich, and M. Westoby. 2003. Least-cost input mixtures of water and nitrogen for photosynthesis. *American Naturalist* 161(1): 98–111.

Yang, C., D. Shi, and D. Wang. 2008. Comparative effects of salt and alkali stresses on growth, osmotic adjustment and ionic balance of an alkali-resistant halophyte *Suaeda glauca* (Bge.). *Plant Growth Regulation* 56: 179–190.

Yuan, Fang, Zhongtao Feng, Yunquan Deng, Bingying Leng, and Baoshan Wang. 2017. How do salt glands develop and secrete salt in recretohalophytes? From structures to genes. *Journal of Plant Physiology & Pathology* 5(5): 121.

Yuan, Fang, Bingying Leng, and Baoshan Wang. 2016. Progress in studying salt secretion from the salt glands in recretohalophytes: how do plants secrete salt? *Frontiers in Plant Science* 7: 977.

Yun, Kassandra B. M., Sonja Koster, Allison Rutter, and Barbara A. Zeeb. 2019. Haloconduction as a remediation strategy: capture and quantification of salts excreted by recretohalophytes. *Science of the Total Environment* 685(1): 827–835.

Zapata, F., and R. N. Roy. 2004. Phosphorus in the soil-plant system. In: Zapata, F., and R. N. Roy (eds.), *Use of Phosphate Rocks for Sustainable Agriculture*. FAO Fertilizer and Plant Nutrition Bulletin 13. Food and Agriculture Organization of the United Nations: 1–10.

Zhang, J.-L., and H. Shi. 2014. Physiological and molecular mechanisms of plant salt tolerance. *Photosynthesis Research* 115(1): 1–22.

Zhang, Mingjun, Scott C. Lenaghan, Lijin Xia, Lixin Dong, Wei He, William R. Henson, and Xudong Fan. 2010. Nanofibers and nanoparticles from the insect-capturing adhesive of the Sundew (*Drosera*) for cell attachment. *Journal of Nanobiotechnology* 8: 20.

Zhu, J. K. 2001. Plant salt tolerance. *Trends in Plant Science* 6(2): 66–71.

Illustration Credits

T = top; B = bottom; L= left; M = middle; R = right

ii Beth Richardson with Kevin Tarner, Georgia Electron Microscopy
v Mason McNair
vi Rachel Hughes
xii Kevin Tarner
7 Phyllis Hughes
17 Kevin Tarner
18 Beth Richardson with Kevin Tarner, Georgia Electron Microscopy
19TL Beth Richardson with Kevin Tarner, Georgia Electron Microscopy
19TR Beth Richardson with Kevin Tarner, Georgia Electron Microscopy
19B Kevin Tarner
21 Rachel Hughes
23 Rachel Hughes
27 Beth Richardson with Kevin Tarner, Georgia Electron Microscopy
29 Rachel Hughes
31 Kevin Tarner
33 Beth Richardson with Kevin Tarner, Georgia Electron Microscopy
34 Kevin Tarner
38 Amityadav8, Creative Commons Attribution-ShareAlike 4.0 International license, https://commons.wikimedia.org/wiki/File:Bamboo_Phytoplasma.jpg
40 Beth Richardson with Kevin Tarner, Georgia Electron Microscopy
42 Kevin Tarner
44T Kevin Tarner and Beth Richardson, Georgia Electron Microscopy
44B Beth Richardson and Kevin Tarner, Georgia Electron Microscopy
47 Kevin Tarner, Georgia Electron Microscopy
48L&R Kevin Tarner, Georgia Electron Microscopy
52L Kevin Tarner, Georgia Electron Microscopy
52R Kevin Tarner
53 Beth Richardson with Kevin Tarner, Georgia Electron Microscopy
54 Kevin Tarner, Georgia Electron Microscopy
56 Rachel Hughes, Beth Richardson and Kevin Tarner, Georgia Electron Microscopy
57TL Rachel Hughes
57TR Beth Richardson with Kevin Tarner, Georgia Electron Microscopy
57B Beth Richardson with Kevin Tarner, Georgia Electron Microscopy
59 Beth Richardson with Kevin Tarner, Georgia Electron Microscopy
60 Rachel Hughes
62 Kevin Tarner
63 Beth Richardson with Kevin Tarner, Georgia Electron Microscopy
65 Rachel Hughes
66 Kevin Tarner, Georgia Electron Microscopy
68 Rachel Hughes
70L&R Kevin Tarner, taken at the University of Georgia Trial Gardens
72 Kevin Tarner
77 Kevin Tarner
79 Rachel Hughes
81 Kevin Tarner
83 Kevin Tarner
85 Rachel Hughes
87 Rachel Hughes
88 Primejyothi, Creative Commons Attribution-ShareAlike 3.0 Unported license, https://commons.wikimedia.org/wiki/File:RubberTapping_5172.jpg
90 Kevin Tarner

94 Kevin Tarner, Georgia Electron Microscopy
96 Kevin Tarner, Georgia Electron Microscopy
98 Kevin Tarner
99 Rachel Hughes
101 Kevin Tarner
104 Rachel Hughes
107L Mason McNair
107R Rachel Hughes
112 Rachel Hughes
115 Rachel Hughes
117L Kevin Tarner
117R Kevin Tarner, Georgia Electron Microscopy
119T Kevin Tarner
119B Beth Richardson and Kevin Tarner, Georgia Electron Microscopy
120 Kevin Tarner
122 Rachel Hughes and Kevin Tarner
124 Alexander Klepnev (CC BY 4.0), https://en.wikipedia.org/wiki/Seed#/media/File:%D0%A0%D0%B0%D0%B7%D0%BD%D0%BE%D0%BE%D0%B1%D1%80%D0%B0%D0%B7%D0%B8%D0%B5_%D1%81%D0%B5%D0%BC%D1%8F%D0%BD.jpg
125 Kevin Tarner
126 Jacob W. Frank, National Park Service, public domain, U.S. government work
127 Kevin Tarner
128 Kevin Tarner
129 Rachel Hughes
135T Kevin Tarner and Beth Richardson, Georgia Electron Microscopy, University of Georgia
135B Berkshire Community College Bioscience Image Library, public domain, https://upload.wikimedia.org/wikipedia/commons/f/f6/Gymnosperm_Stem_One_Year_Pinus_%2835924400930%29.jpg
137 Berkshire Community College Bioscience Image Library, public domain, https://www.flickr.com/photos/146824358@N03/34991095873/in/photostream/
140 Rachel Hughes
142 Kevin Tarner (all photos)
145 Berkshire Community College Bioscience Image Library, public domain, http://blogs.berkshirecc.edu/bccoer/plant-morphology-angiosperm-stems/
149T Kevin Tarner and Beth Richardson, Georgia Electron Microscopy, University of Georgia
149B Phyllis Hughes
152 Amityadav8, Creative Commons Attribution-ShareAlike 4.0 International license, https://commons.wikimedia.org/wiki/File:Bamboo_Phytoplasma.jpg
157 Phyllis Hughes
166 Kevin Tarner
167 Kevin Tarner, Georgia Electron Microscopy
171L&R Kevin Tarner
173 Mason McNair
177 Joseph Meyer, Konversations-Lexikon (1897), public domain
179TL William Morton Wheeler, *Ants: Their Structure, Development and Behavior* (New York: Columbia University Press, 1910), public domain
179TR Eurico Zimbres, public domain, https://commons.wikimedia.org/wiki/File:Cecropia_glazioui.jpg
179B Kevin Tarner
182 Kevin Tarner
183 Judy Gallagher, Attribution 2.0 Generic (CC BY 2.0), https://www.flickr.com/photos/52450054@N04/8505045055
185 Hans Stuessi, Creative Commons Attribution-ShareAlike 4.0 International license, https://commons.wikimedia.org/wiki/File:Compilation_of_seeds_with_elaiosomes.jpg
186L Kevin Tarner
186R Beth Richardson with Kevin Tarner, Georgia Electron Microscopy
187L Shirankallu Sathwik Shastry (username Kateelkshetra), Creative Commons Attribution-Share Alike 3.0 Unported license, https://commons.wikimedia.org/wiki/File:Perfect_Camouflage_(Caterpillar_on_teakwood_branch).jpg
187R Rachel Hughes
188 Judy Gallagher, https://www.flickr.com/photos/52450054@N04/23417536094/
190T&B Rachel Hughes
193 Rachel Hughes
194 Peter Coxhead, Creative Commons CC0 1.0 Universal Public Domain Dedication, https://commons.wikimedia.org/wiki/File:Lithops_aucampiae_var._koelmanii_120502.jpg
195TL Rachel Hughes
195R Matt Lavin, https://flickr.com/photos/35478170@N08/48833060192
195BL Rachel Hughes
196TL Rachel Hughes
196R Rachel Hughes
196BL Kevin Tarner
197 *World Telegram* staff photographer, Library of Congress, public domain

198L Kira Hoffmann, https://pixabay.com/photos/wheat-wheat-field-cornfield-summer-1556698/
198R LSDSL, Creative Commons Attribution-ShareAlike 3.0 Unported license, https://upload.wikimedia.org/wikipedia/commons/thumb/7/79/Ear_of_rye.jpg/640px-Ear_of_rye.jpg
203 Nature Shutterbug, Attribution 2.0 Generic (CC BY 2.0), https://www.flickr.com/photos/deinandra/328674427
204 Kevin Tarner, plant grown in the wonderful University of Georgia Trial Gardens by Brandon Coker
207 Rachel Hughes
209TL Kevin Tarner
209TR Rachel Hughes
209BR Kevin Tarner
210 Frank Kovalchek, Creative Commons Attribution 2.0 Generic license, https://commons.wikimedia.org/wiki/File:Fall_colors_near_the_Eagle_Lake_trailhead.jpg
214 Kevin Tarner
215 Kevin Tarner
217T Kevin Tarner
217B Rachel Hughes
220 James Fynmore, public domain, Adirondack Museum
221 Sven Osborn, public domain, https://en.wikipedia.org/wiki/Grauballe_Man#/media/File:Grauballemannen3.jpg
228 Kevin Tarner
230 Codex Mendoza, public domain
234 Kevin Tarner
235 Kevin Tarner
237 Kevin Tarner
238 Kevin Tarner
245 Kevin Tarner
246 Kevin Tarner
247 Kevin Tarner
248 Kevin Tarner
250 Kevin Tarner
251 Bob Walker, Wikimedia Commons
253 Kevin Tarner
254 Rachel Hughes
256 pixabay.com, Creative Commons license
259 Michael Hermann, Creative Commons Attribution-ShareAlike 3.0 Unported license
260 Srimesh, Wikimedia Commons, Creative Commons license
261 Kevin Tarner
263L Wilhelm Brandt, M. Gürke, F. E. Köhler, G. Pabst, G. Schellenberg, and Max Vogtherr, *Köhler's Medizinal-Pflanzen*, https://commons.wikimedia.org/wiki/File:K%C3%B6hler%27s_Medizinal-Pflanzen_in_naturgetreuen_Abbildungen_mit_kurz_erl%C3%A4uterndem_Texte_(Plate_189)_(8232823234).jpg
263R Rachel Hughes
267T pxfuel.com, Creative Commons Zero-CC0, https://www.pxfuel.com/en/free-photo-osjcf
267B Rachel Hughes
268 U.S. Air Force, public domain (U.S. government work), https://www.afhistory.af.mil/News/photos/igphoto/2000445232/
269 NOAA George E. Marsh Album, theb1365, Historic C&GS Collection, public domain (U.S. government work), https://commons.wikimedia.org/wiki/File:Dust_Storm_Texas_1935.jpg
273T&B Rachel Hughes
275 Kevin Tarner
276L Oliver Macdonald, Creative Commons Attribution-ShareAlike 3.0 Unported license, https://commons.wikimedia.org/wiki/File:Alopecurus_myosuroides_in_Barley1.jpg
276R Jan Kops, *Flora Batava*, vol. 7 (1830), public domain
278 Rachel Hughes
280 Mason McNair
282 Cameron Strandberg, https://commons.wikimedia.org/wiki/File:Fire-Forest.jpg
286 Tanya Hart (CC BY-SA 2.0), https://www.flickr.com/photos/arripay/30460590826/
287 www.peakpx.com, public domain, https://www.peakpx.com/636149/redwood-yosemite-giant-trees-tree-trunk-tree
288 Kevin Tarner
290 Mason McNair
291 Mark Muir, USFS government work, public domain, https://commons.wikimedia.org/wiki/File:AspenOverview0172.jpg
292 Cas Liber, https://upload.wikimedia.org/wikipedia/commons/8/82/LambertiaformLCNP.jpg
294L Rachel Hughes
294R Beth Richardson and Kevin Tarner, Georgia Electron Microscopy
296 Natalie Maguire, Attribution-ShareAlike 2.0 Generic (CC BY-SA 2.0), https://www.flickr.com/photos/natalietracy/40994413982
297T Rachel Hughes
297B Matt Lavin, Attribution-ShareAlike 2.0 Generic (CC BY-SA 2.0), https://www.flickr.com/photos/plant_diversity/36766044673

298 Andrew Massyn, Creative Commons Attribution-ShareAlike 3.0 Unported license, https://commons.wikimedia.org/wiki/File:Syncarpha_vestita_flowers.jpg
301 Ken Lund, Attribution-ShareAlike 2.0 Generic (CC BY-SA 2.0), https://www.flickr.com/photos/kenlund/8602586507
303 Rachel Hughes
305 Kevin Tarner
306L&R Rachel Hughes
307 Kevin Tarner
308 Rachel Hughes
309L Kevin Tarner
309M Rachel Hughes and Kevin Tarner
309R Rachel Hughes
310L Kevin Tarner
310R Rachel Hughes
311 Berkshire Community College Bioscience Image Library, public domain, https://www.flickr.com/photos/146824358@N03/24010130808/
312L Kevin Tarner
312R Beth Richardson with Kevin Tarner, Georgia Electron Microscopy
319 Else-Marie de Leeuw, unsplash.com
321 Rachel Hughes
322 R adept, wikipedia.org
323L SB_Johnny, Wikimedia Commons
323R James St. John, https://commons.wikimedia.org/wiki/File:Rhododendron_maximum_(Fox_Creek,_Grayson_County,_Virginia,_USA)_5_(30421473015).jpg
324L Lars Falkdalen Lindahl, https://commons.wikimedia.org/wiki/File:Young_pine_trees_covered_in_heavy_snow.jpg
324R Rachel Hughes
325 Rachel Hughes
326 Roger Griffith, https://upload.wikimedia.org/wikipedia/commons/b/be/Sitka_Branch_in_the_snow.jpg
327 Jsayre64, Wikimedia Commons
329 PxHere, CC0, https://pxhere.com/en/photo/1354257
331 Mason McNair
332T Rachel Hughes
332BL Grand Canyon National Park (CC BY 2.0), https://www.flickr.com/photos/grand_canyon_nps/12199085015
332BR Dcrjsr, Creative Commons Attribution-ShareAlike 3.0 Unported license, https://commons.wikimedia.org/wiki/File:Pinyon_pine_Pinus_monophylla_in_sagebrush.jpg
333 Rachel Hughes
334 Kevin Tarner
335 Rachel Hughes
336 Kenneth P. James, CC 4.0 license, https://commons.wikimedia.org/wiki/File:FlagFormedRedSpruceDollySodsWilderness.jpg
338 Dimitris Vetsikas, CC0, https://pixabay.com/photos/tree-palm-tropical-wind-storm-3191872/
339 Phyllis Hughes
340 MCSN Michael Achterling / U.S. Navy, https://upload.wikimedia.org/wikipedia/commons/d/d5/Flickr_-_Official_U.S._Navy_Imagery_-_USS_Constitution_sets_sail._%281%29.jpg
342 Rachel Hughes, Bonaventure Cemetery, Savannah, Ga.
343L Rachel Hughes
343R Rachel Hughes, St. Simons Island, Ga.
347 Mason McNair
349 Scot Nelson
350T Basile Morin, https://commons.wikimedia.org/wiki/File:Tortuous_dirt_road_in_wet_paddy_fields_and_green_hills_in_Vang_Vieng,_Laos.jpg
350B Siggy Nowak, https://pixabay.com/photos/mangroves-tropical-water-plant-sea-3788104/
353L Fayette A. Reynolds, Berkshire Community College Bioscience Image Library, public domain, https://www.flickr.com/photos/146824358@N03/35570465200
353R Berkshire Community College Bioscience Image Library, public domain, https://www.flickr.com/photos/146824358@N03/33958474393/
354L Rachel Hughes
354R pxhere.com, public domain, https://pxhere.com/en/photo/1201528
355 Kevin Tarner, Georgia Electron Microscopy
356T Mason McNair
356B USGS government work, https://commons.wikimedia.org/wiki/File:Duckweed_Pond_(15317740476).jpg
357 Rachel Hughes
358 Rachel Hughes
359 Kevin Tarner
361 Keith Weller, USDA, public domain, https://commons.wikimedia.org/wiki/File:Cranberry_harvest_in_New_Jersey.jpg
366 Rachel Hughes
367 Rachel Hughes
369 Kevin Tarner
371L Rachel Hughes
371R Kevin Tarner

372TL Rachel Hughes
372BR Rachel Hughes and Kevin Tarner
373L André-Ph. D. Picard, https://commons.wikimedia.org/wiki/File:Mycorhizes-01.jpg
373R Kevin Tarner
375 Mason McNair
376 Mason McNair
378 Mason McNair
379 Mason McNair
380 Mason McNair
381 Mason McNair
382T Kevin Tarner
382B Mason McNair
384TL Stefan Lefnaer, https://commons.wikimedia.org/wiki/File:Utricularia_australis_sl1.jpg
384TR Michal Rubeš, Creative Commons Attribution-ShareAlike 3.0 Czech Republic license, https://commons.wikimedia.org/wiki/File:Utricularia_aurea_8_Darwiniana.jpg
384B Mason McNair
386 Maxpixel.com, public domain, https://www.maxpixel.net/Flat-Salt-Flats-Nature-Landscape-Sunrise-Travel-2379007
387T Steve Hillebrand, U.S. Fish and Wildlife Service, https://commons.wikimedia.org/wiki/File:Mangrove_plant_roots_provides_an_island_in_the_water.jpg
387B Rachel Hughes
389 Kevin Tarner
391 Rachel Hughes
392 Krzysztof Ziarnek, Creative Commons Attribution-ShareAlike 4.0 International license, https://commons.wikimedia.org/wiki/File:Mesembryanthemum_crystallinum_kz6.jpg
393TL Rachel Hughes
393TR Rachel Hughes
393B Metropolitan Museum of Art, Theodore M. Davis Collection, Bequest of Theodore M. Davis, 1915, https://commons.wikimedia.org/wiki/File:Winged_Scarab_Amulet_MET_30.8.1082a-c_EGDP012497.jpg
396 Rachel Hughes
397 Udo Schmidt, Attribution-ShareAlike 2.0 Generic (CC BY-SA 2.0), https://www.flickr.com/photos/coleoptera-us/9376703412
399 Bananaflo (CC BY 2.5), https://en.wikipedia.org/wiki/Nickel_mining_in_New_Caledonia#/media/File:River_South_New_Caledonia.jpg
471L&R Kevin Tarner

Index

About the Authors

Kevin "the Plant Man" Tarner

Kevin "the Plant Man" Tarner boasts a green thumb forged through a career of dedicated horticultural expertise within both scientific and commercial controlled environment agricultural growing arenas. His passion for cultivating botanical wonders has blossomed into a career marked by a deep understanding of plant biology and appreciation for the natural world. A sought-after garden speaker and horticultural instructor, Kevin enjoys sharing his knowledge and enthusiasm for plants.

When he isn't puttering in his gardens and greenhouses, Kevin loves spending time with his wife and pets, being out in nature, taking photographs, tinkering with mechanical things, yoyoing, woodworking, and playing tabletop games. He is a top competitive expert at his favorite board game, Santorini. Kevin's inaugural literary venture, *Hidden World: The Survival Systems of Plants*, invites readers into a captivating exploration of the intricate survival strategies employed by the botanical realm.

Rachel Hughes

Rachel Hughes has devoted her life to educating people about the environment. Growing up on a wooded granite outcrop, Rachel developed an acute appreciation for life on this planet. Her parents fostered this passion through numerous nature hikes, plant rescues, and community activism. Once in school, she spent most of her time embracing both the sciences and the social sciences, hoping to understand human behavior and the world around her. She spent her college years studying entomology, sociology, and natural resources, obtaining a graduate degree in the latter. Her employment experience has included laboratory and teaching work, mostly at the university level. Currently, she is a university lecturer specializing in ecology and environmental science. She loves spending time in the sun, playing board games with her husband, and spoiling her pets. *Hidden World: The Survival Systems of Plants* is Rachel's first book.